Igor M. Rouzine

Mathematical Modeling of Evolution

De Gruyter Series in Mathematics and Life Sciences

Edited by
Anna Marciniak-Czochra, Heidelberg University, Germany
Benoît Perthame, Sorbonne-Université, France
Jean-Philippe Vert, Mines ParisTech, France

Volume 8/2

Igor M. Rouzine

Mathematical Modeling of Evolution

Volume 2: Fitness Landscape, Red Queen, Evolutionary Enigmas, and Applications to Virology

DE GRUYTER

Mathematics Subject Classification 2020
Primary: 92XX; Secondary: 60XX

Author
Dr. Igor M. Rouzine
Sechenov Institute of Evolutionary Physiology and Biochemistry
Russian Academy of Sciences
Prospekt Thorez 44
194223 St Petersburg
Russian Federation
igor.rouzine@iephb.ru

ISBN 978-3-11-069731-5
e-ISBN (PDF) 978-3-11-069738-4
e-ISBN (EPUB) 978-3-11-069746-9
ISSN 2195-5530

Library of Congress Control Number: 2023937362

Bibliographic information published by the Deutsche Nationalbibliothek
The Deutsche Nationalbibliothek lists this publication in the Deutsche Nationalbibliografie; detailed bibliographic data are available on the internet at http://dnb.dnb.de.

Typesetting: Integra Software Services Pvt. Ltd.
Printing and binding: CPI books GmbH, Leck

www.degruyter.com

Preface

The book represents the continuation of my previous book *Mathematical Models of Evolution. Volume I* on the general theory of evolution in the setting of one-locus and multi-locus models. This volume contains mathematical models tailored to answer specific biological questions arising in virology. Chapter 1 demonstrates how mathematical models help to estimate basic evolutionary parameters of adapting populations not measurable directly by inferring them from genetic samples. Chapter 2 describes an evolutionary approach to infer full fitness landscape. Chapter 3 unravels the mystery of several biological properties of unclear evolutionary purpose. Chapter 4 dwells on the manifestations of Red Queen effect, including the adaptation of a pathogen in the face of the changing immune response and defective interference variants. The book will benefit all readers interested in evolution theory and its applications, especially quantitatively inclined virologists.

Igor M. Rouzine

https://doi.org/10.1515/9783110697384-202

Contents

Chapter 1
Inference of the acting factors of evolution and basic evolutionary parameters from sequence data

Genetic evolution of a virus in an animal host is a result of interplay between a multitude of different factors (Wright 1931; Fisher 1990; Kimura 1994): random mutation, natural selection including epistasis (Chapter 2), random genetic drift, linkage, recombination, transmission between individuals, and spread between infected organs. Transmission between individuals and infection sites creates genetic bottlenecks, which create founder effects due to random sampling from the infecting population. This complexity necessitates the use of mathematical and computational models, which allows to study the system in a reduced, simplified form. The main challenge facing mathematical modeling of real systems is that it is not known in advance which biological factors, among the many existing, impact the system behavior and should be included into the model. The validity of approximations cannot be tested until the analysis or computation is over and depends on the questions asked and parameters inferred.

A modeler, thus, faces a time paradox, as follows. On the one hand, approximations must be introduced in the beginning to make the model tractable. On other hand, discovering an appropriate set of approximations ("the model") is the final aim of such research. A dynamic interplay between theory and experiment, in a continuous loop, offers the resolution of the time paradox (Rouzine and Coffin 1999a). A modeler selects a few basic experimental observations and develops the simplest model to interpret them. Once the first match between data and model predictions has been obtained, the modeler looks for the part in experimental data that contradict the model predictions. When such contrary evidence is found, the modeler changes the initial assumptions, one by one, and finds out which ones removes the disagreement. The loop is repeated, until all data of interest are matched to the predictions. The result is considered a working model subject to further experimental verification or falsification.

In this chapter, this strategy is applied to elucidate the acting factors and main parameters of HIV evolution. The goal is to demonstrate, on three examples, how the models of population genetics can help to determine the dominant evolutionary factors acting on a population and estimate their parameters from genomic data.

https://doi.org/10.1515/9783110697384-001

1.1 Mechanism of HIV diversity and the estimate of average selection coefficient

To study the mechanism of HIV evolution, Rouzine and Coffin (1999c) used 213 sequences of HIV protease protein isolated from 11 HIV-infected individuals not treated previously with protease inhibitors (Lech et al. 1996). The diverse sites were interpreted as sites undergoing slow adaptation during individual infection. The model of slow adaptation is found to be adequate to explain the existence of diverse sites in the protease gene. To predict the average frequency of variable sites in individuals, transmission between individuals has to be taken into consideration. Initially, the best-fit sequence is assumed to be the same between individuals, and the stochastic transmission bottleneck is included in the model. With infection from a single source, such a model could potentially explain the high frequency of variable sites observed in untreated infected individuals. However, this theoretical result turns out to be highly sensitive to the assumption about the absence of coinfection from different sources, and hence, not robust. After excluding several apparent possibilities, a plausible explanation supported by immunological data is found. In the presence of the active immune response, the best-fit sequence of virus differs strongly between individuals due to individual variation in major histocompatibility complex class I (MHC-I) subtypes. Emergence of a variant unrecognized by the immune system (Section 4.2) resets the signs of selection coefficients at many other sites resulting in the gradual growth of compensatory mutations (Sections 2.1–2.3). This section follows this work.

Compared to other viruses, human immunodeficiency virus (HIV) shows a very rate of evolution and high levels of genetic diversity (Wain-Hobson 1993). In infected individuals, HIV establishes a persistent infection with hundreds of genomic sites that vary among and within patients. Mutations abrogating virus recognition by the immune system interfere with vaccine development (Wolfs et al. 1990; Burns and Desrosiers 1994; Moore et al. 1996; Wolinsky et al. 1996), while other mutations impede the efficacy of replication inhibitors (Mayers et al. 1992; Cleland et al. 1996).

From another angle, genomic data contain crucial information about biological factors acting on the virus. The presence of a functional immune response, in addition to the immunological experiments, can be inferred from the genetic diversity data. The dominance of nonsynonymous substitutions in the envelope protein reveals the presence of the selection for diversity acting on this surface protein (Simmonds et al. 1990; Burns and Desrosiers 1991; Pang et al. 1991), caused by the antibody response and by the adaptation of HIV to different cell types.

Two basic types of natural selection can be considered, as follows. Purifying selection exists due to the fact that a certain genetic variant (wild type), has a higher average number of progeny than other variants (higher fitness). The selection coefficient, s, is defined as the relative fitness minus 1, i.e., to the relative difference in the virus growth rates. If this type of selection were the only evolutionary factor acting on a population, the population would eventually become genetically uniform in the better-fit variant (survival of the fittest).

Another type of natural selection is the selection for diversity, which comes in two flavors. The first emerges due to the changes in external conditions, such as the immune response (Sections 4.1 and 4.2). Another flavor is the effect of ecological niche, when different cell tissues favor different best-fit genomic sequences. Under this selection, the population assumes and maintains diverse genetic composition.

The level of genetic diversity, the type of natural selection, and the evolution rate differ between HIV genes. The envelope gene is the fastest-evolving and the most diverse gene, with the variation of 3–5% within an individual (Balfe et al. 1990; Wolfs et al. 1990; Lamers et al. 1993) and 8–17% between individuals from the same location (Balfe et al. 1990; Kuiken et al. 1993; McCutchan et al. 1996). A smaller but still high level of diversity within an individual was reported for the capsid protein, 1–2% (Yoshimura et al. 1996) and for the polymerase and protease proteins, 0.4–1% (Najera et al. 1995; Lech et al. 1996).

In the protease, genetic variation is dominated by synonymous and chemically conservative substitutions (Lech et al. 1996), which implies a dominant role for purifying (directional) selection. In contrast, envelope evolution is dominated by nonsynonymous substitutions indicating the presence of selection for diversity. The focus of this section is on the protease gene and purifying selection. The results are presented qualitatively in Sections 1.1.1–1.1.6 and paralleled by mathematical derivations in Section 1.1.8.

1.1.1 Sequence diversity in protease gene: data analysis

A database of 265 protease gene sequences was obtained by Lech et al. (1996) from patients not treated previously with protease inhibitors. The 265 clones of proviral DNA were isolated from cells of 13 individuals and subjected to 70 cycles of nested PCR amplification. For each individual, the consensus sequence was determined separately, by noting the most frequent variant for each base. It differed between patients. The overall consensus for the entire set of individuals was found to be identical to the subtype B consensus in the LANL database, which is 297 bases long and begins with the conserved sequence CCTCAgATCACTCTT (PQITL), where "g" denotes a variable silent base.

The data were filtered, as follows.

(i) Consensus sequences in patients numbered 06 and 12 had stop codons indicating defective virus; samples from these individuals were excluded from analysis.
(ii) For the remaining 11 individuals, 6% of their sequences contained deletions, insertions, or stop codons; these sequences were also filtered out, since the focus of the work was on point mutations that do not render the virus unable to replicate. The remaining data set consisted of 213 sequences of proviral DNA.

(iii) 25% of the substitutions were transversions with respect to the consensus, defined as mutations that substitute a nucleotide by a nucleotide with a strongly different biochemical structure, such as A → C or A → T. Transversions were ignored, replacing them with the consensus sequence. As a result, each base had only two alleles: either A/G or C/T. A model taking into account transversions would require including two more parameters, the forward and back mutation rates for transversions, that are small and were known, at that time, with poor accuracy. The parsimony principle prohibits such complications, without much gain in information.

(iv) To exclude potential errors during PCR rounds, alleles present in a single copy in the entire database were ignored. The "sporadic" mutations emerged with a frequency of 0.06% per base, consistent with the PCR error rate of 1 per 10^5 bases per cycle (Tindall and Kunkel 1988; Barnes 1992).

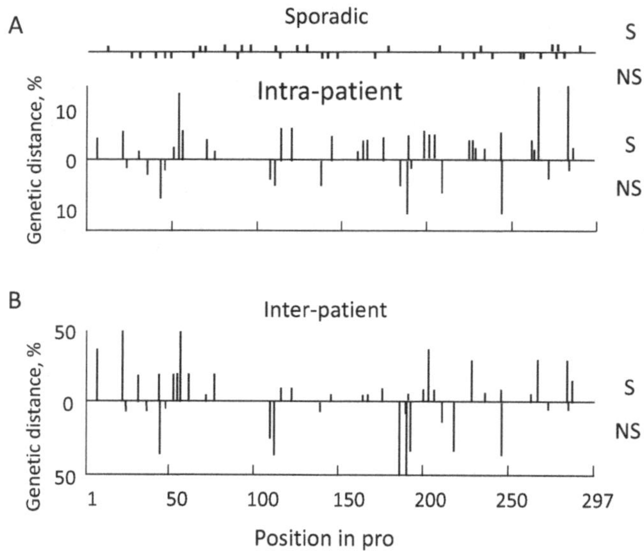

Fig. 1.1: Intrapatient (A) and interpatient (B) genetic distances, averaged over patients, at different positions in the HIV protease gene from data (Lech et al. 1996).
The upper and lower histograms in each figure correspond to synonymous and nonsynonymous sites, respectively. Dots on the upper horizontal line in panel a show the positions of sporadic mutations seen only once in the data set. Based on Rouzine and Coffin (1999c).

In the filtered database, each base was classified as either "variable" or "conserved." For each variable base and each individual, the frequency of minority alleles with respect to the common consensus was calculated. For each individual and each base, the *intrapatient* genetic distance (heterozygosity) was calculated as the proportion of randomly sampled sequence pairs that differ at that base. The *interpatient* genetic distance was calculated in the same way, but for pairs sampled from different individuals.

Both genetic distances can be expressed in terms of the minority allele frequency with respect to any reference sequence, such as the database consensus sequence or the best-fit sequence. Let us denote by f_{ki} the allelic frequency at site i for individual k. Then, the intrapatient distance for each patient, T_{ki}^{intra}, and the interpatient genetic distances for a pair of individuals, $T_{k_1k_2i}^{\text{inter}}$, can be written as

$$\begin{aligned} T_{ki}^{\text{intra}} &= 2f_{ki}(1-f_{ki}) \\ T_{k_1k_2i}^{\text{inter}} &= f_{k_1i}(1-f_{k_2i}) + f_{k_2i}(1-f_{k_1i}) \end{aligned} \tag{1.1}$$

By the definition, the intrapatient and interpatient genetic distances vary between 0 and 1/2 and 0 and 1, respectively. The intrapatient distance is maximal when a population has 50:50 composition in two alleles, $T_{ki}^{\text{intra}} = 1/2$, $f_{ki} = 1/2$. As one can show, the interpatient distance is equal or larger than the average of the two intrapatient distances. The intrapatient distance is maximal, $T_{k_1k_2i}^{\text{inter}} = 1$, when the two virus populations are uniform and composed of opposite alleles, either $f_{k_1i} = 0$, $f_{k_2i} = 1$ or vice versa, $f_{k_1i} = 1$, $f_{k_2i} = 0$.

Note that both genetic distances are independent of the choice of the reference sequence: the right-hand sides of eqs. (1.1) are invariant to the replacement $f_{ki} \to 1 - f_{ki}$. This invariance makes the intrapatient genetic distance a much better parameter for the comparison of theory with experiment than the substitution frequency. Indeed, as argued below, the best-fit variant and the consensus variant do not necessarily coincide.

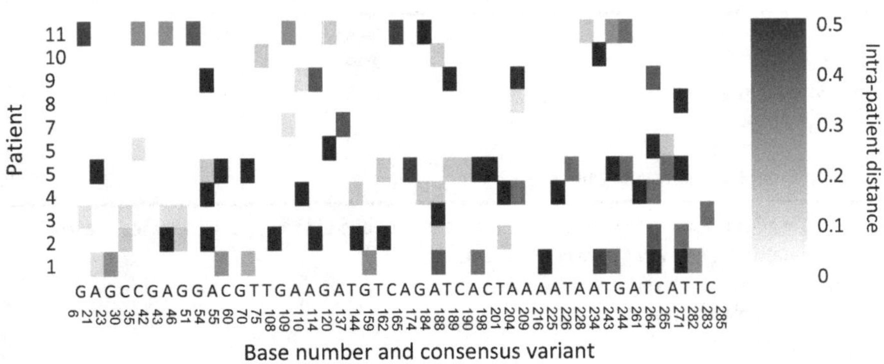

Fig. 1.2: Gray-scale diagram of intrapatient genetic distance, T_{ki}^{intra}, in different patients at different variable sites in the HIV protease gene.
The intrapatient genetic distance is indicated by the degree of shading, as shown on the scale on the right. Letters and numbers under the diagram show consensus nucleotides and positions in the protease gene, respectively. Based on Rouzine and Coffin (1999c).

Both types of distances averaged over either individuals or their pairs, respectively, are shown for different sites (Fig. 1.1). The intrapatient distance is shown for individual patients as well (greyscale diagram in Fig. 1.2). Genetic distance was also averaged over three groups of sites, for each individual separately: (i) sites that are variable in the individual, (ii) sites that are variable in any individual, and (iii) all sites of the protease gene (Tab. 1.1). Because multiple substitutions within the same codon are very rare, all mutations can be classified into synonymous (not changing the amino acid) or nonsynonymous (changing the amino acid), with separate treatment of these two classes (Tab. 1.1). When two minority alleles were found in the same codon of the same sequence, they were counted as belonging to two different sequences.

The main conclusions from data analysis are, as follows:

(i) Genetic polymorphism within an individual is focused on rare sites: 47 sites of the 297 sites total.
(ii) Variable site locations are shifted between individuals: an average variable site appears in only ~16% of individuals.
(iii) Variable sites are highly diverse: $\langle T_{ki}^{\text{intra}} \rangle = 0.27$ per variable site, with 0.5 being the possible maximum.
(iv) Strikingly, synonymous and nonsynonymous variable sites have similar diversity.
(v) However, diverse synonymous sites are twice as frequent as nonsynonymous sites.

Tab. 1.1: Genetic diversity in the protease gene of HIV based on data in Lech et al. (1996).

Mutation type	Average intrapatient distance, T (%)[a]			No. of variable sites
	Individual variable site[b]	Any variable site[c]	Any site[d]	
Synonymous	29.5	4.7	0.23	31
Nonsynonymous	21.9	4.3	0.72	16
Synonymous and nonsynonymous	27.5	4.5	0.05	47
Sporadic synonymous[e]			0.07	16
Sporadic nonsynonymous[e]			0.12	23
Sporadic synonymous and nonsynonymous[e]			0.23	39

[a]Intrapatient genetic distance in the protease, *T*, averaged over one of three groups of bases in each infected individual and then averaged over individuals.
[b]Bases that are variable in a given individual only.
[c]Bases that are variable in any individual.
[d]All bases in the protease.
[e]Substitutions present in a single copy per database and excluded from the first three rows as PCR error suspects. Note the difference in relative number of synonymous sites between sporadic and nonsporadic substitutions.

1.1.2 Model 1: deterministic adaptation of virus in an individual

Below a family of four models is analyzed. In all models, evolution is treated as a deterministic process, with purifying selection and mutation as the only evolutionary factors, while random nature of mutation and random genetic drift are neglected (Rouzine 2020b). The preponderance of synonymous substitutions (Section 1.1.1) implies the dominant role of purifying selection as opposed to selection for diversity. A genomic test presented in Section 1.2 demonstrates a large population size and a small role for stochastic effects in a chronic untreated HIV infection. Linkage effects and epistasis are neglected as well (they are the focus of Chapter 2), and evolution is considered at a single site, independently on the other sites. The best-fit virus is assumed to be the same for all individuals. This and the other assumptions are discussed in detail in Section 1.1.7.

The goal of this study is to explain three observations, as follows:

(i) the high level of diversity at individual variable sites,
(ii) the same average diversity for synonymous and nonsynonymous variable sites,
(iii) the difference and overlap in the genomic location of variable sites between infected individuals.

The first explanation of these facts is, as follows: weakly deleterious ("mutant") alleles at a site of interest are generated in some individuals by random mutation and passed down the chain of transmission (Fig. 1.3). Due to random sampling during transmission, the infecting viral sequence may or may not contain a mutated allele at the site. If a person is infected with a mutant virus (person A in Fig. 1.3), the virus gradually adapts and reverts to the wild type. During the adaptation process, when the proportions of mutant and wild type in a population are comparable, the site is very diverse. To find out if this model explains the fact successfully, the process will be studied as two parts: the evolution in an individual (Model 1), and the evolution with transmission between persons (Model 2). The output of Model 1 is used as the input for Model 2.

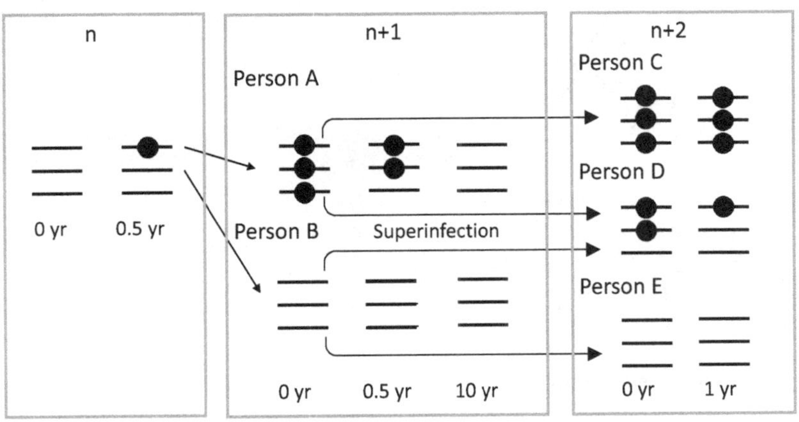

Fig. 1.3: Model of evolution of a nucleotide along the chain of infected individuals under purifying selection.
The numbers at the top denote cycles of transmission. Circles denote mutants. A mutant base appears spontaneously in person *n*, who infects person A with the mutant and person B with the wild-type variant. Person A passes the mutant down the chain to person C, after which his/her own virus population slowly reverts to the wild type. In person D, who is stably coinfected with a pair of sequences polymorphous at that base, selection clears the mutant virus rapidly. Based on Rouzine and Coffin (1999c).

Each variable site can have one of two variants, either A/C or G/T (Section 1.1). Variant fitness (reproduction number) is defined as the average number of cells infected by one infected cell. One of the two alleles is better-fit and called "wild type." A site will be characterized by three parameters: the relative fitness difference between the two alleles, termed "the selection coefficient" and denoted s, and the forward and reverse mutation rates, μ_r and μ_f. The estimates of the HIV mutation rate are $4 \cdot 10^{-5}$ for A → G and C → T substitutions, and 5–10 times lower for the reverse substitutions (Mansky and Temin 1995). The selection coefficient, s, varies strongly among different bases and will be treated as a fitting parameter. For the mathematical definition of parameters, see Section 1.1.8.

To complete the model, one has to add the initial conditions regarding the initial diversity of population. It is known that <1% of sexual contacts result in a systemic HIV infection (Gray et al. 2001; Wawer et al. 2005; Fraser et al. 2007), which implies a very small inoculum seeding the infection. The role of this bottleneck will be analyzed in detail in Section 3.2. Not surprisingly, the early virus population (days after infection) is usually genetically uniform, implying transmission of one, rarely two sequences. In this section, a single-sequence transmission is assumed. Multiple-sequence transmission will be considered below (Sections 1.1.4, 1.1.8.2, and 1.1.8.3).

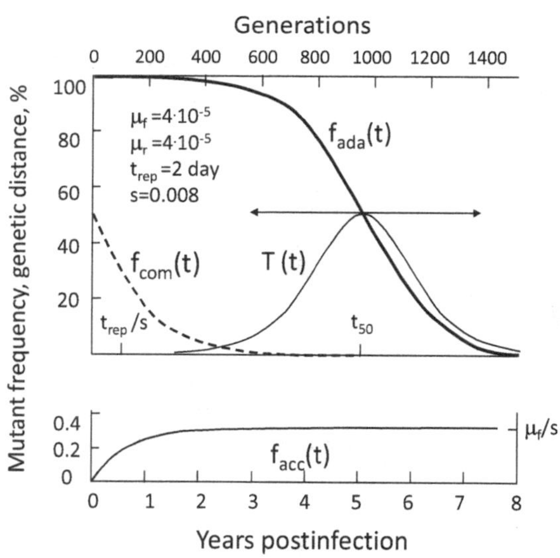

Fig. 1.4: Time dependence of the mutant frequency at different initial populations: purely mutant (adaptation, thick line in upper panel), purely wild type (accumulation, lower panel), or 50% mutant (growth competition, dashed line).
The thinner curve shows the intrapatient genetic distance $T(t)$ during adaptation. The horizontal bar is the time interval during which the base would be classified as variable (5% to 95%). Parameters are shown: the forward (wild type → mutant) and reverse mutation rates, μ_f and μ_r, corresponding to either A or C wild type; the replication cycle time, t_{rep}, and selection coefficient, s. The adaptation half-time, t_{50}, is given by the equation $t_{50} = (t_{rep}/s)\log(s/\mu_r)$. Based on Rouzine and Coffin (1999c).

Suppose, the infecting population is 100% mutant. Because persistent HIV infection involves many cycles of cell infection, and the population size is large (Section 1.2), wild-type variants can emerge frequently due to mutation and gradually grow until the population becomes almost entirely wild type (Fig. 1.4, upper panel). The end population will be in a steady-state population with a small proportion of mutants, due to the balance between selection and mutations (Wright 1931; Coffin 1995). The time to 50% composition denoted t_{50} is inversely proportional to s and directly proportional to the time per infection cycle (Section 1.1.8.2). The cycle time is approximately equal to the average productive period of an infected cell before it dies, which was estimated in the early drug-response studies to be about $t_{rep} \approx 2$ days (Ho et al. 1995; Wei et al. 1995; Perelson et al. 1996; Haase 1999). Within the time interval $t = t_{50} \pm 2/s$, where time t is measured in generations of infected cells, the population will be diverse above $T \sim 0.015$. Because an average untreated infection lasts 10 years, the mean sampling time is in the middle, $\langle t \rangle t_{rep} \approx 5$ years (Smith et al. 1997). The result of fitting in Section 1.1.8.1 for the case when A or C is wild type predicts the average selection coefficient, $s = 0.008$. For individuals sampled earlier or later, s will be higher or lower, respectively. When G or T is wild type, mutation rate is higher, and fitting yields $s = 0.006$. [After the original work by Rouzine and Coffin (1999c) came out, the infected cell generation interval was re-estimated to be shorter, $t_{rep} \sim 1$ day, which leads to $s = 0.003$.]

The model of virus adaptation naturally explains most experimental features discussed above:

(i) An average diverse site is very diverse, because it is in the process of adaptation.
(ii) Variable bases are rare in genome and vary across individuals, because they have small s, because an individual may or may not be infected with a mutant virus, and because most of patients are sampled outside of the narrow window near t_{50} (Fig. 1.4).
(iii) Synonymous and nonsynonymous diverse sites have approximately equal diversity, because they are preselected based on similarly small s.
(iv) But synonymous variable sites are more abundant (Tab. 1.1), because they do tend to have smaller s.

1.1.3 Model 2: chain of single-clone transmission

The remaining task is to explain (match) the frequency with which a variable site appears in individuals, 16% (Tab. 1.1). For this goal, the evolution along the chain of transmission must be included into the model (Fig. 1.3). Suppose that each person infects the next person at a fixed time t^* since his or her own infection. Infection occurs with a single virus sequence, which is sampled randomly and can be either mutant or wild type at the variable site of interest. Then, the frequency of variable sites in the individual number $n+1$ in the transmission chain is equal to the probability f_n^* that the virus transmitted from person number n is mutant.

As the derivation given in Section 1.1.8.2 predicts, the mutant probability f_n^* grows with every transmission n, until it saturates at a plateau value f_∞^*, which represents the transmission steady state (Fig. 1.5A). If the transmission time t^* is shorter than the adaptation half-time, $t^* < t_{50}$, the plateau value f_∞^* is much higher than the steady-state mutant frequency in an individual μ/s. If, in addition, the forward mutation rate is much higher than the reverse mutation rate, f_∞^* can be close to 100%.

Thus, if a single virus is transmitted, any site with a small selection coefficient will eventually become highly diverse. Qualitatively speaking, the strong accumulation of mutants along transmission chain is the combined effect of the stochastic transmission bottleneck, which makes the initial population genetically uniform at

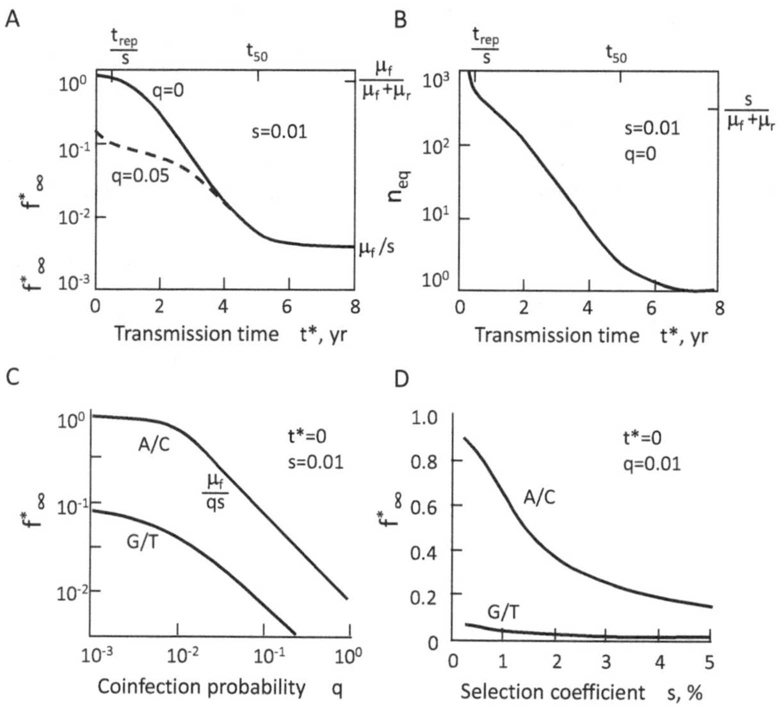

Fig. 1.5: Predictions of the chain transmission model. (A) Probability of a mutant base being in the inoculum at the chain steady state, f_∞^*, as a function of the transmission time, t^*. Solid line, strictly single-clone infection; dashed line, coinfection from two sources with probability $q = 0.05$. (B) Number of cycles required to reach the chain steady state, n_{eq}, as a function of transmission time, t^*. (C) Probability of a mutant base for a very short transmission time, $t^* \to 0$, versus the probability of dual infection, q. Upper and lower curves correspond to wild-type A/C and G/T, respectively. (D) The same probability versus s, for a fixed value $q = 0.01$. Parameters $t_{rep} = 2$ days; $\mu_f = 4 \cdot 10^{-5}$, $\mu_r = 4 \cdot 10^{-6}$ for wild-type A/C and vice versa for G/T. Based on Rouzine and Coffin (1999c).

the site of interest, and the short time between transmissions that does not give enough time for natural selection to act before the next transmission event.

The number of transmissions needed to reach the inoculum steady state and f_∞^* are linearly proportional to each other (Fig. 1.5A, B; Section 1.1.8.2). Even though the average transmission interval is short, $t^* t_{\text{rep}} < 1$ year, hundreds of transmission cycles are required to explain high $f_\infty^* \sim 0.1$. Thus, within this model, the accumulation of mutants has occurred long before the start of the HIV pandemic, which includes the human endemic and enzootic infection in the primate host.

1.1.4 Model 3: coinfection from independent sources

As the derivation in Section 1.1.8.3 demonstrates, above predictions critically depend on the assumption that each individual is always infected from a single source. Even very rare coinfections from independent sources can prevent the accumulation of mutant alleles and hence reduce the variable site density.

As mentioned above, the factor suppressing natural selection in single-clone transmission is that the initial virus is genetically uniform, and that it is passed quickly further down the chain (person C in Fig. 1.3). Suppose now, an individual is infected, within a short time and before further transmission, with two virus variants coming from two different infected individuals, one mutant and another wild type (patient D in Fig. 1.3). This super-infection scenario is supported by the observations in simians and humans during the first weeks postinfection (Zhang et al. 1993; Wyand et al. 1996). Then, both genetic variants are well-represented in the initial population, and natural selection between the two viruses starts to act immediately, without waiting for new mutations to occur [see curve $f_{\text{com}}(t)$ in Fig. 1.4]. As a result, mutant virus will get almost extinct in the host before transmitting further. Such a rare coinfection event can interrupt a long chain of mutant allele transmissions. Thus, even sparse coinfections can stop mutants from accumulating.

The dampening effect of rare coinfections on genetic diversity is especially strong for short t^*, i.e., exactly when the probability of mutant in inoculum f_∞^* is predicted to be very high (Fig. 1.5A). Even 5% of coinfections, $q = 0.05$, decrease the value of f_∞^* by an order of magnitude relative to pure single-virus transmission (Fig. 1.5C). Transmission of multiple virus variants from the same source, as expected, does not greatly affect the accumulation of mutants in the transmission chain (Section 1.1.8.4). To conclude this subsection, the topology of epidemiological routes (tree-like or with loops) is very important for virus evolution.

1.1.5 Probability of coinfection estimated for HIV

Thus, the predicted diversity in the protease gene critically depends on the coinfection probability q, which must be estimated from independent experimental data. Due to the difference in mutation rates, the probability of mutant base in inoculum f_∞^* is predicted to be much less for the sites with wild-type G or T than for A or C (Fig. 1.5C). If sampling time in patients were fixed, f_∞^* would be equal to the observed frequency in individuals, 16% (Tab. 1.1). Because the sampling time varies broadly between patients, and each base is expected to be diverse in a narrow time window $t = t_{50} \pm 2/s$, one expects f_∞^* to be larger than 16%.

The estimate of q is obtained in Section 1.1.8.5, under the assumptions, as follows.

The variation in virus replication time t_{rep} is relatively small and can be neglected (Ho et al. 1995). The time to AIDS is uniformly distributed between 2 and 14 years (Smith et al. 1997), and the test time is assumed to change between 1.3 and 8.7 years. Because the final estimate is robust to the distribution density of s, it is assumed to be uniform in the interval $s \in [0.005, 0.05]$ (a more realistic distribution will be presented in Section 2.4). Under these assumptions, the observed experimental frequency of variables sites in individuals, 16%, requires $q < 0.01$ (Section 1.1.8.5). This corresponds to $f_\infty^* \sim 1$ for wild-type A or C and $f_\infty^* \sim 0.1$ for wild-type G or T (Fig. 1.5C).

Thus, the effect of even rare coinfections on the predicted frequency of variable bases is very strong: less than 1% of coinfections from independent sources should occur to explain the observed virus diversity. Only 10% of coinfections would decrease the predicted number variable sites by the factor of 10 (cf. Fig. 1.5C). In fact, in the countries where no single HIV subtype dominates, coinfections with different HIV subtypes occur in 10 to 15% of cases (Pfutzner et al. 1992; Pieniazek et al. 1995), not counting the cases of intersubtype recombination, where coinfection took place in the past (Groenink et al. 1992; Sabino et al. 1994). [The effect of superinfection protection in SIV-infected animals is observed, but is only partial (Wyand et al. 1996).] Assuming that the two subtypes are present in a population in equal quantities, one obtains $q = 0.2 - 0.3$ or even more, which is 20 to 30 times higher than the predicted ceiling $q < 0.01$. The coinfection rate observed in high-risk groups is much higher than 1%, which buries the entire idea of deleterious alleles accumulating due to transmission bottleneck. Another explanation is due.

1.1.6 Model 4: individual variation in wild type due to MHC subtypes

The probable explanation is the existence of the immune response and antigenic evolution (Sections 4.1 and 4.2). At the time the original work was produced, the role of the immune response in HIV-infected individuals was hotly debated between immunologists and virologists. Rouzine and Coffin (1999c) took the side of immunologists and assumed that that the cytotoxic T-cell (CTL) response contributes substantially to virus control. Right after the work has been published, this view has been confirmed in direct experiments on CD8 T-cell depletion (Kuroda et al. 1999; Schmitz et al. 1999; Wong et al. 2010).

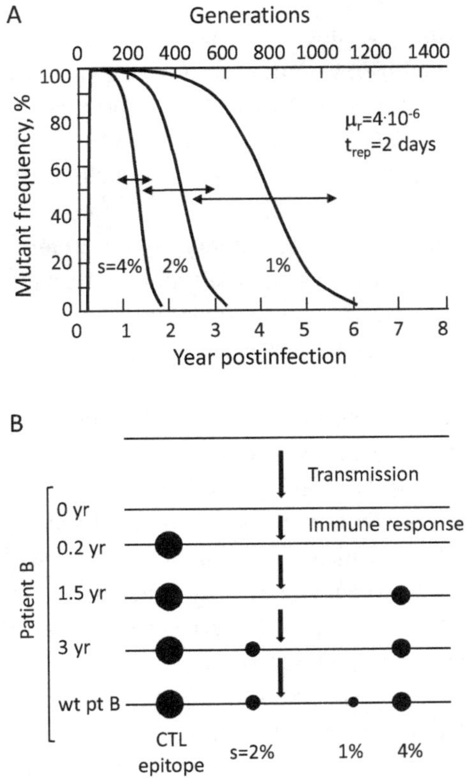

Fig. 1.6: A working model of evolution in *pro*. A quick initial switch of dominant epitopes redefines wild type for a number of sites, which then evolve towards the new wild type at speeds inversely proportional to their selection coefficients. (A) Time dependence of mutant frequency at three sites linked to an epitope. The three selection coefficients after the epitope switch are shown near corresponding curves. (B) Evolution of an individual consensus sequence at the three sites. pt, patient. Based on Rouzine and Coffin (1999c).

HIV-specific lymphocytes, including CD8 T cells and CD4 T cells, recognize short stretches of 8–9 amino acids (epitopes), located everywhere in the genome (Murphy 2011). By analogy with the well-studied viruses, such as LCMV and influenza, Rouzine and Coffin (1999c) assumed that protease rapidly accumulates antigenic escape mutations within the CTL epitopes, abrogating their recognition by the virus, the immune memory cells ensuring that these changes become permanent. Once an antigenic escape mutant is fixed, the best-fit configurations for a number of other bases are defined due to epistatic interactions (Chapter 2). These sites now become effectively 100% mutant and start to adapt toward the new wild type, as described by Model 1, under a constant selection. Some of the substitutions are synonymous due to natural selection acting on the mRNA or transcription levels. About 10–15 years later, the model was confirmed by detailed observations of the antigenic escape in CTL epitopes (Goonetilleke et al. 2009; Ganusov et al. 2011; Song et al. 2012; Liu et al. 2013; Batorsky et al. 2014). The dynamics of such escape is considered in Section 4.2.

The location of CTL epitopes in genome are determined by individual immunogenetics, specifically, by the six HLA subtypes (Murphy 2011). This fact explains why the overlap of variable sites among patients is only partial. Indeed, it occurs among patients that share some (but not all) of HLA subtypes. An allele, which is wild type for one person (the site does not fall within an epitope), is mutant for another (the site falls within an epitope). In such a model, the frequency of variable sites reflects the degree of the HLA subtype overlap and the frequency of epitopes for each HLA subtype. In total, 20–30 epitopes are active in each patient (Goonetilleke et al. 2009; Ganusov et al. 2011; Song et al. 2012; Liu et al. 2013; Batorsky et al. 2014).

Most antigenic escape mutations in epitopes are fixed rapidly, within several months after infection. These fixed mutations change the sign of selection coefficients for many compensatory sites outside of epitopes, where adaptation occurs with different rates (Fig. 1.6). Fixation of compensatory sites with a larger degree of compensation (larger s) occurs sooner, and those with a smaller s later. During the period of months to years, the average s of variable sites decreases in time, slowing down the average tempo of evolution (Section 2.4). A gradual shift from nonsynonymous to synonymous variable sites take places, because the second tend to have smaller s. At the same time, the average intrapatient distance of any variable site stays at the same high value, as in Tab. 1.1. The envelope gene of HIV, which is under antibody response, shows the opposite effect: the synonymous-to-nonsynonymous ratio is decreasing at early stages of infection (Burns and Desrosiers 1991; Pang et al. 1991). For the detailed dynamics of antigenic escape from antibody response, see Section 4.1.

The proposed model of evolution of protease as an afterglow of the immune escape is indirectly corroborated by compensatory mutations emerging after the fixation of primary mutations conferring resistance to drug treatment. After 32 to 60 weeks of treatment by indinavir, a protease inhibitor, Condra et al. (1996) observed that individual's consensus of protease changed at 10.4 sites per patient, of which a quarter were synonymous. The higher fraction of synonymous variable sites compared to the drug-naive patients (66%, Tab. 1.1) is due to the difference in the average testing time, 1 year and 5 years post-infection. Most nonsynonymous sites in the drug-naive patients have already finished adaptation by the testing time.

1.1.7 Discussion

Thus, the distribution of alleles, in a relatively conserved gene of HIV, between and within patients, is shown to be consistent with the reversion of deleterious mutations with small selection coefficients $s \sim 0.005$. The origin of these deleterious alleles was investigated using a family of models. The initial idea was that these deleterious alleles as originating from stochastic bottlenecks of transmission is shown to be valid, at least, in theory. This mechanism would be similar to the accumulation of deleterious mutations in vesicular stomatitis virus passaged in culture (Steinhauer et al. 1989; Duarte et al. 1992; Clarke et al. 1993; Elena et al. 1996). [Another explanation of these laboratory experiments is Muller's ratchet effect (Duarte et al. 1992), in which the accumulation of deleterious alleles is faster than in the one-locus model due to linkage effects (Muller 1932; Fisher 1958; Felsenstein 1974), see Chapter II in Rouzine (2020b)].

However, the transmission bottleneck model fails to explain the observed overlap of variable sites between patients, when realistic amounts of coinfection from epidemiologically distant sources are included. Hence, the transmission bottleneck model was rejected, and the origin of the deleterious alleles is attributed to mutations in CD8 T-cell epitopes. In turn, epitope mutations flip the signs of selection coefficients of the sites that compensate for the fitness cost of the epitope mutations and located outside of epitopes. This model "antigenic escape and epistasis" naturally explains the both the difference and overlap in diverse sites between individuals and is, importantly, confirmed by later experiments and modeling (Goonetilleke et al. 2009; Ganusov et al. 2011; Song et al. 2012; Liu et al. 2013; Batorsky et al. 2014). Various approximations of models are discussed in Section 1.1.8.6.

To summarize, the distribution of polymorphisms of the HIV protease gene across sites and individuals is explained by rare sites adapting under constant selection. Most of these sites compensate for the cost of the early "antigenic escape" mutations decreasing the virus recognition by the immune response.

1.1.8 Mathematical derivations

The remaining subsections serve as a mathematical appendix.

1.1.8.1 Virus evolution in an individual

Deterministic equations for the two-allele model have the form of ODE

$$\frac{dn_1}{dt} = (1-\mu_r)\kappa_1 n_1 + \mu_f \kappa_2 n_2 - \frac{n_1}{t_{\text{rep}}} \tag{1.2}$$

$$\frac{dn_2}{dt} = (1-\mu_f)\kappa_2 n_2 + \mu_r \kappa_1 n_1 - \frac{n_2}{t_{\text{rep}}} \tag{1.3}$$

Here n_1 and n_2 are the numbers of less-fit (mutant) and better-fit (wild-type) alleles, respectively; κ_1 and κ_2 are their replication coefficients; t_{rep} is the replication cycle time; μ_f and μ_r are the forward and reverse mutation probabilities (rates), respectively. Average time of cell death (replication cycle) t_{rep} is assumed to be the same for the two variants; cell death is a random Poisson process (for the long-term evolution under consideration, neither assumption is important). By the definition, the selection coefficient s is the relative fitness difference, $s = (\kappa_2/\kappa_1) - 1$, which is constant and meets the double inequality $\mu_{f(r)} \ll s \ll 1$. The ratio κ_2/κ_1 is the standard definition of the relative fitness measured experimentally from the exponential growth rates.

Equations (1.2) and (1.3) can be reduced to a single equation by introducing the mutant share in population (mutant frequency) $f = n_1/(n_1+n_2)$. Differentiating f in time with the use of eqs. (1.2) and (1.3), one gets

$$\frac{df}{dt'} = -sf(1-f) + \mu_f(1-f) - \mu_r f + O(\mu s) \tag{1.4}$$

where $O(\mu s)$ is a second-order correction to be neglected, and $t' \equiv \kappa_1 t$ is dimensionless time in units of replication cycles, $1/\kappa_1 = t_{\mathrm{rep}}$. Below, notation t' is replaced with t keeping these units in mind. Solving eq. (1.4) for the general initial condition $f(0) = f_0$ yields

$$f(t|f_0) = \tilde{\mu}_f + \frac{f_0\left(1+\tilde{\mu}_f+\tilde{\mu}_r\right) - \tilde{\mu}_f}{f_0 - \tilde{\mu}_f + (1-f_0+\tilde{\mu}_r)e^{-st}} + O\left[(\tilde{\mu}_{f(r)}/s)^2\right] \tag{1.5}$$

with the new notation $\tilde{\mu}_f \equiv \mu_f/s$, $\tilde{\mu}_r \equiv \mu_r/s$.

The cases of mutant reversion (virus adaptation) and accumulation of mutants are given by initial conditions $f_{\mathrm{rev}}(t) = f(t|1)$ and $f_{\mathrm{acc}}(t) = f(t|0)$. For a realistic set of parameters, the two functions are plotted in Fig. 1.4. The adaptation half-time, t_{50}, is defined by the condition $f_{\mathrm{rev}}(t_{50}) = 1/2$. Using eq. (1.5) with $f_0 = 1$, one gets $t_{50} = \left(t_{\mathrm{rep}}/s\right)\log(s/\mu_r)$. In the limit of large t, such that $t - t_{50} \gg t_{\mathrm{rep}}/s$, the mutant frequency $f(t)$ converges to the steady-state value $f_{\mathrm{ss}} = \tilde{\mu}_f = \mu_f/s \ll 1$.

1.1.8.2 Chain of single-clone transmission

In Model 2, each infected individual infects the next person in the transmission chain with a single virus variant at time t^* after his own infection. The probability that mutant variant infects individual n is denoted f_n^*. The average value of the mutant frequency in person n at time t is given by the branching process expression

$$f_n(t) = f_n^* f_{\mathrm{rev}}(t) + \left(1-f_n^*\right) f_{\mathrm{acc}}(t) \tag{1.6}$$

For self-consistency, f_n^* must be equal to the average mutant frequency in the source individual at time t^* post-infection

$$f_n^* = f_{n-1}(t^*) \tag{1.7}$$

Together, eqs. (1.6) and (1.7) fully describe the evolution along the transmission chain. The solution depends on the relation between t^* and t_{50}, as follows. If transmission occurs sufficiently late, so that $t^* - t_{50} \gg t_{\mathrm{rep}}/s$, the probability f_n^* can be shown to converge, after several transmissions, to the small steady-state value in a host, f_{ss}. However, if transmission occurs prior to the adaptation time, so that $t^* < t_{50}$ & $|t^* - t_{50}| \gg t_{rep}/s$, the probability f_n^* changes slowly with n, and eqs. (1.6) and (1.7) can be simplified to

$$\frac{df_n^*}{dn} = -\frac{f_n^* - f_\infty^*}{n_{eq}} \tag{1.8}$$

where

$$f_\infty^* \equiv \frac{f_{acc}(t^*)}{1 - f_{rev}(t^*) + f_{acc}(t^*)} \tag{1.9}$$

$$n_{eq} \equiv \frac{1}{1 - f_{rev}(t^*) + f_{acc}(t^*)} \tag{1.10}$$

Solving eq. (1.8) at the general initial condition $f^*(0) = f_0^*$, one gets

$$f_n^* = f_\infty^* \left(1 - e^{-\frac{n}{n_{eq}}}\right) + f_0^* e^{-\frac{n}{n_{eq}}} \tag{1.11}$$

Equation (1.11) demonstrates that the probability of a mutant base converges, on the scale of n_{eq} transmission events, to the "inoculum steady-state value" f_∞^* given by eq. (1.9). The dependence of both f_∞^* and n_{eq} on the transmission time t^* is exponential (Fig. 1.5A and B).

In the case of moderately short transmission times, such that $t_{rep}/s \ll t^* < t_{50}$ & $|t^* - t_{50}| \gg t_{rep}/s$, the steady-state values for the inoculum, f_∞^*, and in the host, f_{ss}, are simply related

$$f_\infty^* \approx n_{eq} f_{ss} = \frac{n_{eq}\mu_f}{s}$$

$$f_\infty^* = \frac{\mu_f}{\mu_f + \mu_r}$$

Although $f_{ss} = \mu_f/s$ is usually very small, the value n_{eq} in eq. (1.10) is very large due to $f_{rev}(t^*) \approx 1$, so that f_∞^* can be on the order of 1.

1.1.8.3 Coinfection from independent sources

Suppose now, that a pair of virus genomes can be transmitted from two different persons. The probability of this event is denoted q. The probability of single-genome transmission is $1 - q$. Then, eq. (1.6) is replaced by a more complex expression

$$f_n(t) = \left[(1-q)f_n^* + q(f_n^*)^2\right] f_{rev}(t) + \left[(1-q)(1-f_n^*) + q(1-f_n^*)^2\right] f_{acc}(t) + 2qf_n^*(1-f_n^*)f_{com}(t) \tag{1.12}$$

where $f_{com}(t) \equiv f(t|1/2)$ in eq. (1.5) is the mutant frequency in the growth competition of two strains (Fig. 1.4). The value of f_n^* meets the same consistency eq. (1.7). In eq. (1.12), the two infection sources are assumed to be statistically independent (epidemiologically distant). The inoculum steady-state value f_∞^* is found from the quadratic equation

$$q(f_{acc} + f_{rev} - 2f_{com})(f_\infty^*)^2 - \left[(1+q)f_{acc} + (1-q)(1-f_{rev}) + 2q\left(\frac{1}{2} - f_{com}\right)\right] f_\infty^* + f_{acc} = 0 \quad (1.13)$$

where f_{acc}, f_{rev}, and f_{com} are all evaluated at $t = t^*$, and the smaller of the two roots is chosen, because the larger one is above 1.

The value f_∞^* as a function of transmission time t^* is plotted at the fixed coinfection probability $q = 0.05$ in Fig. 1.5A. The value f_∞^* is plotted as a function of q in the limit $t^* \to 0$ in Fig. 1.5C. At not too-small values of q, such that $\mu_f + \mu_r \ll q$, one has

$$f_\infty^*(t^* = 0) \approx \frac{2\mu_f}{qs}$$

At sufficiently small coinfection probability, $q \sim 1/n_{eq}$, this expression matches the formula $f_\infty^* = n_{eq} f_{ss}$ obtained above for a single-clone transmission.

This derivation is based on the assumption that the virus population after the coinfection consists of equal amounts of the two variants. The original work by Rouzine and Coffin (1999c) also analyzed the case of a random initial composition uniformly distributed between 0 and 100% and found out that this change of the model does not perturb f_∞^* at $q \to 0$. At moderate values of q, its effect is to increase f_∞^* by the factor of 1.5 (the upper curve in Fig. 1.5C moves up), which does not change any conclusions.

1.1.8.4 Coinfection from the same source

Above, coinfection occurred from two statistically independent sources. Let is consider the opposite case and assume that, if two virus variants are transmitted, they come from the same source. (Obviously, if the two sources are different but epidemiologically close, the situation is essentially the same.) In this case, eq. (1.12) is replaced with two separate equations for the average values of the mutant frequency, $f_n(t)$ and of its square, $y_n(t) = \langle [f_n(t)]^2 \rangle$ that have a form

$$\begin{aligned} f_n(t) &= \left[(1-q)f_n^* + q\, y_n^*\right] f_{rev}(t) + \left[(1-q)(1-f_n^*) + q(1 - 2f_n^* + y_n^*)\right] f_{acc}(t) \\ &\quad + 2q(f_n^* - y_n^*) f_{com}(t) \\ y_n(t) &= \left[(1-q)f_n^* + q\, y_n^*\right] f_{rev}^2(t) + \left[(1-q)(1-f_n^*) + q(1 - 2f_n^* + y_n^*)\right] f_{acc}^2(t) \\ &\quad + 2q(f_n^* - y_n^*) f_{com}^2(t) \end{aligned} \quad (1.14)$$

where both f_n^* and y_n^* are defined by the same consistency condition eq. (1.7). The inoculum steady-state values f_∞^* and y_∞^* are calculated from the steady-state conditions

$f_n^* = f_{n-1}^* = f_\infty^*$ and $y_n^* = y_{n-1}^* = y_\infty^*$. In the most important case $t^* \to 0$, in which f_∞^* is the largest is maximal, the dependence on q cancels out, and one arrives at the same result as in the case of single-clone transmission

$$f_\infty^* = y_\infty^* = \frac{\mu_f}{\mu_f + \mu_r}, \quad t^* \to 0 \tag{1.15}$$

1.1.8.5 Estimation of the value of q for HIV

The average observable mutant frequency denoted f_∞^{exp} is given by

$$f_\infty^{\text{exp}} = \frac{\int ds\, \varphi(s) f_\infty^*(s) g(s)}{\int ds\, \varphi(s)} \tag{1.16}$$

where $\varphi(s)$ is the distribution density of selection coefficients, and $f_\infty^*(s)$ was derived above (Fig. 1.5C and D). The limits of the integrals in s not shown in eq. (1.16) are the boundaries of the interval in s where the site is variable and will be specified below. Finally, the factor $g(s)$ in eq. (1.16) is the fraction of individuals tested within the time interval in which a site with selection coefficient s is observably variable (Fig. 1.4)

$$g(s) = \int_{t_{\text{rep}}(L-\alpha)/s}^{t_{\text{rep}}(L+\alpha)/s} dt\, \phi(t) \tag{1.17}$$

where $\phi(t)$ is the distribution density of the testing time among individuals, $L \equiv \log\left(\frac{s}{\mu_r}\right) = 7.8$ for $s \sim 0,01$, and $\alpha \sim 1$ depends on the interval in f considered "variable." At an average sample size of 20 genome sequences in Tab. 1.1, a site with $0.05 < f < 0.95$ is observably diverse, which yields $\alpha \approx 3$; note that the dependence of sample size is slow (logarithmic). Assuming that testing time distribution $\phi(t)$ is uniform in the interval $t_{\text{rep}} t =$ [1.25, 8.75] years, and using $L \gg 1$, eq. (1.17) yields

$$g(s) \approx \begin{bmatrix} 0.53\dfrac{\bar{s}}{s} & \text{if } 0.57\bar{s} < s < 4\bar{s} \\ 0 & \text{otherwise} \end{bmatrix} \tag{1.18}$$

Here $\bar{s}$ is defined by the equation

$$\frac{t_{\text{rep}} L}{\bar{s}} = \bar{t}\ t_{\text{rep}} = 5\,\text{years}$$

The interval in s in eq. (1.18) determines the integration limit in eq. (1.16). Approximating the distribution density $\varphi(s)$ by a constant in this interval of s, and substituting $f_\infty^*(s)$ from eq. (1.13), one obtains that eq. (1.16) fits the observed value of the variable site fraction in an individual, $f_\infty^{\text{exp}} = 0.16$ (Tab. 1.1), at $q = 0.0085$. [Here, the wild-type allele is assumed to be A or C, because they are predicted to dominate diversity (Fig. 1.5C).]

1.1.8.6 Approximations

Ignoring transversions: A double transversion may resemble a transition and increase with the transition number count. Because single transversions contribute 25% to the net diversity, this error is on the order of $0.25^2 = 0.0625$.

Neglecting stochastic effects in steady state: Stochastic effects include random mutation times and random genetic drift. The linkage disequilibrium test in Section 1.2 demonstrates that the stochastic effects on the evolution of separate sites are relatively small, because the population size of infected cells is very large (Rouzine and Coffin 1999b).

The same wild-type sequence for all cells and individuals in the absence of the immune response (Models 2 and 3): HIV adapts to new cell types and changes, such as naive CD4 T cells, while replicating in memory CD4 T cells. However, a change comparable to the 16% in Tab. 1.1 can be caused only by a very sharp change of selection conditions: (i) anti-viral drug, protease inhibitor (Condra et al. 1996), and (ii) the immune response in a host (or a vaccine), hence the choice of Model 4.

"Well-stirred pot" approximation: Virus population has been approximated by a well-mixed pot of different alleles. In a real host, HIV is distributed in a complex way between and within different lymphoid organs, and within these organs, infected cells are nonuniformly distributed into islands (Haase et al. 1996; Reinhart et al. 1998; Zhang et al. 1998; Haase 1999). At the same time, visualization of HIV variants in spleen by selective labeling shows that most of these islands are shared by different variants implying good local mixing (Reinhart et al. 1998). Also, rapid dynamics of virions (Ho et al. 1995; Moore et al. 1996; Perelson et al. 1996) implies that a significant part of new infections is due to far-travelling virus particles (Rouzine and Coffin 1999a).

Independent sources of coinfection: Model 3 assumed that typical sources of coinfection are epidemiologically far apart and hence statistically independent. Whether two typical individuals in a population are epidemiologically close or distant depends on the coinfection frequency q and on the population size. Coinfections randomize the distribution of virus variants between hosts and make them more statistically independent. If q is much less than 1, a mutant allele will be passed in a long transmission chain without reverting. Then, two hosts sampled from a population are more likely to be correlated. If, however, $q = 0.1 - 0.2$, as actually observed in multi-clade countries, two typical animals are statistically independent even in a small population. Therefore, epidemiological proximity does not help to explain the high mutant frequency in individuals Tab. 1.1; one still must assume that q is very small.

Absence of group selection: Miralles et al. (1997) discussed the possibility of group selection, as opposed to the selection at the level of individual genomes. However, there is no evidence for group selection in RNA viruses.

Neglecting selection for diversity: In the protease gene in late chronic infection, synonymous and chemically conservative substitutions dominate the nonsynonymous nonconservative changes by the factor of four compared to the random ratio. This why selection for diversity is neglected. The early infection when most of epitope mutations are fixed due to the rising immune response is analyzed in Section 4.2.

Neglecting linkage and recombination: All the models above assumed that evolution of separate loci is independent. In fact, this is not the case, because they are inherited as a set. In the 1990s when the original work (Rouzine and Coffin 1999c) was published, the mathematical techniques for including multi-locus linkage did not exist as yet. Such technique was developed later (Rouzine et al. 2003), which triggered a series of work by different groups and created a new subfield of population genetics [see for review (Rouzine 2020b)]. Linkage effects are considered in Chapter 2.

Neglecting stochastic effects: The above models are deterministic, although stochastic effects are created by the randomness of mutation and random genetic drift. that. The classical one-locus theory demonstrates that the impact of stochastic effects depends on the population size and its genetic diversity [see for review Rouzine et al. (2001)]. The population size of HIV and the validity of deterministic approximation are analyzed in Section 1.2.

1.2 Estimate of the effective population size

Random mutation, natural selection, and random genetic drift due to progeny number variation (neglected in Section 1.1) represent the main actors in a one-locus model of evolution (Fig. 1.7A). As it is well known in population genetics, see reviews in Rouzine et al. (2001) or Rouzine (2022), the relative importance of these factors varies between three broad intervals of the population size (number of individuals in a population), N, with the approximate boundaries given by the inverse selection coefficient $1/s$ and the inverse mutation rate $1/\mu$, as follows:

(i) In the largest populations, $N \gg 1/\mu$, mutation and selection dominate, and random genetic drift can be neglected. Because many mutation events are averaged out, mutation acts as a deterministic factor, and the entire dynamics is almost deterministic, with some small fluctuations. This scenario was considered in Section 1.1.

(ii) In the smallest populations, $N \ll 1/s$, natural selection can be neglected and stochastic neutral theory applies (Kimura 1994). Diverse populations evolve due to random genetic drift. Mutation events are very rare, random events important only in genetically uniform populations. Their role is to create new alleles.

(iii) The broad intermediate interval $1/s \ll N \ll 1/\mu$ has mixed properties depending on its diversity: random drift controls weakly diverse populations, more diverse populations are controlled by selection, and uniform populations stay uniform until a rare mutation event (Rouzine et al. 2001).

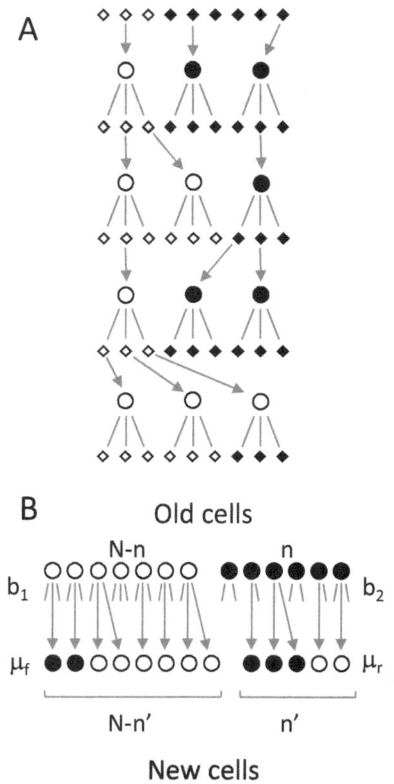

Fig. 1.7: The three factors of evolution of the stochastic one-locus two-allele model.
(A) Random drift of genetic composition because of sampling of infecting virions. Circles denote productively infected cells; small diamonds represent free virus particles. Two genetic variants of the virus are shown as black or white. (B) A virus population model including the factors of evolution: random drift, selection, and mutation. Two consecutive generations of infected cells are shown. Lines radiating from circles denote virions produced by infected cells, some of which (shown by arrows) infect new cells. A cell infected with mutant virus (black circle) leaves fewer infectious progeny than the wild type (white circle). Based on Rouzine and Coffin (1999b).

Whether the genetic diversity of HIV in an average untreated individual is controlled by Darwinian forces, i.e., random mutation and natural selection, or stochastic effects, was a topic of hot debates back in mid-1990s. Although the population size, defined as the number of virus-producing infected cells, was measured in tissue to be $N = 10^7–10^8$ virus-producing cells (Haase 1999) and, hence, much greater than the inverse mutation rate, $\mu = 0.4\ 10^{-5}$ to $4\ 10^{-5}$ per base (Mansky and Temin 1995), which is consistent with deterministic evolution, there could be quite a few scenarios in which the effective size of HIV population N_{eff} is much smaller than the total size N (Leigh-Brown 1997). For example, not all RNA-producing cells may produce virus that can reach other target cells fast enough (Fig. 1.8). Because of this complication, the estimates of N_{eff} obtained by different methods varied broadly between 100–1000 (Leigh-Brown 1997) to 10^8 (Coffin 1995), and this debate lasted for two decades.

The challenge with estimating effective population size N_{eff} is that the method must be robust to the assumptions of the model. The number of possible models, some of which are discussed in this book, is infinite. Therefore, Rouzine and Coffin (1999b) followed in this section decided to find a striking effect that would reveal a low population size regardless of the assumptions. The effect of clonal interference was predicted by Fisher (1930) and Muller (1932) who demonstrated mathematically that, in a small

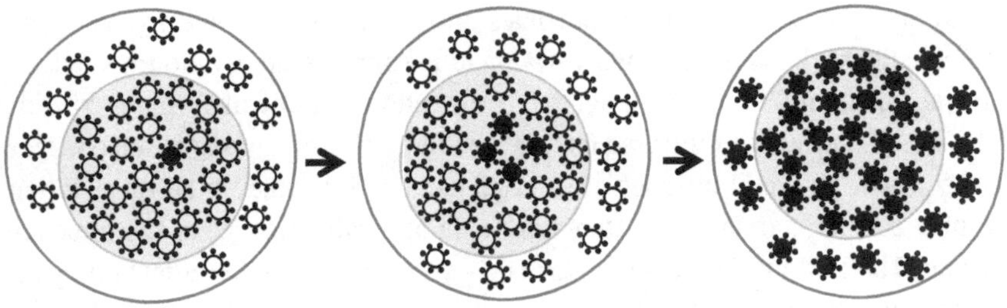

Fig. 1.8: The effective population size may be much smaller than the census population size. Beneficial viral mutants (black) arise in the effective virus subpopulation (N_{eff}, gray circle) and spread gradually to the entire census population (white circle).

population, beneficial mutations are fixed in a population one at a time, due to mutual interference. Sexual reproduction and recombination were thus proposed to emerge in the course of the evolution of species to counteract clonal interference and thus accelerate adaptation (Fisher 1930; Muller 1932; Maynard Smith 1971; Felsenstein 1974). This important prediction was supported by later studies, which calculated the speed of adaptation in asexual and sexual populations using two-locus (Otto and Barton 1997; Gerrish and Lenski 1998) and multi-locus models (Rouzine et al. 2003; Rouzine and Coffin 2005; Desai and Fisher 2007; Rouzine and Coffin 2007; Hallatschek 2011; Neher et al. 2013; Weissman and Hallatschek 2014). At one time point, this effect can be observed as strong "linkage disequilibrium" (LD) between pairs of different loci: the frequencies of the four genetic variants at two loci (haplotypes) are not multiplicative over the one-locus frequencies. In a small population, alleles do not segregate independently, because selection at different loci does not act independently (Lewontin 1964; Hill and Robertson 1968).

Below the one-locus model (Section 1.1) is generalized to include stochastic effects, and the basic stochastic evolution theory is illustrated by Monte-Carlo simulation. Then, the stochastic model is expanded to a two-locus scenario. An LD test based on the least-represented haplotype is applied to a published database of HIV sequences for the protease gene (Lech et al. 1996) and the envelope gene (Holmes et al. 1992). In the end, the analysis demonstrates an effective population size of 10^5 infected cells or more, which implies the quasi-deterministic regime. That the linkage equilibrium test is robust to model details is confirmed by repeating it for a selectively neutral model, a model with recombination, and a model with epistasis. The estimate of Rouzine and Coffin (1999b) was later confirmed by other methods (Frost et al. 2000; Maldarelli et al. 2013; Pennings et al. 2014).

1.2.1 One-locus model of stochastic evolution

The initial model is the same as in Sections 1.1, but now it includes the stochastic factor of genetic drift, as follows. A virus population is represented by N productively infected cells (Fig. 1.7B) infected by either the wild-type virus or the mutant virus. Each cell produces a number of infectious virus particles and then dies. The average progeny number differs slightly between the two variants (alleles) by relative amount s, which creates natural selection (Fig. 1.7B). To account for random genetic drift, the N virions infecting the next generation of infected cells is randomly chosen among a much larger of produced virions. The total number of infected cells is fixed and equal to N. When infecting a new cell, a genome can mutate with a small probability, μ, to the opposite genetic variant. The HIV mutation rate is $\mu = (0.4–4)10^{-5}$ per base per infection cycle, depending on the nucleotide pair and the direction of the substitution (Mansky and Temin 1995). Straightforward generalization of the same model for two loci and four variants (haplotypes) is given in Section 1.2.3.

The frequency of less-fit allele in the population changes slightly between consecutive generations due to the combined effect of (*i*) natural selection, (*ii*) random drift due to the random choice of infecting virions, and (*iii*) random mutation. Due to the last two factors, the evolutionary dynamics of the system, in general, cannot be described by deterministic ODEs, eqs. (1.2) and (1.3). Hence, the evolutionary dynamics cannot be predicted with certainty, even if one knows the initial conditions with infinite accuracy. Yet, it is possible and useful to calculate the probability of finding the mutant frequency within an interval of values $[f, f + df]$ at a given time t. The simplest way to study evolution is to simulate the process described above using a pseudorandom generator (Monte-Carlo simulation). A more general, analytic technique is the Kolmogorov equation (Fisher 1958; Kimura 1994) representing a particular case of Markov process. Kolmogorov equation is also a case of Fokker-Plank equation used in statistical physics. It can be solved in different particular situations and tested by Monte-Carlo simulation [see Rouzine et al. (2001) for review]. This section is based on computer simulation.

1.2.2 Three regimes of evolution

As already mentioned in introduction, the leading factors of evolution differ between three broad intervals of population size N, with boundaries $1/s$ and $1/\mu$ (Rouzine et al. 2001). For example, selection is negligible if $N \ll 1/s$ (the neutral limit). Random genetic drift and the randomness of mutation times are small corrections if $N \gg 1/\mu \sim 10^5$ (the deterministic limit). Between the two limits, there is a broad interval of population sizes, $1/s \ll N \ll 1/\mu$, with mixed properties. In this regime, random drift dominates the dynamics of weakly polymorphous populations, whereas highly polymorphous populations are quasi-deterministic, with natural selection as the dominant force. The characteristic copy number called “stochastic threshold” is, again, $1/s$.

Given initial conditions, a typical stochastic trajectory of the mutant frequency can be simulated following the rules of the model specified above. Eventually, the population arrives at a dynamic steady state where its statistical properties no longer change in time. The way this transition occurs depends on the interval of N. Let us consider three kinds of initial conditions: (*i*) 100% wild type, (*ii*) 100% mutant, and (*iii*) 50%–50%. The respective *gedanken* experiments defined by these initial conditions are the accumulation of deleterious mutants, adaptation (fixation of an advantageous allele), and the growth competition experiment.

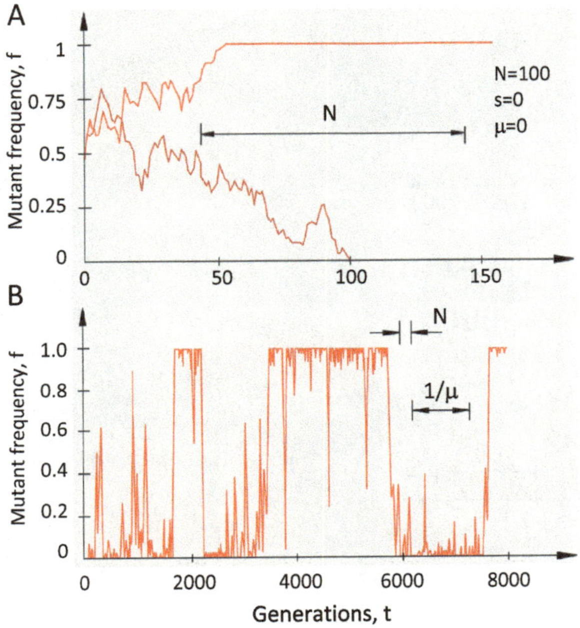

Fig. 1.9: Time dependence of mutant frequency in the neutral limit, $N \ll 1/s$, obtained by Monte-Carlo simulation.
(A) Two representative Monte-Carlo runs in the growth competition experiment, $f(0) = 1/2$, and (B) the accumulation/adaptation experiment on a long-time scale, $f(0) = 0$. The latter time dependence was obtained at $\mu = 10^{-3}$ and $N = 50$ and then rescaled along the horizontal axis to correspond to the values of μ and N shown. Based on Rouzine and Coffin (1999b).

In the deterministic limit, $N \gg 1/\mu$, the dynamics in the three "experiments" can be calculated from the deterministic eqs. (1.2) and (1.3) (Fig. 1.4). The steady-state mutant frequency is equal to μ/s. The dynamics of the accumulation of mutants and the steady-state are confirmed by direct simulation in this regime (Fig. 1.10C). The time of the transition to the steady-state is given by $1/s$ for the accumulation and competition. Adaptation takes a longer time, $t_{50} = (1/s)\log(s/\mu) \gg 1$.

In the opposite neutral limit $N \ll 1/s$, when selection is not important, the growth competition looks like a ragged curve that depends strongly on a realization (random run) (Fig. 1.9A). With an equal probability, one competitor is driven to extinction due to random drift at an average time scale $t \sim N$. Then, mutation brings the extinct allele back to the population, but then becomes extinct again. After many attempts, the allele succeeds in taking over the entire population, and the other allele becomes extinct, and so on. The resulting time dependence switches back and forth between the two monomorphic states at random moments of time (Fig. 1.9B).

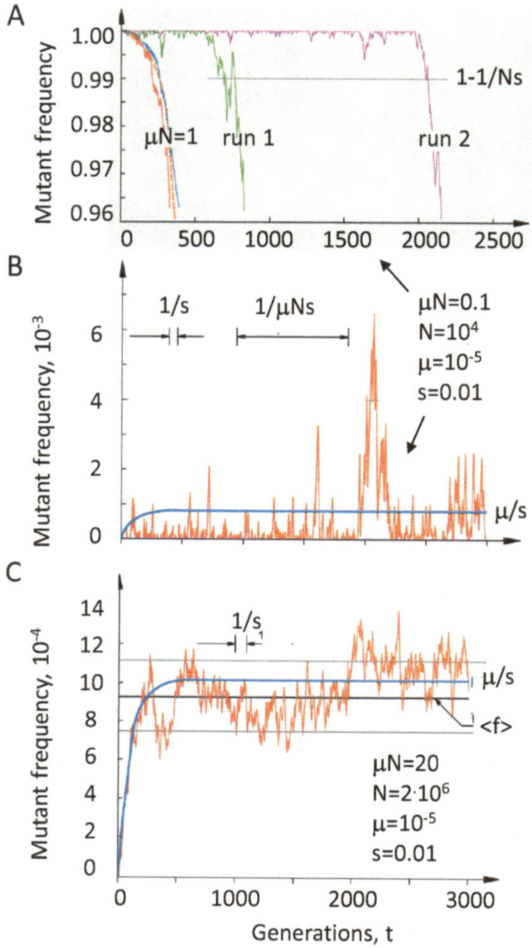

Fig. 1.10: Simulated dependence of the mutant frequency on time at population sizes $N \gg 1/s$. (A) The beginning part of the adaptation experiment in the selection–drift regime ($1/s \ll N \ll 1/\mu$). Two representative Monte-Carlo runs are shown (purple and green curves). Two runs at $N = 1/\mu$ are shown for comparison (blue and red curves). They are very close the deterministic prediction (dashed curve). (B) The accumulation experiment in the selection–drift regime. (C) The accumulation experiment in the quasi-deterministic regime, $N \gg 1/\mu$. (A-C) Smooth curves correspond to the limit of infinite N. Based on Rouzine and Coffin (1999b).

The intermediate interval, $1/s \ll N \ll 1/\mu$, has features common with both extreme regimes (Fig. 1.10). The growth competition simulation does not differ much from the deterministic case (Fig. 1.4), except for small fluctuations. Fixation of a beneficial allele, however, is strongly delayed compared to deterministic time t_{50}, and the random delay obeys Poisson statistics with the average delay time ~ $1/(\mu Ns)$ (Fig. 1.10A). The delay is caused by the waiting time to produce a single copy of beneficial allele, $1/(\mu N)$, and by the small probability s of the new subpopulation reaching the stochastic threshold (Rouzine and Coffin 1999c). There exists an approximate critical size $1/s$, above which random drift is weaker than natural selection. If and once the clone reaches that size, it will grow rapidly and in an "almost deterministic" fashion (Fig. 1.10A). Similar time and clone size scales appear in the mutant accumulation experiment (Fig. 1.10B), except that $1/s$ is the typical maximum number of deleterious mutants. As in the neutral regime at small population sizes, the population remains genetically uniform most of the time.

1.2.3 Two-locus model and linkage disequilibrium test

As is clear from the above examples, the role of stochastic factors depends on the population size, N. To estimate its effective value in an average untreated patient, the next task is to conduct an LD test using pairs of diverse loci and a two-locus model, as follows.

Early in infection, the HIV population is uniform or almost uniform genetically due to the transmission bottleneck and early selection (Holmes et al. 1992; Zhang et al. 1993; Delwart et al. 1994; Liu et al. 1997). The highly diverse sites are those that are adapting (Section 1.1) (Rouzine and Coffin 1999c). Let us choose two diverse sites and classify all genomes in the population into four bins (haplotypes): *ab*, *Ab*, *aB*, and *AB*, where the lowercase and uppercase letters denote less-fit and better-fit alleles, respectively. During the process of adaption, the initial population comprises uniform haplotype *ab*. The final population is uniformly *AB*. The haplotypes *Ab*, *aB* exist only during the transition. When the population is small, $N\mu \ll 1$, and given a limited sample size, one of the four haplotype classes will be empty, because adaptation at each site starts at a random time (Fig. 1.10A).

The simulated haplotype frequencies are simulated in the two-locus model, as a function of time, when both selection coefficients are the same. In the deterministic limit $N\mu \gg 1$, all four haplotypes are well-represented in the same time interval (Fig. 1.11A).

If however, the population is small, $N\mu \ll 1$, the two loci adapt at different random times, even for equal selection coefficients (Fig. 1.11B and 1.10A). Almost simultaneous adaptation can occur with a small probability in one of two scenarios, as follows. In *Scenario I*, two mutations at two sites generate two haplotypes, *Ab* and *aB*. By chance, the size of both clones reach the stochastic threshold $1/s$ at approximately the same time. Then, a third mutation within one of the two clones generates a new haplotype *AB* that subsequently is lucky to reach the stochastic threshold (Fig. 1.11C). In *Scenario II*, a mutation in *ab* creates either *Ab* or *aB* haplotype, which is lucky to grow above the threshold. By chance, very soon after it happens, a second mutation within that new clone generates haplotype *AB* (Fig. 1.11D). In either scenario, there are

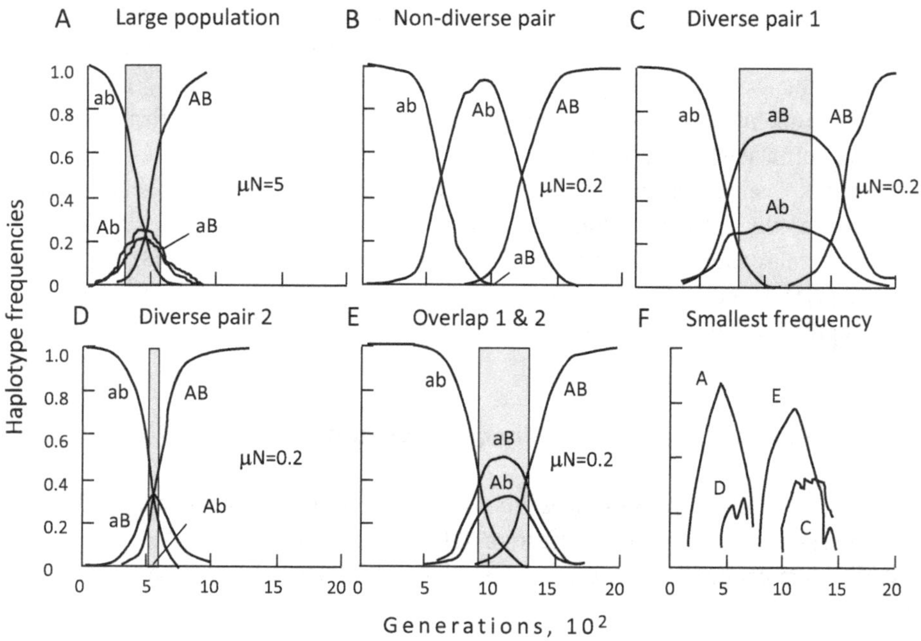

Fig. 1.11: Computer simulation of adaptation at two sites in selection–drift regime ($1/s \ll N \ll 1/\mu$). The four curves in A–E are frequencies of four haplotypes (shown). *A* and *B* denote the first and second site, *A* and *a* denote wild type and mutant. The selection coefficient, *s*, is equal for the two sites. Parameter values: all panels are obtained at $s = 0.1$, $N = 5000$, and rescaled along the time axis to correspond to $s = 0.01$. The mutation rate $\mu = 10^{-3}$ in panel A and 4×10^{-5} in B–F. Runs like that in panel B are the most frequent, pattern C is less frequent, and patterns D and E are the least frequent. Gray shading shows the time interval in which both sites of a pair have a mutant frequency in the interval 25–75%. Panel F shows the time dependence of the smallest haplotype frequency for the four runs in cases A and C–E. Based on Rouzine and Coffin (1999b).

no more than three well-represented haplotypes, at any time point (Fig. 1.11C and D). The fourth haplotype can be well-represented only if *Scenarios I* and *II* overlap, which has a very small probability (Fig. 1.11E).

To test the sequences of the protease gene from untreated patients (Lech et al. 1996) for this effect of "missing haplotype," all site pairs with the mutant frequency in the interval 25–75% were found. For each pair, four haplotypes were counted according to consensus or anticonsensus variant at each base. (The test does not depend on whether the consensus is best fit or not.) In three pairs of the four studied, all four haplotypes were present (Tab. 1.2). The test was repeated on the envelope protein sequences from untreated patients (Holmes et al. 1992). Once again, only two pairs out of the six total were missing a haplotype (Tab. 1.2). Therefore, the effective population of HIV is sufficiently large to be in the deterministic regime (or at the border).

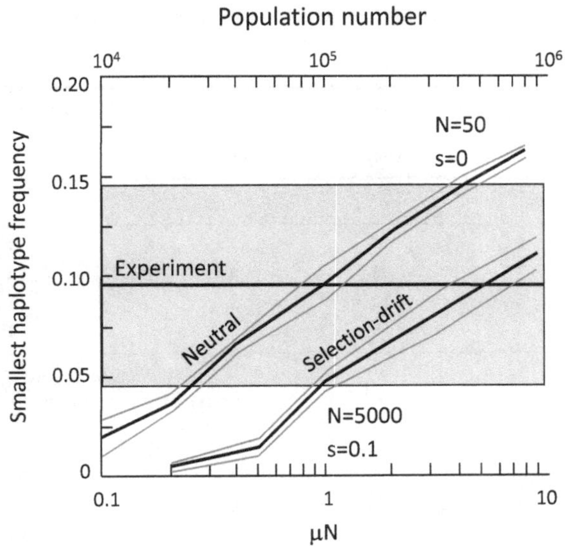

Fig. 1.12: Predicted dependence of the average frequency of the least-represented haplotype on the population number.
Crossover from the deterministic, $N \gg 1/\mu$, to a stochastic, $N \ll 1/s$, regime is shown for two values of the selection coefficient: $s = 0.1$, and $s < 10^{-5}$. Only simulation runs in which the genetic composition at each site is in the interval 25–75% in a time interval (cf. panels C–E in Fig. 1.11) are used for averaging, with different runs weighed accordingly to the length of this time interval (i.e., same selection criterion as in experiment). Thick lines are the average; thin lines show the 95% – confidence region. Dependence on μN was obtained by varying μ at a fixed population size: $N = 5\ 10^3$ for the selection–drift regime, and $N = 50$ for the neutral regime. Population numbers shown in the upper horizontal axis correspond to the fixed mutation rate $\mu = 10^{-5}$. The thick horizontal line and shaded band are the experimental average and the 95%-confidence region obtained from data in Tab. 1.2 based on Rouzine and Coffin (1999b).

To obtain a quantitative bound on the population size from below, the average frequency of the least-abundant haplotype was calculated at different population sizes N and compared with the observed data (Tab. 1.2). The time dependence of the least-represented haplotype frequency for different scenarios between representative Monte-Carlo runs (Fig. 1.11 A–E) is plotted in Fig. 1.11F. Then, this quantity was averaged over 100 simulation runs at different N (Fig. 1.11 and legend).

The corresponding experimental estimate was obtained by combining data on the protease and envelope genes (Tab. 1.2). The value $\mu = 10^{-5}$, which is the log average between the rate A → G or C → T and the rate of opposite transitions, was used (Mansky and Temin 1995). Plotting the result against population size, the population size is certainly ($P = 0.05$) larger than 9×10^4 infected cells, with the most likely value larger than 5×10^5 infected cells (Fig. 1.12).

Tab. 1.2: Distribution of sequences among four haplotypes for a few highly diverse pairs of sites in the HIV genome.

Site numbers	Pair				Sample size
	a–a	a–c	c–a	c–c	
Protease gene					
21, 174	1	6	6	1	14
21, 201	4	3	3	4	14
114, 209	0	6	14	3	23
174, 201	2	5	5	2	14
Envelope gene					
−17, −2	3	4	3	5	15
−2, 3	0	6	4	5	15
−17, 3	1	6	3	5	15
3, 10	0	4	5	6	15
−2, 10	2	4	3	6	15
−17, 10	2	5	3	5	15

Three samples (of 14, 23, and 15 sequences) were isolated from three HIV-infected individuals. The protease and envelope (V3 region) sequences were obtained from (Lech et al. 1996) and (Holmes et al. 1992), respectively. The GenBank accession nos. for (Holmes et al. 1992) were M84240–M84314. Letters "a" and "c" in the first line denote consensus and anticonsensus, with respect to the sample consensus. Site numbers in the first column for *pro* are standard nucleotide positions; for *env*, they are codon positions counted from the GPG crown of the V3 loop (the first G is numbered 0).

1.2.4 Estimation of the effect of recombination

HIV has recombination (Hu and Temin 1990; Robertson et al. 1995), and the missing fourth haplotype could be produced by recombination even if the population is small. Recombination redistributes already-existing alleles among sequences. To generate a diverse locus pair, two mutations have to generate haplotypes Ab and aB within the ab population (Fig. 1.11C). Then, recombination can generate haplotype AB from haplotypes Ab and aB. All four haplotypes will be present simultaneously, only if this event happens early enough, before clone ab becomes extinct (Fig. 1.11E). Effectively, in the two-locus model, recombination is equivalent to a second mutation, $Ab \to AB$ or $aB \to AB$. Therefore, the plots shown in Fig. 1.11 describe the case of recombination as well, except that the mutation rate μ should be replaced with its effective value including recombination

$$\mu_{\text{eff}} = \mu + \frac{1}{8}\langle 4f_{Ab}f_{aB}\rangle \rho L f_{\text{coinf}} \tag{1.19}$$

where ρ is the recombination rate per site per replication cycle, L is the distance between the two sites along the genome, f_{Ab} and f_{aB} are single mutant frequencies, f_{coinf}

is the fraction of coinfected cells among productively infected cells, $\langle\ldots\rangle$ denotes averaging over realizations, and 1/8 is the combined probability of the pair of proviruses in the cell being *Ab* and *aB*, a heterozygous pair of RNA genomes packed in a virion, and the resulting recombinant being *AB* rather than *ab*.

Parameters in eq. (1.19) are estimated, as follows. The average recombination rate is $\rho = 4 \cdot 10^{-5}$ per base per replication cycle (Hu and Temin 1990). The average pair distance in Tab. 1.2 is $L = 71$. Parameter f_{coinf} was unknown at the time of the original work and was estimated indirectly, as follows.

T-cell turnover rate in human and simians shows the daily replacement of 2–5% of T cells in a chronically infected individual with the current CD4 cell count 200/μl blood, which is 10^9 cells per day (Hu and Temin 1990; Mohri et al. 1998). The average number of virus-producing cells is 4 10^7 cells (Haase et al. 1996). The average life span of a virus-producing cell was estimated 1–2 days (Ho et al. 1995; Wei et al. 1995; Haase 1999). Thus, a cell has less than 3% chance to be infected. Assuming that infected cells are randomly chosen, and that the time a cell remains permissive for virus after infection ~0.5 day until CD4 receptor deregulation, one obtains that the fraction of double-infected cells among virus-producing cells does not exceed $f_{\mathrm{coinf}} = ((1\text{–}2)\,\mathrm{day}/0.5\,\mathrm{day})(3\%/2) = (3\text{–}6)\%$, even if superinfection resistance is absent. Substituting the values of ρ, L, and f_{coinf} into eq. (1.19) and estimating the average $\langle 4f_{Ab}f_{aB}\rangle$ from the scenario in Fig. 1.11C as 0.5, recombination increases the effective mutation rate by less than a factor of two. Therefore, recombination does not significantly alter the estimates of the effective population size. A more accurate quantitation of doubly infected cells in Section 1.3 from genetic data shows that f_{coinf} is even smaller than estimated above.

1.2.5 Robustness to approximations

Above consideration was based on four approximations whose importance for the results needs to be tested, as follows:

(i) Selection coefficients at two sites, s_A and s_B, was assumed to be equal. As Rouzine and Coffin (1999b) checked numerically at $\mu N = 1$, the difference between the two values by the factor of 2 makes a variable pair of loci much less likely to be observed (in 4.6% of 1,800 simulation runs against 66% of 50 runs at $s_A = s_B$), because they adapt at different times. At the same time, the least haplotype frequency averaged over such runs changes little compared to the case $s_A = s_B$ (0.051 ± 0.013 against 0.044 ± 0.009).

(ii) Epistasis also contributes to linkage disequilibrium, leading to distortion of the estimated population size. Hence, the test was modified to include either strong positive epistasis, where the fitness difference between *AB* and *Ab*/*aB* is twice larger than that between *Ab*/*aB* and *ab*, or strong negative epistasis, where the fitness of *Ab*/*aB* and *AB* are the same. Both modifications were found to increase predicted LD and make the least-represented haplotype at $\mu N = 1$ smaller, which only increases the final estimate of N by an order of magnitude. (Effects of epistasis are considered in Chapter 2.)

(iii) The initial virus population was assumed to be 100% less fit at both bases. In principle, a small admixture of *Ab* and *aB* could later be amplified by natural selection. A simulation $\mu N = 1$ shows that the average frequency of the least-represented haplotype either remains about the same or even decreases, depending on the relative initial quantities of *Ab* and *aB*: 0.047 ± 0.018 at 0% and 0%; 0.061 ± 0.018 at 1% and 5%; and 0.029 ± 0.005 at 2% and 2%.

(iv) An HIV population is assumed to be a "well-stirred pot" of infected cells and virions. In principle, the population comprised weakly connected subpopulations (Frost et al. 2001) could have several overlapping adaptation processes, which could explain the presence of all four haplotypes, even if population size of each subpopulation is small. Two lines of evidence demonstrate that this is not the case. Firstly, virions travel far and rapidly from their producing cells (Ho et al. 1995; Moore et al. 1996; Perelson et al. 1996) ensuring good genetic mixing (Rouzine and Coffin 1999a). Secondly, topography of HIV variants by selective labeling reveals that most of these islands are shared by different variants implying good local mixing (Reinhart et al. 1998). Also, the island pattern likely results from the nonuniform formation of permissive cells in germinal centers, rather than a cell-to-cell infection spread as proposed by Grossman et al. (1998) and Haase (1999).

1.2.6 Discussion

The early work reviewed above demonstrated that the effective population size in an average untreated HIV-infected individual is large, and that the evolution of separate bases is mostly deterministic. The controversy started by Leigh-Brown (1997) who applied a "neutrality test" (Tajima 1989) to the sequence samples of the envelope gene (Holmes et al. 1992). The results suggested that the average HIV population in untreated patients is very small, $N_{\text{eff}} = 100\text{–}1000$, and proposed that HIV evolution is selectively neutral. The reason for the discrepancy between the two early tests is, as follows. Firstly, the "neutrality test" (Tajima 1989) assumes that selection is neutral and then checks whether the average number of diverse sites in a sample of sequences grows, roughly, as the logarithm of the sample size, within a predicted statistical interval (Waterson 1975). The general problems with this test are that the prediction was not shown to be statistically distinct from the predictions of alternative models, and that the predicted dependence on the sample size is very slow, while the predicted statistical error is quite large, not much smaller than the average. The estimate of the effective population size from the number of segregating sites by Leigh-Brown (1997) was also based on the *a priori* assumption that the neutral model applies. Natural selection decreases the number of diverse sites to the few sites with a small s, which effect was misinterpreted as a small population size.

The estimate of, at least, 10^5–10^6 infected cells per untreated patient made in 1999 was confirmed later by other groups who used other methods, such as delays in fixation of beneficial alleles (Frost et al. 2000) and the clonal structure of genome regions linked to these alleles ("soft sweeps" versus "hard sweeps") (Pennings et al. 2014; Rouzine, Coffin et al. 2014). Nonlinear accumulation of genetic diversity is also indicative of a nonneutral evolution with a large population size (Maldarelli et al. 2013).

The conclusion of this section about a relatively weak role of stochastic effects in the steady-state HIV population is restricted to the evolution of separate bases, whenever such a model applies. Due to the multi-locus nature of HIV evolution and linkage effects, stochastic effects are very important at realistic population sizes, at the high-fitness front of the traveling wave (Rouzine et al. 2003; Rouzine 2020b). Also, the viral transmission is essentially stochastic due to the random sampling of inoculum (Sections 1.1 and 3.1) and variation in HLA subtypes between individuals (Sections 1.1 and 4.2). Genetic bottleneck created by antiviral therapy is another source of stochasticity affecting the chances of HIV eradication (Rouzine, Razooky et al. 2014).

1.3 Estimate of recombination rate

In this section, the effective recombination rate (frequency of coinfected cells) r in untreated patients is estimated from two measures of linkage disequilibrium by comparing genomes simulated in a Monte-Carlo code to real HIV polymerase gene sequences. The result of inference is $r \approx 1\%$ per genome per generation. Despite the small value of r, the predicted adaptation rate is higher by the factor of two than in the absence of recombination, which demonstrates the importance of rare recombination in partly sexual evolution.

The models reviewed in Sections 1.1 and 1.2 reflected the knowledge of the twentieth century. By the start of the third millennium, new methods of genome sequencing were invented, and more comprehensive studies were carried out. These studies confirmed the previous predictions and offered new important insights. In keeping with the predictions of the models in Section 1.1, the pattern of genetic diversity within an untreated HIV-infected individual changes as infection progresses, as follows. In the first weeks of infection, before the host immune response is in full power, mutations are found once per sequence sample per site and accumulate with the neutral mutation rate, $\mu = (2\text{–}3)\ 10^{-5}$/base/day Keele et al. (2008); Kearney et al. (2009); Salazar-Gonzalez, et al. (2009). This patter, however, is not maintained forever (Maldarelli et al. 2013). Several months later, minority alleles concentrate in a few highly diverse ($> 5\%$) sites, about 15 sites per individual (Goonetilleke et al. 2009). Almost all substitutions are nonsynonymous and occur within the immunologically important regions, CD8 T lymphocyte (CTL) epitopes. A year later, in the chronic phase of infection, the diversity pattern changes again, as described in Section 1.1. Highly diverse sites expand to 200–300 sites, far beyond the number of active CTL epitopes (20–30), and many are synonymous. The high diversity is maintained for a long time, and most of these sites evolve very slowly, over a period of years. Sometimes, late immune escape mutations are observed (Asquith et al. 2006).

Meanwhile, the models of HIV evolution progressed from deterministic models assuming a single site and two loci (Rouzine and Coffin 1999c) or neutral evolution (Leigh-Brown 1997), to models including random genetic drift (Rouzine and Coffin 1999b; Frost et al. 2000; Frost et al. 2001; Rouzine et al. 2001) to the more advanced models including linkage and recombination of many polymorphic sites (Rouzine 2020b).

The key evolutionary factor missing in the models of the twentieth century was many-locus linkage effects. Experiments (Gerrish and Lenski 1998; Rice 2002) and early few-locus models (Muller 1932; Fisher 1958; Hill and Robertson 1968; Felsenstein 1974; Otto and Barton 1997) demonstrated that the evolution of different loci, due to common ancestry, is entangled. Competition between clones containing beneficial mutations at different sites slows down adaptation. The "background selection" limits the fixation of new alleles to the few high-fitness genomes in a population (Good et al. 2012) and requires their surfing in the presence of recombination, according to Rouzine and Coffin (2007); Neher et al. (2010). The magnitude of all these linkage effects increases with the number of adapting sites (Maynard Smith 1971; Rouzine et al. 2003).

The quantitative effect of linkage on the evolution speed and genetic diversity for an asexual population was derived in the general form by several groups of theoretical physicists (Rouzine et al. 2003; Desai and Fisher 2007; Brunet et al. 2008; Rouzine et al. 2008; Hallatschek 2011; Good et al. 2012) who used the techniques of statistical physics. These teams described the evolving population as a traveling wave moving in fitness coordinate (Tsimring et al. 1996; Kessler et al. 1997), with the speed controlled by stochastic effects at the high-fitness front. Genealogical properties were obtained analytically as well by Brunet et al. (2007); Walczak et al. (2012); Neher and Hallatschek (2013). In keeping with the strong role of linkage, the series of models predicted a strong role for rare recombination events in evolution of a long genome (Rouzine and Coffin 2005, 2007; Neher et al. 2010; Rouzine and Coffin 2010; Neher et al. 2013).

These theoretical advances in the field of population genetics turned out to be of great practical help for HIV research. Like many other viruses, HIV has a mechanism for recombination. A cell super-infected by multiple strains of HIV can produce virions containing heterologous pairs of RNA genomes. Upon entry into a new cell and the reverse transcription of the genomic RNA into a proviral DNA, the polymerase protein switches between the two templates several times, creating a recombinant DNA provirus. What was unknown, however, is what percent of infected cells is dually infected by genetically remote strains and, hence, contributes to the effective recombination rate, r.

To estimate this parameter from HIV sequence data, Batorsky et al. (2011) whose work is reviewed in this section, used computer simulation based on a model including mutation, the natural selection with a variable selection coefficient, random genetic drift, linkage, and recombination. Using the simulated sequences and sequence data from HIV-positive individuals, as well as two sensitive measures of linkage disequilibrium, they estimated the value of r in a typical individual.

1.3.1 Methods

1.3.1.1 Model of population

The model, implemented as a Wright–Fisher algorithm (Batorsky et al. 2011) https://github.com/irouzine/Strong-linkage-in-sex, represents a generalization of the two-locus model in Section 1.2 for a large number of loci. A population is represented by N genomes with L binary sites, which correspond to HIV proviruses integrated into cell DNA. Each site carries either a less-fit or better-fit allele, denoted 0 and 1, respectively. For HIV in steady-state, a generation is equal to a day (Markowitz et al. 2003; Brandin et al. 2006)].

At each generation, all individuals die and are replaced with their identical progeny, so that the total number of genomes stays fixed. The progeny number of a genome is random, with the average given by the relative fitness, which depends on the number of beneficial alleles. Strictly speaking, the multinomial distribution for the progeny number is the most appropriate choice for the progeny distribution. It could be implemented by a broken-stick algorithm (Chapter 2). A simpler choice used below is the cell division model, where the progeny number is either 0 or 2, with different probabilities, which creates the same average natural selection and the same genetic drift as the multinomial distribution.

If no mutation or recombination event took place, a progeny genome is identical to the parent. Genome comprising less-fit alleles 00000 . . . has the smallest relative fitness chosen to be 1. Mutation rate per site is $\mu = 3 \times 10^{-5}$(Mansky and Temin 1995). A beneficial mutation event at a site i increases fitness by the factor of $\exp(s_i)$, where s_i is the selection coefficient. Epistasis is absent from the model; this factor is discussed in Chapter 2. Deleterious mutation events are ignored, which is a fair approximation when far from the steady state (mutation-selection balance, Section 1.1). The values of s_i are sampled according to a distribution, and their values are fixed before simulations begin. Two forms of distribution are considered, the exponential and the power law, $(1/s_0)\exp(-s/s_0)$ and $a{s_0}^a/(s+s_0)^{a+1}$, respectively, where s_0 and a are input parameters of the model. The main focus will be on the exponential distribution, because it is often observed in experiment, for the reasons discussed later in this book (Section 2.4). In keeping with the discussion and results of Section 1.1, the initial population is assumed to be uniformly less-fit at each site.

Before the random sampling of progeny, rN pairs of genomes (rN coinfected cells) are randomly selected from the population. Each pair of genomes undergoes recombination crossovers at M randomly selected genomic sites, creating two child genomes with alternating parental segments. One of the two recombinants replaces one parent. The effective recombination rate, r, represents the probability of outcrossing equal to the fraction of coinfection among infected cells.

For the exponential form of the distribution of s, there are 6 model parameters: N, L, μ, r, M, s_0. For the power law distribution, there is an extra parameter, a. Three of these parameters are known from experiment: $\mu = 3 \times 10^{-5}$ (Mansky and Temin 1995), $M \approx 10$ (Levy et al. 2004); and from the data in Batorsky et al. (2011), $L \approx 2,000$. The fourth parameter is the effective population $N = 10^5$–10^6 (Rouzine and Coffin 1999b; Frost et al. 2000) (Section 1.2). Remaining parameters r and s_0 are inferred from fitting single-time sequence data obtained by the experimental team in Batorsky et al. (2011), as explained in the following sections.

1.3.1.2 Linkage disequilibrium measures

Two statistics of linkage disequilibrium (LD) were used to estimate the two parameters, as follows. In simulation and patient data, one selected the "very diverse" pairs of sites such that $0.25 < f < 0.75$ for both sites. The first statistics of LD was defined as $1 - \langle f_{AB}^{\mathrm{LRH}} / (f_A f_B) \rangle$, where f_{AB}^{LRH} is the least-represented haplotype frequency for sites A and B, $f_A f_B$ is the product of the respective one-site frequencies, and $\langle \cdots \rangle$ denotes averaging over "very diverse" pairs. The second statistics of LD was the fraction of "very diverse" pairs for which $f_{AB}^{\mathrm{LRH}} < 0.04$.

1.3.1.3 Patient data

Nine sets of HIV polymerase gene sequences were sampled from four untreated patients, at 2 to 3 time points for each patient. The total sequence number per sample changed between 12 and 42, with an average of 25. Plasma samples were obtained from of adult chronically infected patients enrolled in clinical studies at the NIH Clinical Center at Bethesda, Maryland, USA. All patients gave informed consent in the written form. Single-genome sequencing was used to obtain sequences, as described by Palmer et al. (2005). The method has no assay-related recombination. Sequence length varied between 1,200 and 1,250 nucleotides. Software Clustal X (DNASTAR-MEGALIGN or MEGA) was used for alignment. Compared to the polymerase consensus of 1,100 nucleotides, each sequence was converted into a binary string. Because the above LD measures are not dependent on which allele is beneficial, the results are invariant to the consensus sequence. The sample sizes, the numbers of diverse sites, and the two measures of LD are given in Tab. 1.3.

1.3.2 Simulation of HIV adaptation

Figure 1.13 shows observable quantities of evolution for a set of values of $r = 0.01$ and $s_0 = 0.005$ estimated in the next section from data. As predicted by asexual and sexual

Tab. 1.3: Observable quantities calculated from HIV *pol* sequences.[1]

Patient Sequence[2]	Sample size	No. of diverse sites	% in simulation[3]	Obs. diverse sites $f > 4\%$	% in simulation	Very div. sites $f > 25\%$	% in simulation	Avg. LD of very diverse pairs	95% CI over pairs	Fraction of very div. pairs with 4th hapl. <4%
Patient 1										
1	20	63	0.28	63	0.28	11	0.05	0.43	0.09	0.09
2	30	72	0.32	28	0.12	14	0.06	0.75	0.05	0.6
3	17	49	0.22	49	0.22	11	0.05	0.8	0.05	0.31
Patient 2										
1	30	46	0.2	9	0.04	4	0.02	0.75	0.33	0.33
2	12	21	0.09	21	0.09	5	0.02	0.7	0.28	0.3
Patient 3										
1	18	39	0.17	39	0.17	8	0.04	0.55	0.14	0.18
2	43	71	0.31	36	0.16	5	0.02	0.53	0.25	0.2
Patient 4										
1	38	76	0.34	44	0.19	13	0.06	0.67	0.09	0.44
2	22	70	0.31	70	0.31	11	0.05	0.37	0.09	0.07
Average ±SD	26	56±19	0.25±0.08	40±19	0.16±0.09	9±4	0.04±0.02	0.62±0.15		0.28±0.17

[1]Results shown are compared with simulation predictions to estimate model parameters (Fig. 1.14).
[2]2–3 time points are shown for each patient.
[3]To compare with simulation, where L variable sites are modeled, the number of diverse sites observed in the polymerase is extrapolated to the entire genome by dividing the number of diverse sites by the total number of sites in the *pol* sequences (1,100 nt) multiplying by the length of HIV genome (9700 nt) and dividing by the length HIV genome considered variable (2,000 nt).

models (Rouzine et al. 2003; Rouzine and Coffin 2005; Desai and Fisher 2007; Brunet et al. 2008; Neher et al. 2010; Rouzine and Coffin 2010; Hallatschek 2011), the fitness distribution (Fig. 1.13A) represents a solitary wave of a fixed profile traveling towards higher fitness. The average adaptation rate (the time derivative of log fitness) is 0.0023/day, which is larger than the asexual value at $r = 0$ by the factor of 2. Thus, only 1% of recombination accelerates adaptation twice.

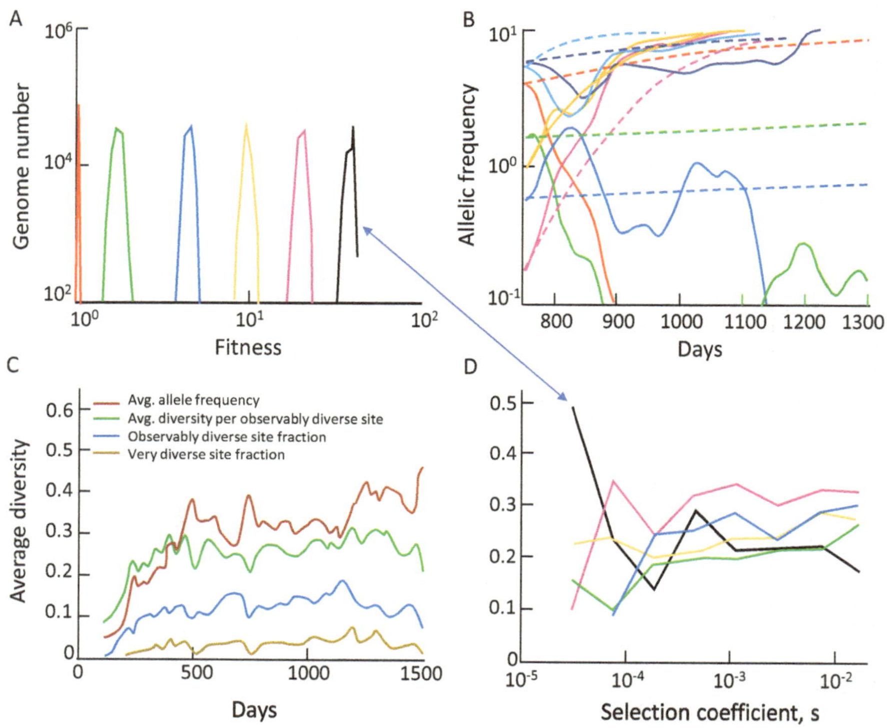

Fig. 1.13: Simulation of viral adaptation during a chronic HIV infection.
(A) The population distribution in fitness moves as a wave. Distribution is shown at 300 day intervals: day 1 (red), day 300 (green), day 600 (blue), day 900 (yellow), day 1200 (magenta), day 1500 (black).
(B) Dynamics of beneficial alleles at high frequencies. Trajectories shown for 8 randomly selected sites with frequency of beneficial alleles per site, $f > 0.1$ at day 750: in simulation (solid), 1-site model prediction (dashed). The values of s at the sites shown: 0.02 (cyan), 0.004 (black), 0.004 again (red), $4.5 \cdot 10^{-4}$ (green), 0.01 (yellow), 0.01 (blue), 0.01 (magenta). (C) Quantities characterizing average diversity of the population. Fraction of "observably diverse" sites, defined as having more than 0.04 of minority allele (blue), fraction of "very diverse" sites, defined as having more than 0.25 of minority allele (brown), average frequency of beneficial alleles per site $\langle f \rangle$ (dark red), genetic diversity $\langle 2f(1-f) \rangle$ averaged over "observably diverse" sites only (green). (D) Average diversity of "observably diverse" sites classified according to their selection coefficient s for 10 equidistant time points at 300-day intervals, the same times and colors as in A. Parameters in (A–D) are estimated from the HIV polymerase gene sequences in a representative patient (Fig. 1.14): $r = 0.01$, $s_0 = 0.005$, $M = 10$, $L = 2,000$, $N = 10^5$, $\mu = 3 \times 10^{-5}$. Based on Batorsky et al. (2011).

Beneficial allelic frequencies of a few sites, $f_i(t)$, are shown as a function of time t in Fig. 1.13B. The deterministic dynamics described by the single-site model (Section 1.1) is shown by the dashed lines. All trajectories have a strong stochastic component. Two kinds of behavior are observed. At some sites, $f_i(t)$ fluctuates around the monotonic trajectory predicted by the single-site model. At other sites, beneficial alleles fall to a low level in 50–200 generations, where they are maintained only by new mutation events.

Four observable diversity measures in Fig. 1.13C include

(i) the average frequency per site $\langle f \rangle$,
(ii) the fraction of sites i with minority allele frequency min $[f_i(t), 1-f_i(t)] > 0.04$ termed "observably diverse (in a sample of 25),"
(iii) the same with more than 0.25 of the minority allele termed "very diverse," and
(iv) intrapatient genetic distance $T = 2\langle f(1-f)\rangle$ averaged over "observably diverse" sites.

After about 500 days, these quantities attain a steady state, with some fluctuations. The average height of their plateaus is not far from the single-site model prediction with a broadly variable s (Rouzine and Coffin 1999c) (Section 1.1). This is true both for the exponential and power-law distribution of s (with $a > 1$). When the recombination rate changes by two orders of magnitude, $0.01 < r < 1$, the average diversity stays between 0.25 and 0.33, close to the observed value in patients (Tab. 1.1). In contrast to the average diversity of observably diverse sites, their fraction in genome is sensitive to both r and s_0 and can be used to estimate s_0 from sequence data (see next subsection).

The average diversity of "observably diverse" sites grouped according to their selection coefficient s is shown in Fig. 1.13D for five time points. The dependences on s are weak, showing that sites with large s and small s contribute roughly equally to net diversity at any times. This feature cannot be explained by an independent-locus model (Section 1.1) and points to strong linkage effects at small recombination rates. In the independent-locus limit, the dependence in Fig. 1.13D would have a sharp maximum in s moving in time towards smaller s. The weak dependence on s indicates that alleles with small s are linked to alleles with large s and that selection takes place at the genome level. (This situation corresponds to the beginning of adaptation. As the population becomes closer to the best-fit sequence, the dynamics at sites with different s differs, see Section 2.4 for the detailed analysis at $r = 0$.)

1.3.3 Estimation of recombination rate and average selection coefficient

The observables shown in Fig. 1.13 describe the evolutionary process, for the sake of general information. However, they are not suitable for the main goal of estimating the effective recombination (coinfection) rate r. Usually, r is estimated from longitudinal samples demonstrating a genetic crossover (Neher and Leitner 2010). The original method proposed here has the advantage of using a one-time sample only.

The method relies on two related measures of linkage disequilibrium (LD) sensitive to recombination rate r, as follows:

(i) The least-represented haplotype frequency (Section 1.2) averaged over pairs of "very diverse" sites and normalized to its predicted zero-LD value.

(ii) The frequency of "very diverse" site pairs that miss a haplotype in the sample. Nine sequence sets of HIV polymerase gene, which takes 11% of HIV genome, were isolated from 4 untreated patients, with 2 to 3 time points per patient (Tab. 1.3; Section 1.3.1.3). The LD values calculated from patient data were fit to the two measures of LD evaluated from a 11% length of the simulated genome, for different values of r.

Using the best-fitting value of s_0 (below), the two simulated measures of LD plotted as a function of the recombination rate r, were compared to the range of LD evaluated from patient data (Tab. 1.3, Fig. 1.14A and B). The fast decrease of both LD measures with r demonstrates their sensitivity to r. The interval of r that matches data is 0.05–0.15, with the center at $r = 0.01$.

The estimate is sensitive to the crossover number per genome M. The above estimates assumed $M = 10$ (Levy et al. 2004). Decreasing M by the factor of 3 increases predicted LD, which increases the estimate of outcrossing rate r by the factor of 2 to 3 (Fig. 1.14A and B).

The value for s_0 used in Fig. 1.14A and B was estimated by matching the net change in virus fitness predicted by simulation, as a function of s_0, to the experimental value (Fig. 1.14D). The experimental fitness gain was estimated from the additional information, as follows. A large class of the models of HIV dynamics comprising the populations of susceptible, infected, and immune cells predict a steady state (Nowak and Bangham 1996; Muller et al. 2001; Wodarz 2001; Rouzine et al. 2006; Sergeev, Batorsky and Rouzine 2010; Rouzine et al. 2015; Rouzine 2022). In all these models, the infected-cell number is pinned near the activation threshold of the immune cells, and the susceptible cell number is inversely proportional to the average viral fitness. This prediction is common for the models, in which the immune response controls HIV. The last fact has been proven experimentally in many studies (Kuroda et al. 1999; Schmitz et al. 1999).

Hence, if virus fitness slowly increases in the process of adaptation, the steady-state CD4 count decreases slowly, which can be linked to the progression to AIDS (Section 3.1) (Rouzine 2020a). Before the fully developed AIDS, the steady state CD4 T-cell count drops from ~ 500 to 5–50 cells/ml blood. This decrease corresponds to a fitness increase between 10 and 100, which amount specifies the middle estimate $s_0 = 0.005$ for both distributions of s considered here, the exponential and the power law ($a = 2$).

Another method to estimate s_0 is to match the simulated fractions of "observably diverse" and "very diverse" sites to the patient data for the polymerase gene (Fig. 1.14C), whence s_0 is greater than 0.003. From either method s, one obtains $s_0 = 0.003 - 0.005$.

The main assumption of the above simulation is that the initial population is uniformly less-fit. Adding some beneficial alleles at the start weakly affected the estimates of r and s_0. Examination showed that, at small recombination rates ~1%, most

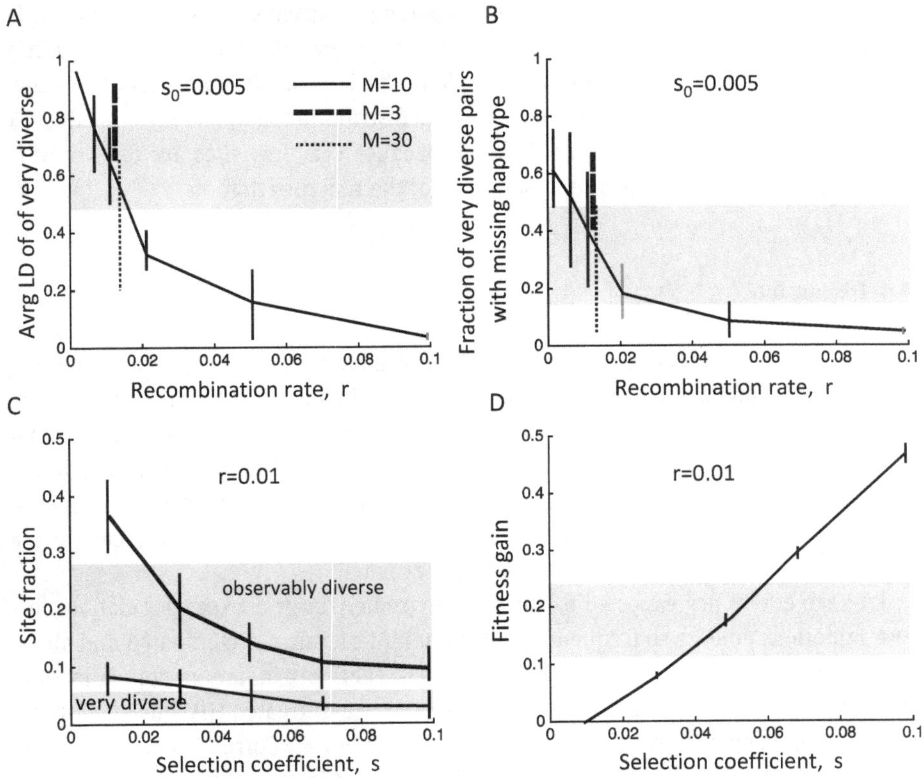

Fig. 1.14: Estimation of model parameters for chronic HIV infection.
Recombination rate, r, and average selection coefficient, s_0, are estimated by comparing four observable quantities in data with their predicted values in simulation. Grey regions indicate the mean ± 1 standard deviation of quantities calculated from data from the *pol* gene in 4 patients (Tab. 1.3). In simulation, each quantity is calculated at day 1,500 for 11% of the genome. An average over 16 random runs is shown, with the error bars showing the standard deviation over runs. Results are shown for a single population size, $N = 10^5$, and 3 values of the crossover number, $M = 3$ (dashed line), 10 (solid line), and 30 (dotted line). Results for different M are shifted slightly horizontally for clarity. (A) Linkage Disequilibrium (LD) of "very diverse" pairs for $s_0 = 0.005$. LD is defined as $1 - \langle f_{AB}^{\mathrm{LRH}} / f_A f_B \rangle$, where f_{AB}^{LRH} is the frequency of the least-represented haplotype for a pair of sites A and B, $f_A f_B$ is the product of one-site frequencies or the value of f_{AB}^{LRH} is in the absence of linkage for $s_0 = 0.005$. (B) Another measure of LD: fraction of very diverse pairs with the frequency of the least represented haplotype below 0.04 for $s_0 = 0.005$. (C) Fraction of sites that are "observably diverse" (fat line) and "very diverse" (thin line) for $r = 0.01$. (D) Population average fitness for $r = 0.01$. Other fixed parameters are shown in the legend to Fig. 1.13. Based on Batorsky et al. (2011).

preexisting alleles rapidly become extinct due to random drift. At larger $r > 5\%$ and at the values of other parameters relevant for HIV, preexisting alleles are very important and dominate evolution (Rouzine and Coffin 2005; Gheorghiu-Svirschevski, et al. 2007; Rouzine and Coffin 2007, 2010).

The presented results focused on the exponential distribution of s, because it is frequently observed in experiment (Imhof and Schlotterer 2001; Kassen and Bataillon 2006; Acevedo et al. 2014; Stern et al. 2014; Wrenbeck et al. 2017) (Section 2.4). However, similar estimates follow for the power-law distribution, $as_0^a/(s+s_0)^{a+1}$, with $a=2$ or larger. For $a<1$, data cannot be fit at all, because very few sites are diverse; it is highly unlikely that this case is representative of the real distribution.

1.3.4 Discussion

Thus, the incidence of recombination in an average untreated HIV-positive individual is $r \approx 1\%$. This value is too low to compensate for the linkage effects. In particular, the adaptation rate is smaller than that predicted by the single-site model by a factor of 4 (Batorsky et al. 2011), and large LD exists (Fig. 1.14A and B). Nevertheless, the adaptation rate is higher, and LD is smaller, than in the case of asexual reproduction, by the factor of 2. This result emphasizes the importance of sexual reproduction, even if the fraction of individuals that participate in it is very small.

Linkage effects are expected to vanish in extremely large asexual populations, because mutations emerge so frequently, that clonal interference is eliminated and allelic correlations are averaged out (Rouzine et al. 2003). Therefore, a limited population size is a necessary for linkage effects to exist; for an average untreated HIV infection, it was estimated between 10^5 and 10^8 infected cells (Rouzine and Coffin 1999b; Frost et al. 2000; Frost et al. 2001; Pennings et al. 2014; Rouzine, Coffin et al. 2014). Although finite population size N is important in a multi-locus system, all the observables there depend only logarithmically on N (Rouzine 2020b). Changing the population size from $N=10^5$ to 10^8 elevates the estimate of r by less than 50% (results not shown).

Similar estimates of the two parameters, $r \approx 0.02$ and $s_0 \approx 0.005$, were obtained by Neher and Leitner (2010). To estimate r and s_0, these authors used two simplified, essentially different models. To estimate r, they assumed selectively neutral evolution of partly linked locus pairs and fit the dynamics of the appearance of the initially missing haplotype to data. To estimate s_0, they assumed the deterministic evolution of a site under positive selection and unlinked from the other sites.

The method used by Neher and Leitner (2010) has the advantage of being less laborious. The main advantage of this study is that the two parameters are estimated from the same, full model without using two different models with mutually excluding approximations. Another benefit is that the present method works with single-time samples. The third, experimental advantage is that the sequences were obtained using SGS method less prone to PCR-derived mutation and recombination than the one used by Neher and Leitner (2010).

The present model made some simplifications as well:

(i) Deleterious mutation events were ignored, assuming that the system is far from equilibrium.

(ii) The estimates for the polymerase gene were extrapolated to the entire genome.
(iii) Constant purifying (directional) selection is assumed. Some rare but important sites, such as antibody or CTL epitopes require separate treatment (Chapter 4).
(iv) Epistasis was neglected. Epistasis exists for RNA viruses (Burch et al. 2003; Weinreich et al. 2005; Ayme et al. 2007) including HIV (Bonhoeffer et al. 2004; Poon et al. 2010) and may contribute to the observed values of LD, thus underestimating the recombination rate. (In Section 1.2, epistasis was accounted for when estimating N. Epistatic signature in genomic data is studied in detail in the next chapter.)

The obtained estimate of r contradicts an early claim that all infected cells are superinfected several times (Jung et al. 2002). The experimental team of this study (Batorsky et al. 2011) found that less than 10% of infected cells have a second provirus (data not shown). The effective coinfection rate is even lower due to the proximity effects of virus infection causing similarity between the coinfecting sequences.

To conclude this section, matching predictions of a Monte-Carlo model of multilocus adaptation to the RNA sequence data from HIV-infected individuals, the estimates of the effective recombination rate and the average selection coefficient were obtained for HIV-infected individuals. The results emphasize the importance of linkage and recombination in lentiviral evolution.

Chapter 2
Inference of fitness landscape from sequence data

Fitness landscape is the dependence of fitness, defined as the average progeny number, on genomic sequence. It includes fitness changes caused by separate mutations (selection coefficients), as well as their interaction (epistasis). Fitness landscape can be viewed as a particular case of a phenotype–genotype map, with fitness as the trait of interest. While viruses or bacteria allow to measure fitness by site-directed mutagenesis, in most cases, fitness is very difficult to measure directly. The goal of this chapter is to describe a method of fitness landscape inference from a large set of genomic sequences.

2.1 Universal evolutionary footprint of epistasis

In an organism, a region of the genome, of a protein, of a messenger RNA often interacts biochemically with the other regions. This interaction, epistasis, is an omnipresent property of biological networks (Cordell 2002; Cordell 2009; Zuk et al. 2012; Wei et al. 2014). Epistasis plays an important role in evolution of populations and heritability of traits. Taking epistatic contributions into account improves predictions of phenotype in various species including chicken (Carlborg et al. 2006; Alvarez-Castro et al. 2012), yeast (Segrè et al. 2004; Brem et al. 2005; Schuldiner et al. 2005), viruses and bacteria (Nijhuis et al. 1999; Levin et al. 2000; Piana et al. 2002; Handel et al. 2006; Cong et al. 2007), and diverse plants (Dudley and Johnson 2009; Hu et al. 2011; Wang et al. 2012). Genetic studies of human diseases report epistatic interactions for a large number of pairs of loci (Wan et al. 2010; Lippert et al. 2013) suggesting a major role for epistasis (Combarros, van Duijn et al. 2009; Bell et al. 2011; Kölsch et al. 2012; Lu et al. 2012; Bullock et al. 2013; Génin et al. 2013; Rhinn et al. 2013; Zhu et al. 2013).

Although many epistatic interactions were proposed, few of them were confirmed (Combarros, Cortina-Borja et al. 2009; Wei et al. 2014). This is despite the fact that dozens of statistical methods were developed to infer locus interactions (McKinney and Pajewski 2011; Ritchie 2011; Zhang et al. 2011; Huang et al. 2013; Pang et al. 2013). They include regression analysis (García-Magariños et al. 2009; Chen et al. 2011), Bayesian methods (Zhang and Liu 2007; Tang et al. 2009; Zhang et al. 2011), and haplotype statistics (Hoh and Ott 2003; Ueki and Cordell 2012). All of these methods infer epistasis from the statistical association of alleles at pairs of loci. Despite the multitude of methods, detecting epistasis remains a serious challenge, and few reports of epistasis are reproducible and validated experimentally (Cordell 2009; Wei et al. 2014).

One reason for this problem is a small number of epistatic pairs among all possible pairs. Another reason is strong statistical noise discussed in Section 2.2. In addition to these two issues, a separate measure that could directly quantitate the strength of epistasis without entanglement of the other parameters of a population was absent until

https://doi.org/10.1515/9783110697384-002

recently. Before the work by Pedruzzi et al. (2018) discussed in this section, the detection methods relied on diverse statistical markers designed to recognize interacting pairs without finding the exact strength of epistasis (Cordell 2009; Wei et al. 2014).

A crucial biological scenario that requires the quantitative knowledge of epistasis is a viral population, which adapts after a sudden change in external conditions. Another example is an evolutionary bottleneck: an example from virology is the viral transmission to a new host, the spread to a different organ, or coping with a new therapeutic agent. Often, an adapting organism or virus passes through intermediate genetic variants with reduced fitness, termed "fitness valleys" (Levin et al. 2000; Piana et al. 2002; Weissman et al. 2009; Gonzalez-Ortega et al. 2011), as it happens when a pathogen evolves resistance to drugs or the immune response (Sections 4.1 and 4.2). Compensatory mutations mentioned in Section 1.1 emerge and are fixed expediting the development of drug resistance (Nijhuis et al. 1999; Levin et al. 2000; Piana et al. 2002; Handel et al. 2006; Cong et al. 2007; Gonzalez-Ortega et al. 2011; Noviello et al. 2011).

Rather than interacting by direct biochemical binding, epistatic loci may represent allosteric sites (Noviello et al. 2011), which cannot be inferred based on structural studies. A recent example is a pair of mutations E138K and M184I conferring resistance against four FDA-approved drugs in Phase 3 clinical trials resulting in HIV treatment failure. The pair 150L and A71V is also associated with drug resistance of HIV (Meher and Wang 2012; Yu et al. 2015). To predict these epistatic pairs from a genetic database, it is important to understand how to identify their fingerprint at the genetic level.

This section following Pedruzzi et al. (2018) introduces and tests, analytically and by simulation, a measure of epistasis that depends only on relative epistatic strength and topology, but not on the state of the population and other model parameters. An analytic argument based on the Wright–Fisher model of asexual populations is developed to study the haplotype distribution in the process of adaptation. The focus is on the case of positive epistasis in a population evolving after a sudden change of environment (the fitness valley situation). The principal finding is the existence of a quasi-equilibrium between epistasis and disorder caused by stochastic factors of random mutation and genetic drift in the regime of slow adaptation characteristic for the regime of traveling wave (Rouzine 2020b). In the general case of a pair of loci linked to a long genome, this quasi-equilibrium, not to be confused with the quasi-linkage equilibrium of (Kimura 1965), results in a universal relationship ("universal footprint") between haplotype frequencies of a pair that depends only on the relative strength of epistatic interaction network with respect to the average selection coefficient, but not any other parameters of the system. The universal footprint is generalized for any topology of epistatic network, several examples of which are considered explicitly.

2.1.1 Model of stochastic evolution with epistasis

The model is a variation on the model described in Section 1.3 following Batorsky et al. (2011), whose upgraded code is found in https://github.com/irouzine/Strong-linkage-in-sex. The difference is that recombination is now absent and epistasis is added. Consider a population of N binary sequences $\{K_i\}$, for example, 0111010101, where each evolving site (locus) numbered $i = 1, 2, \ldots, L$ is either a nucleotide or amino acid position. A case of two possible alleles is considered: $K_i = 0$ or $K_i = 1$. In real life, the binary assumption is valid in moderate-term evolution, where only two variants are observed during a period of time for most of loci. The genome is assumed to be long, $L \gg 1$. Evolution of the population is simulated in discrete time as a Wright–Fisher process, as follows. The evolutionary factors included are constant natural selection, random mutation with rate per genome μL, and random genetic drift (Rouzine et al. 2001; Nielsen and Slatkin 2013). Recombination is absent. Each generation, all individuals are replaced with their progeny, whose numbers are random and calculated with the use of the broken-stick algorithm (multinomial distribution) takes into account both natural selection and genetic drift, as follows. Specifically, an interval $[0, N]$ is broken into N intervals with the length proportional to the average progeny numbers of the corresponding individual genomes, and N random points are generated over the entire interval. The number of points falling into interval i are the number of genome i. Thus, the total population stays constant between generations and equal to N. The average progeny number (fitness) of each genome e^W is given by

$$W = \sum_{i=1}^{L} s_i K_i + \sum_{i<j}^{L} s_{ij} K_i K_j \tag{2.1}$$

$$s_{ij} \equiv E_{ij} \left(|s_i| + |s_j| \right) T_{ij} \tag{2.2}$$

The first term on the right-hand side of eq. (2.1) corresponds to noninteracting mutations, with selection coefficient s_i for each site i.

According to the theorem of population genetics, different loci evolve independently in infinite population if the contributions of mutations to the log fitness are additive. Interactions between loci, at the level of gene regulation and transcription, as well as mRNA or protein folding and interaction, create nonadditive effects to log fitness. The second term in eq. (2.1) describes pairwise interactions of loci with magnitudes s_{ij} given by eq. (2.2); higher-order interactions are neglected. An alternative "mean-field" model of global epistasis where epistatic effects are averaged out over a large number of pairs is investigated by Good and Desai (2015) and Jerison and Desai (2015).

Factor E_{ij} is the relative strength of interaction between sites i and j, while the topology matrix $T = T_{ij}$ indicates an interacting pair by $T_{ij} = 1$ and a noninteracting pair by $T_{ij} = 0$. For the simplest case of a site pair with $s_i < 0, s_j < 0$ not interacting with any other site, the value $E_{ij} = 1$ corresponds to the full mutual compensation of the two deleterious mutations, $W = 0$. By the definition, the sign of E_{ij} is the same as the sign of the interaction term in fitness s_{ij}, regardless of the signs of selection coefficients s_i, s_j. If the interaction term increases fitness, one has $E > 0$ (positive epistasis), and if it decreases fitness, one has $E < 0$ (negative epistasis).

2.1.2 The footprint of epistasis for a single interacting pair in a long genome

The task is of this section is to develop a method to estimate epistasis E for a locus pair from available sequence samples that would be least-affected by the presence of the other loci and model parameters. The focus will be on the regime of negative selection, $s < 0$, $|s| \ll 1$ with positive epistasis, $E > 0$. Thus, the effect of two interacting alleles together is less deleterious than the sum of their effects if they occurred independently. In other words, alleles are mutually compensatory mutations.

For starters, consider an interacting pair of sites with negative selection coefficients $-s_1$, $-s_2$ and epistatic strength E in a long genome with a total log-fitness W (Fig. 2.1). The epistatic interaction of the pair with the other sites elsewhere in the genome is neglected. The approximation is lifted in Section 2.1.5, where the effect of more complex epistatic clusters is considered.

The population is assumed to be in the process of multi-locus adaptation with strong clonal interference. In this case, various models demonstrate that the distribution density of fitness W across individual genomes represents a traveling wave, which moves slowly toward higher fitness and is relatively narrow at any realistic population size [see Rouzine (2020b) for review]. In the present context of our work, one can consider it the same for all genomes. There is no need to consider the wave

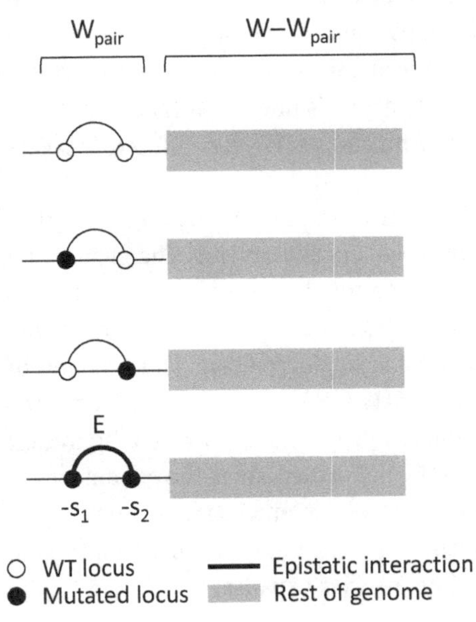

Fig. 2.1: A pair of interacting sites in a long genome.
Open and filled circles: wild type 0 and mutated allele 1. Thick line: existing interaction. Black line: potential interaction. Grey box: the rest of genome. Relative strength of epistasis E is defined in eq. (2.2). Based on Pedruzzi et al. (2018).

explicitly. Analysis given below will rely on the fact that the wave is relatively narrow and relatively slow, because the speed of movement is limited by rare mutation events adding new genomes at the high-fitness front (Rouzine et al. 2003).

The task is to predict the allelic association established during such a slow adaptation. For this purpose, all individual genomes in a population are classified according to the haplotype sequence of the pair of interest: 00, 01, 10 and 11 (Fig. 2.1). Each haplotype class is diverse at the other sites of genome. The pair contributes W_{ij} to the logarithm of the total genome fitness W, which depends on the haplotype sequence as given by

$$\begin{aligned} W_{00} &= 0 \\ W_{01} &= -s_1 \\ W_{10} &= -s_2 \\ W_{11} &= -(s_1+s_2)(1-E) \end{aligned} \tag{2.3}$$

Next, the evolution of a population occurs under competing factors, as follows. Stochastic factors, such random mutation and random genetic drift due to fluctuations of progeny number, increase disorder. The standard measure of disorder is entropy S, which is defined as the logarithm of the number of the equally probable configurations, $S = \log N_{\text{conf}}$. The opposing force is the deterministic effect of natural selection, which decreases disorder and increases Malthusian fitness (average progeny number). As already mentioned, the fitness distribution in the multiple-locus regime is a narrow peak at $W_{\max}(t)$, which is changing very slowly in time t (Rouzine et al. 2003; Desai and Fisher 2007; Brunet et al. 2008; Rouzine et al. 2008). Hence, at any given moment of time, entropy can be assumed to be maximal under the condition that fitness of all genomes is approximately the same and equal to $W \approx W_{\max}(t)$. The principal hypothesis made here is that entropy calculated over the ensemble of sequences adjusts sufficiently quickly and approach its current maximum. This hypothesis of quasi-equilibrium is tested below by computer simulation. Under this assumption, the maximum value of $S(t)$ depends only on $W(t)$ and not on t explicitly, as given by $S(t) = S[W(t)]$. Examples of function $S(W)$ for different types of epistatic network are presented in Tab. 2.1.

Consider all sequences in a haplotype class, for example, 10 (Fig. 2.1). The other sites of the genome outside of this site pair can be genetically diverse. The probability of haplotype, f_{10}, by the definition, is the ratio of the number of sequence configurations in the class, denoted $\exp(S_{10})$, to the total possible configuration number. According to the hypothesis above, the entropy of each haplotype class is a function of the fitness of the rest of genome, as given by $S_{10} = S(W - W_{10})$. The average frequency at which each haplotype is found in a population is proportional to its configuration number, as given by $f_{10} \propto \exp(S_{10}) = \exp[S(W - W_{10})]$. Further, since the genome is assumed to be long, $L \gg 1$, one can safely assume that W_{10} is much smaller than W, so that the corresponding change in entropy is relatively small and proportional to W_{10}. Hence, one can approximate that change by a linear series expansion

Tab. 2.1: Expressions for fitness $-W/s_0$, the exponential of entropy $\exp(S)$, and haplotype frequencies f_{10}, f_{11} in different cases of epistatic network topology (Fig. 2.6). In the expressions for entropy, strong inequality $k_i \ll L$ is assumed.

Topology	Fitness	exp(*S*)	Haplotype frequency
Isolated pairs	$k_1 + 2k_2(1-E)$	$C_{L/2}^{k_2} C_L^{k_1}$	$f_{10} = k_1/L$ $f_{11} = 2k_2/L$
Double arches	$k_1 + 2k_2(1-E) + 2k_3(3-4E)$	$C_{L/3}^{k_3} C_{L/3}^{k_1} C_{L/3}^{k_2} 2^{k_2} 3^{k_1}$	$f_{10} = f - f_{11}$ $f_{11} = 3(k_2 + k_3)/(2L)$ $f = (k_1 + 2k_2 + 3k_3)/L$
Triple arches	$k_1 + 2k_2(1-E) + 3k_3(1-2E)$	$C_{L/3}^{k_3} C_{L/3}^{k_1} C_{L/3}^{k_2} 3^{k_2} 3^{k_1}$	$f_{10} = (k_1 + k_2)/L$ $f_{11} = (k_2 + 3k_3)/L$
Chain	$k_1 + \sum_{i=2}^{I} (i - 2(i-1)E) k_i$	$\prod_{i=1}^{I} C_L^{k_i}$	$f_{10} = \sum_{i=1}^{I} k_i/L$ $f_{11} = \sum_{i=2}^{I} (i-1) k_i/L$
Binary tree		$\prod_{i=1}^{I} C_{L_i}^{k_i} (A_i)^{k_i}$	
Double arches unequal	$k_1 + 2k_2(1-E) + 3k_3(1-E) + k_2'(2-E)$	$C_{L/3}^{k_3} C_{L/3}^{k_1} C_{L/3}^{k_2} C_{L/3}^{k_2'} 3^{k_1}$	$f_{10} = f - f_{11}$ $f_{11} = 3(k_2 + k'_2 + k_3)/(2L)$ $f = (k_1 + 2k_2 + 2k'_2 + 3k_3)/L$

$$S(W - W_{10}) \approx S(W) - \beta W_{10} \tag{2.4}$$

where $\beta = dS/dW$. Combining eqs. (2.3) and (2.4) and its analogues for the other three haplotypes, the haplotype frequencies are expressed in terms of s_1, s_2, and E as given by

$$\begin{aligned} f_{10} &= f_{00} e^{-\beta s_1} \\ f_{01} &= f_{00} e^{-\beta s_2} \\ f_{11} &= f_{00} e^{-\beta (s_1 + s_2)(1-E)} \end{aligned} \tag{2.5}$$

where f_{00} is found from the normalization condition $f_{00} + f_{01} + f_{10} + f_{11} = 1$. After excluding β, s_1, s_2 from eq. (2.5), one obtains (Pedruzzi et al. 2018)

$$\frac{f_{11}}{f_{00}} = \left(\frac{f_{10} f_{01}}{f_{00}^2} \right)^{1-E}$$

which is equivalent to

$$E = 1 - \frac{\log(f_{11}/f_{00})}{\log(f_{01} f_{10}/f_{00}^2)} \tag{2.6}$$

Equation (2.6) can be used to estimate the strength of interaction E from a single-time sample of DNA, RNA, or protein sequences. Unlike the existing measures of linkage disequilibrium, this measure of epistasis not only detects the fact of allelic correlation, but also has a direct biological meaning, does not depend on a population state or other model parameters, and has a fair degree of universality, as we show below. Hence, it will be called Universal Footprint of Epistasis (UFE). (A similar method can be used to measure relative selection coefficients s_1, s_2 from a sequence set, in the units of unknown parameter $1/\beta$, see Section 2.4).

It is worth stressing once again that the above analytic result, eq. (2.6), is based on the hypothesis of quasi-equilibrium of "adiabatically slow" process of traveling wave. [This condition is not to be confused with "quasi-linkage-equilibrium" (Kimura 1965; Neher and Shraiman 2011) in very large populations in the presence of recombination not considered here.]

The result and the hypothesis were tested by Monte-Carlo simulation described in Section 2.1.1 for a long genome comprising isolated epistatic pairs with epistastic strength E. The estimate from eq. (2.6) was compared with the actual value of E set in simulation (Fig. 2.2). The simulation demonstrates that eq. (2.6) represents a fairly accurate estimate of E after ~ $1/s$ generations have passed, which is the time of establishment of the traveling wave (see Section 2.4). Similar result follow in a broad parameter range set including much larger N. In the very long-term, however, when deleterious mutation almost balances natural selection, strong deviations of UFE from E occur (Fig. 2.2). Thus, the estimate in eq. (2.6) does not work near true equilibrium, it applies only in quasi-equilibrium maintained during the process of the slow adaptation at multiple loci.

2.1.3 The long genome of isolated pairs

To consider an explicit example of entropy as a function of fitness, and further verify the validity of eq. (2.6), it is instructive to consider the entire genome comprising isolated epistatic pairs. For the sake of simplicity, all selection coefficients and epistatic strengths are assumed to be the same for all sites and pairs, $s_i = -s_0, E_{ij} = E_0$. The method and the analytic results are described below, and detailed derivation is given in Section 2.1.7. Epistatic networks with more complex topology will be analyzed in Section 2.1.5.

Mutated clusters are grouped by their size and counted: k_1 single mutations and k_2 mutated pairs (Fig. 2.3). Fitness and entropy can be both expressed in terms of these numbers (see Section 2.1.7.1 for the exact expressions). To obtain the most disordered (and hence the most probable) state, k_1 and k_2 are chosen from the condition that entropy S is maximal subject to the restriction that fitness density W/L is fixed. Assuming that mutations in the genome are rare, $k_1 \ll L$ and $k_2 \ll L$, entropy S can be approximated by a function continuous in its arguments, k_1 and k_2. The task is to find the derivatives of $S(k_1k_2)$ in both variables and express the average frequencies

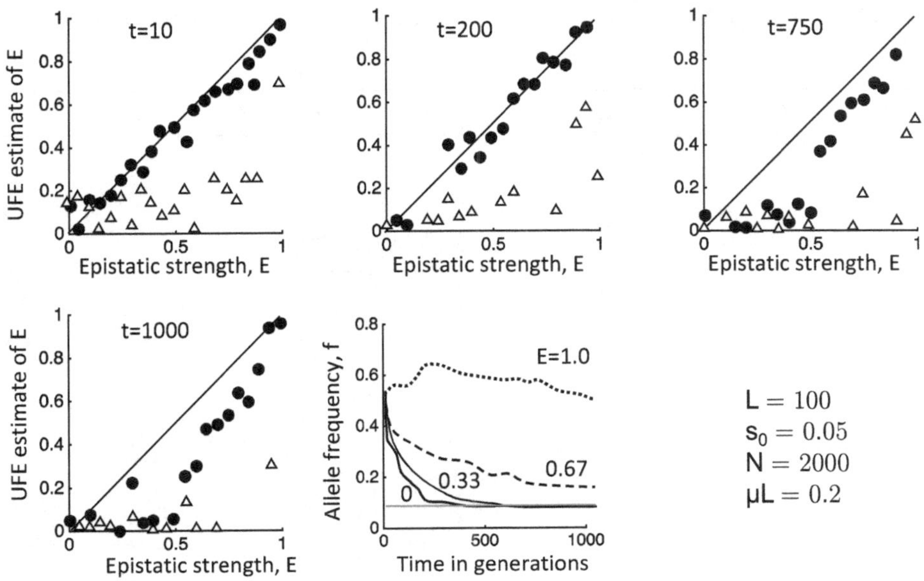

Fig. 2.2: The value of E estimated from UFE relation, eq. (2.6), as a function of the actual value of E. Shown are times in generations. Bottom right: Time evolution of the mutant frequency f. Each dot represents a single Monte-Carlo run. Initial population is randomized with $f = 0.5$. Haplotype frequencies are averaged over sites and pairs. Filled circles: known epistatic pairs. Open triangles: the same number of randomly chosen pairs. Parameter set: $L = 100, s_0 = 0.05, N = 2000, \mu L = 0.2$, one bond per interacting site.

of haplotypes 10 and 00 in terms of k_1 and k_2. From the condition that entropy is maximum for a fixed "fitness frequency" f_0 defined as $f_0 = |W|/(sL)$, and when mutations are rare, $f_{00} \approx 1$, one arrives at a relation between haplotype frequencies identical to eq. (2.6). Finally, one uses the condition $f_{10} + f_{11} = f$ to express the average mutation frequency f in terms of E and f_0. The mathematical details are given in Section 2.1.7.1.

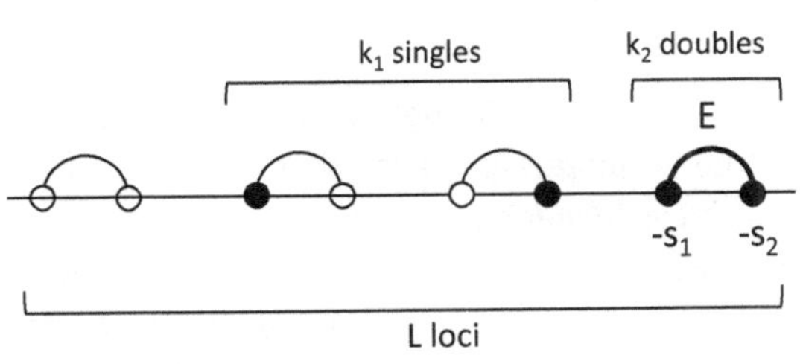

Fig. 2.3: Long genome of interacting pairs.
Interacting pairs with different haplotypes. Based on Pedruzzi et al. (2018).

At the half-compensation point $E = 0.5$, mutated pairs and singles have the same fitness and the same frequency, $f_{11} = f_{10}$. Off this point, one group strongly dominates

Tab. 2.2: Correlation coefficients and critical points for different examples of epistatic network topology (Fig. 2.6).

Name	**Bond #**	E_c	E_{UFE}	**Interval of E**	D_{11}	D_{10}	$f(E)$
Arch	$b_2 = 1$	1	1	$E < \frac{1}{2}$	f^{-2E}	1	f_0
				$\frac{1}{2} < E < 1$	f^{-1}	$f^{\frac{E-1/2}{1-E}}$	$\frac{f_0}{1-E}$
Triple arch	$b_3 = 3$	1/2	1/4	$E < \frac{1}{4}$	f^{-2E}	1	f_0
				$\frac{1}{4} < E < \frac{1}{3}$	f^{1-6E}	1	f_0
				$\frac{1}{3} < E < \frac{1}{2}$	f^{-1}	$f^{\frac{E-1/3}{1/2-E}}$	$\frac{f_0}{1-2E}$
Double arch	$b_3 = 2$	3/4	1/2	$E < \frac{1}{2}$	f^{-2E}	1	f_0
				$\frac{1}{2} < E < \frac{3}{4}$	f^{-1}	$\frac{1}{3} f^{\frac{2E-1}{3-4E}}$	$\frac{f_0}{1-4E/3}$
Chain	$b_i = i - 1$	1/2	$\frac{1}{2} - \frac{1}{\log \frac{1}{f_0}}$	$\frac{1}{2} - E \gg \frac{1}{\log \frac{1}{f_0}}$	f^{-2E}	1	f_0
				$\frac{1}{2} - E \ll \frac{1}{\log \frac{1}{f_0}}$	f^{-1}	$\frac{f_0}{f}$	$\frac{f_0}{(1-2E) \log \frac{1}{f_0(1-2E)}}$
Binary tree	$b_i = i - 1$	1/2	$\frac{1}{2} - \frac{\log 4}{\log \frac{1}{f_0}}$	$E_{\text{UFE}} - E \gg \frac{1}{\log \frac{1}{f_0}}$	f^{-2E}	1	f_0
				$E_{\text{UFE}} - E \ll \frac{1}{\log \frac{1}{f_0}}$	f^{-1}		

Here b_i is the number of bonds for a cluster of i mutations, E_c is the point of full compensation, E_{UFE} is the maximum epistatic strength at which UFE still applies. D_{11} and D_{10} are pairwise correlation coefficients, f is mutant allelic frequency, $f_0 = -W/(sL)$ is the effective "fitness frequency."

another: at $E < 0.5$, singles overcome doubles, and doubles dominate singles at $E > 0.5$. Thus, the presence of epistasis amends the intuitive expectation that double deleterious mutations are much more rare than single deleterious mutations.

In addition to UFE, it is informative to derive a standard measure of linkage disequilibrium, Pearson correlation coefficients $D_{11} = f_{11}/f^2$, $D_{10} = f_{10}/[f(1-f)]$. By the definition, if the allelic frequencies are statistically independent, $D_{11} = D_{10} = 1$. Their values above unit show positive correlation, and the values below unit negative correlation. The correlation coefficients can be expressed in terms of the average mutant frequency f (Tab. 2.2). In turn, f can be expressed in terms of the input parameters E and normalized fitness $f_0 = -W/(sL)$. The analytic results are compared with simulation at the initial frequency value $f(0) = 1/10$ (Fig. 2.4). As expected, the frequency of mutant alleles f, which is small in the absence of epistasis, diverges near the full compensation point $E = 1$ until it reaches the value of 0.5 (Figs. 2.4C and 2.5C; Tab. 2.2). The half is caused by the symmetry between the wild-type and mutant alleles existing at $E = 1$. Correlation coefficient D_{11} sharply peaks at $E = 1/2$ and then decreases linearly with E to the value of 2 at point $E = 1$ (Figs. 2.4A and 2.5A), where $f_{00} = f_{11} = 1/2$. In contrast, coefficient D_{10} at low E stays near 1 and declines rapidly to zero at $E > 1/2$, where the doubles outnumber the singles (Figs. 2.4B and 2.5B). Simulation confirms analytic results. Most importantly, UFE holds in the entire interval of E (Figs. 2.2 and 2.5D).

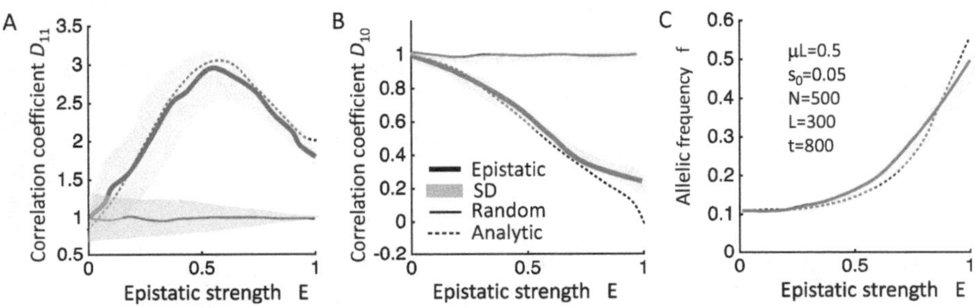

Fig. 2.4: Epistasis causes strong linkage disequilibrium: analytics and simulation.
(A, B, C) Correlation coefficients D_{11}, D_{10} and less-fit allele frequency f after t generations are shown as a function of E. D_{11}, D_{10} are calculated from eq. (2.6) using simulated values of f_{ij} and f averaged over sites and pairs. Solid curves correspond to the average over 300 runs, and the shaded areas show the standard deviation among runs, for epistatic pairs (fat curves) and the same number of random pairs (thin). Dashed curve is the analytic prediction. Parameters: $N = 500$, $s_0 = 0.05$, $L = 300$, $\mu L = 0.5$, $t = 800$, $f_0 \approx 0.1$. Initial population is randomized with $f = 0.5$. Based on Pedruzzi et al. (2018).

2.1.4 Full compensation and UFE interval

The topology of actual epistatic network can be more complex than a set of isolated pairs (Fig. 2.6B–E). Specific examples are discussed in Section 2.1.5, but first, it is useful to derive two important points in E for any topology using the general expression for fitness, eq. (2.1). The first observation is that the point of full compensation is given by

$$E_c = \min_i[i/(2b_i)] \tag{2.7}$$

where b_i is defined as the (average) number of bonds in a cluster of size i. The point $E = E_c$ represents the threshold of full compensation for one of cluster sizes. At $E > E_c$, the genetic stability of the entire population is lost (Tab. 2.2): the critical clusters rapidly expand in time, until mutant frequency is no longer small, and the entire genome has mutated. If mutations have a specific important phenotype, such as resistance of a virus (tumor) to a treatment or the immune response, this is the point where a virus becomes resistant (a tumor starts to grow). Evolutionary escape of viruses from the immune response is considered in Sections 4.1 and 4.2.

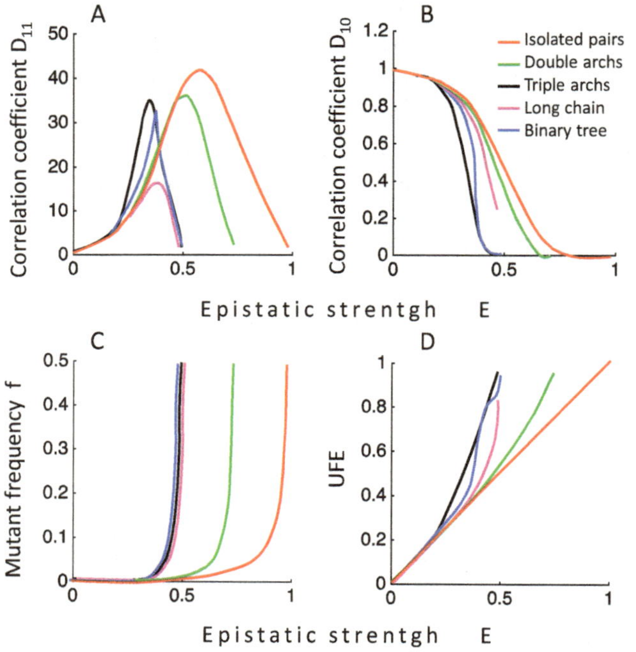

Fig. 2.5: UFE is preserved for different topologies at moderate epistatic strengths. Dependences on E are shown for (A, B) correlation coefficients, (C) mutation frequency f, and (D) UFE on E for the five topologies in Fig. 2.6. UFE is the estimate of E, eq. (2.6), from haplotype frequencies f_{11}, f_{01} derived analytically for each topology (Sections 2.1.7–2.1.12). Parameters: $f_0 = 1/100$. Thus, UFE is exact for the isolated pairs, and overestimates E at large E for other topologies. Asymptotic expressions are given in Tab. 2.2. Based on Pedruzzi et al. (2018).

For isolated pairs, eq. (2.7) gives $E_c = 1$, as it should. In a general topology of epistatic network, the compensation point E_c can be less than 1, because each mutation can be compensated by more than one mutation, depending on the number of bonds b_i.

The second important result is the interval $E < E_{\text{UFE}}$ where UFE given by eq. (2.6) is an accurate estimate of epistatic strength, which upper boundary is given by

$$E_{\text{UFE}} = \min_{i>2}\left[\frac{i-2}{2(b_i-1)}\right] \leq 1/2 \tag{2.8}$$

Because both derivations of eq. (2.6) in Sections 2.1.2 and 2.1.3 neglect triple mutations and higher clusters, $E < E_{\text{UFE}}$ represents the condition that epistatic doubles are much more numerous than triplets and larger clusters, so that these last can be neglected. The case of general topology and $E > E_{\text{UFE}}$ is considered in the next section.

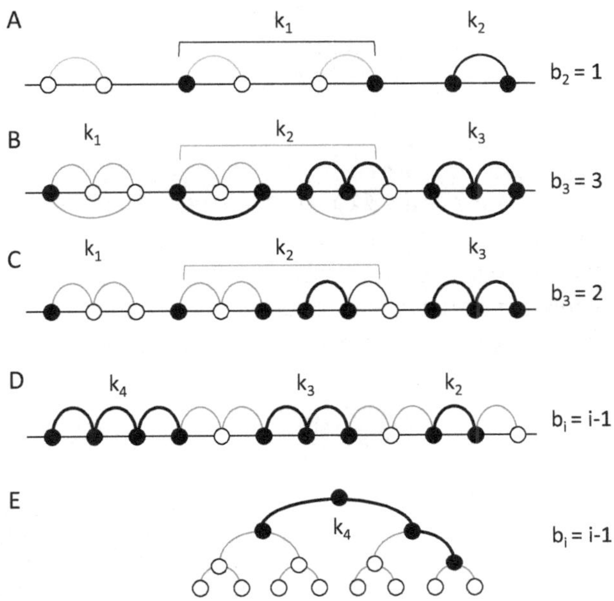

Fig. 2.6: Examples of epistatic network.
Filled and open circles denote mutated and wild-type genomic sites, respectively. Existing and potential epistatic interactions between sites are shown by thick and thin arches, respectively. Thick arches connect clusters of mutated sites. k_i is the number of the cluster with i sites, b_i is the bond number per cluster. Different topologies correspond to A) isolated pairs, B) isolated triple arches, where each site has two epistatic partners, C) double arches, where three sites are involved in two epistatic associations, D) a long chain where each site forms two pairs. Note that (A–D) show connection topology but not the actual site order. (E) Binary tree: possible site order and the equivalent tree structure. Based on Pedruzzi et al. (2018).

2.1.5 Effects of network topology

In this section, several examples of more complex network topology are considered (Fig. 2.6B–E). At large epistatic strengths, epistatic clusters of more than two mutations become numerous. The starting point is to express entropy, fitness, and the haplotype frequencies in terms of the average numbers of mutated clusters of a different size found in a genome. These expressions and the final results are listed in Tabs. 2.1 and 2.2, respectively. The general method and the final results for UFE and correlation coefficients for various topologies are discussed below. Detailed derivations are given in Section 2.1.7.

General method for uniform s and E. Consider a weakly diverse population, $f \ll 1$, and assume that all mutations are deleterious with equal selection coefficient $s_i = -s_0 < 0$, and that epistatic strength is equal for all interacting pairs, $E_{ij} = E > 0$. In this case, a genome can be fully characterized by the set of the numbers of mutated clusters k_i of different size i with b_i bonds (Fig. 2.6). The consideration below is limited to simple topologies in which b_i assumes a single value for each cluster size i. Fitness given by eq. (2.1) can represented as a sum over clusters of different size

$$W \equiv -s_0 f_0 L = -s_0 \sum_{i=1}^{i_{max}} k_i (i - 2Eb_i) \tag{2.9}$$

New notation $f_0 \equiv -W/Ls_0$ has the meaning of the frequency of mutations in the absence of epistasis with the same total fitness W. The bond number b_i for cluster size $i > 2$ depends on the particular topology (Fig. 2.6); for any topology, $b_1 = 0$, $b_2 = 1$. (Please note that Fig. 2.6 shows only the topology of the interactive network and not the actual location of epistatic sites in the genome, which does not affect the results, as long as recombination is absent.)

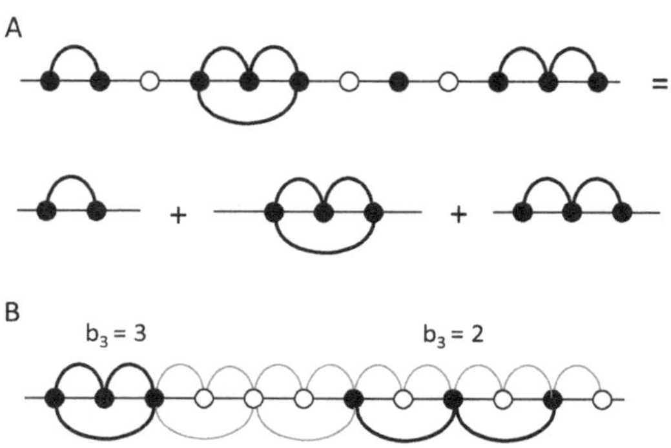

Fig. 2.7: Variations of topology.
(A) A diverse topology consisting of finite subgraphs of epistatic interactions from Fig. 2.6 can be considered as a composition of uniform topologies (i.e. single pairs, double and triple arches), setting aside nonepistatic loci. (B) Example of the complex topology, when two subclusters of the same size may have different number of bonds (not considered in this work). Based on Pedruzzi et al. (2018).

Entropy in quasi-equilibrium. As mentioned above, the values of k_i are determined by the condition that entropy S is maximal subject to the condition that fitness is fixed and given by eq. (2.9). By the definition, entropy is the logarithm of the configuration number

$$e^S = \prod_{i=1}^{i_{max}} C_{L_i}^{k_i} (n_i)^{k_i} \tag{2.10}$$

Here L_i is the number of possible positions for a cluster of size i, and n_i is the number of its possible configurations given position. Topology of epistatic network determines these values. Because deleterious mutations are assumed sparse, as given by

$$f = \frac{1}{L} \sum_i ik_i \ll 1 \tag{2.11}$$

the cluster overlap is neglected. The last condition is violated near the full-compensation point, $E = E_c$, where f diverges (below). Entropy given by eq. (2.10) is maximal with respect to k_i with the fitness restriction from eq. (2.9).

From eqs. (2.9) and (2.11), one gets $f(E=0) = f_0$. The "fitness density" $f_0 \equiv -W/s_0L$ is treated as a given input parameter. At positive E, we have $f(E) > f_0$. The dependence $f(E)/f_0$ is derived from eq. (2.9), for each topology, in Section 2.1.7.

Haplotype frequencies. Haplotype frequencies can be expressed in terms of the numbers of clusters k_i

$$f_{11} = \frac{1}{L_{\text{pair}}} \sum_i k_i b_i \tag{2.12}$$

$$f_{10} = f_{01} = f - f_{11}$$

where $L_{\text{pair}} = \sum_{ij} T_{ij}$ is the total number of interacting pairs in the genome. In turn, the correlation coefficients D_{ij} are expressed in terms of haplotype frequencies given by eq. (2.12)

$$D_{11} = \frac{f_{11}}{f^2}, \quad D_{10} = \frac{f_{10}}{f(1-f)} \tag{2.13}$$

If two sites are statistically independent, by the definition, $D_{11} = D_{10} = 1$. The values above 1 show positive correlation.

Nonepistatic sites and nonuniform network. Noninteracting sites in the genome can be ignored (Fig. 2.7A), because fitness W and entropy S are additive over epistatic and nonepistatic part, and the frequencies f_{11}, f_{10}, f, and f_0 are defined for epistatic sites only. Hence, one can separate the genome into two parts and maximize their entropy values separately. For the same reason, a diverse mixture of disconnected epistatic graphs can be split into independent parts with uniform segments (Fig. 2.7A).

Network topology: triple arches. Consider now a specific topology, the periodic sequence of three-node graphs connected by three bonds (Fig. 2.6B). Entropy and fitness depend on the number of single mutations, k_1, double mutations, k_2, and triple mutations, k_3 (Tab. 2.1). According to the general method, we maximize entropy, eq. (2.10), in these three variables while keeping fitness, eq. (2.09), fixed.

The results derived below (Section 2.1.7) are, as follows. Full compensation point is $E_c = 1/2$ (Fig. 2.5A, B; Tab. 2.2). The dependence of correlation coefficients D_{10}, D_{11} on E is different in three intervals of epistatic strength E, as follows. At weak epistasis $E < 1/4$, the triplets are very few compared to the doubles, as given by $k_3 \ll k_2$. In this interval, UFE in eq. (2.6) is accurate (Fig. 2.5D). At stronger epistasis $1/4 < E < 1/3$, the triplets outcompete the doubles, $k_3 \gg k_2$, so that interacting pairs of mutated loci are found mainly in triplets. As a result, UFE estimate predicts a larger value of E than the actual value (Fig. 2.5D), which also results in a steeper increase of correlator D_{11} with E than for the topology of isolated pairs. At even stronger epistasis, $1/3 < E < 1/2$, triplets outnumber both the doubles and single mutant alleles. When the point of full compensation $E = 1/2$ is approached, the accumulation of triplets causes divergence of average mutant frequency f and a linear decrease in D_{11} (Fig. 2.5A, C). Thus, the presence of extra bonds compared to the topology of isolated pairs generates a positive correction to UFE above the value of E, which has to be taken into account when measuring E from sequence data.

Double arches. To test this conclusion further, a bond from each triple arch is removed producing the double arch topology (Fig. 2.6C). The results of this change from are shown in Fig. 2.5 and Tab. 2.2. The intermediate interval $1/4 < E < 1/3$ disappears. Instead, singles, doubles, and triplets have the same fitness (and, hence, similar abundance) at the same point, $E = 1/2$. Below this point, singlets are the most numerous, while above this point triplets dominate. The upper boundary of the UFE accuracy E_{UFE} moves up from 1/4 to 1/2, and full compensation takes place at a larger value $E_c = 3/4$. The derivation is found in Section 2.1.7.

Long chain. So far, small interacting clusters were considered. A more complex and more realistic epistatic network is a long chain of neighbor-interacting sites (Fig. 2.6D). In this topology, a mutated cluster of any size i smaller than L can exist, and it has $b_i = i - 1$ epistatic bonds. After maximizing entropy at a fixed fitness, one obtains that the frequencies of clusters of different size form a geometric progression (Section 2.1.7). Due to the assumption $f \ll 1$, its denominator is very small, except in the direct vicinity of the full compensation point $E \approx 1/2$. Otherwise, clusters larger than two are negligible. As a result, the UFE formula derived for an interacting pair is valid in the most of the interval of E. Only in a narrow vicinity of compensation point, large clusters become important causing divergence of f and overestimation of E from UFE (Fig. 2.5D, Tab. 2.2).

Binary tree. A tree is a graph where any two nodes can be connected by a single path (Fig. 2.6E). Results for the binary tree and for the chain are similar (Section 2.1.7), because the relationship between the number of bonds and nodes is exactly the same for the two topologies, $b_i = i - 1$. Hence, the critical point, $E_c \approx 1/2$ is also the same as in the chain. The main difference is in entropy, which is larger for the tree than for the chain. A cluster of size i is a subtree with $n_i = (2i)!/[i!(i+1)!]$ configurations, instead of only one, as in the chain. This nuance favors larger clusters, even though their fitness is exactly as in the chain. As a result, near the critical point $E_c \approx 1/2$ where larger clusters become important, both the correlation coefficients and UFE increase more sharply than for the chain topology, and the peak in D_{11} is higher (Fig. 2.5A). As in the chain, estimate $UFE \approx E$ remains accurate until close to full compensation (Fig. 2.5D). Similar results are obtained for a tree with any number of branches, even if random.

Doubles arches with unequal interactions. In the examples above, all epistatic interactions were assumed equal, but in real systems they vary. Hence, it is interesting to know how the result depends on that variation. The sensitivity to this factor can be illustrated in the simplest case of "double arches" (Fig. 2.6C). Suppose the left arch, E, is stronger than the right arch, which has strength $E/2$. The results can be compared between the cases of equal and unequal interaction, as follows (Fig. 2.8, Tab. 2.2, details in Section 2.1.7). In contrast to the case of equal interactions, there are three intervals of E with different behavior instead of two. Full compensation occurs at larger E, at $E_c = 1$, as for isolated pairs. Correlation coefficient D_{10} in the last interval $[2/3, 1]$ changes its behavior: for the symmetric arches, it decreased exponentially, and for the asymmetric arches, it decreases as a power law. UFE now is intermediate between the two values of E, but closer to the larger of the two.

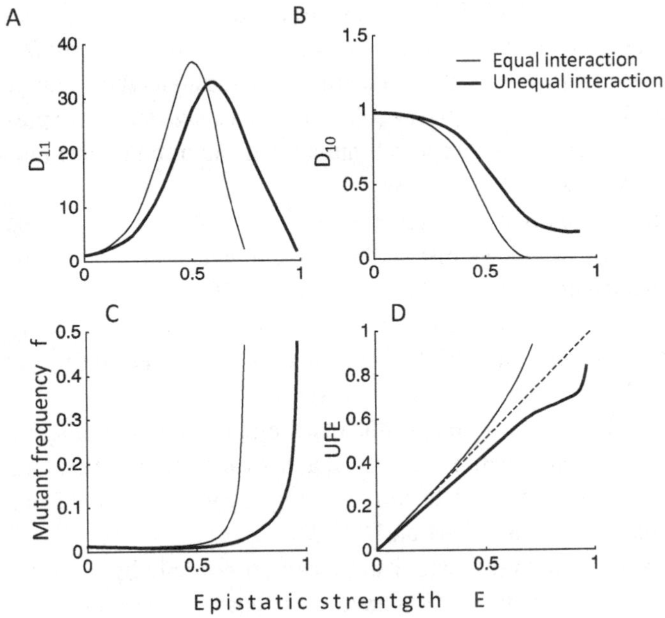

Fig. 2.8: Results for the network of double arches (Fig. 2.6C) with unequal epistatic strength. Y-axis: correlation coefficients, average frequency and UFE. X-axis: epistatic strength E. The two cases are equal epistatic strength (thin curves) and unequal strengths E and $E/2$ (thick curves); $f_0 = 1/100$ is in 61 put parameter. Based on Pedruzzi et al. (2018).

2.1.6 Discussion

Using analysis and Monte-Carlo simulation of adapting asexual populations, a relationship between haplotype frequencies of a pair of sites f_{11}, f_{10}, f_{00} is obtained (UFE). This measure can be used to estimate the strength of relative pairwise interaction E from sequence data independently on other parameters (apart from the topology of the epistatic network). For example, selection coefficients, mutation rate, and population size may be unknown, which fact does not affect the results. At moderate epistatic strengths E, UFE is shown to be independent on topology and equal to E upon sufficient averaging. For the simplest topology of isolated epistatic pairs, $UFE = E$ in the entire interval of E until the full compensation point, which depends on a topology. For more complex networks, the model predicts the point in E above which UFE acquires topology-dependent corrections. These corrections can be compensated when epistatic pairs are determined from genomic sequences, as discussed in Section 2.3. The point of full compensation (genetic instability) in E is derived in the general form for the case of a uniform network; it decreases with the number of interactions per interacting locus. These predictions can be used for identifying, from sequence data, the clusters of compensatory mutations and the genetic stability thresholds, such as precancer cells or drug resistant virus (Nijhuis et al. 1999; Levin et al. 2000; Handel et al. 2006; Cong et al. 2007; Gonzalez-Ortega, et al. 2011).

The results demonstrate the existence, at any time of point of adaptation process, of a quasi-equilibrium between natural selection and entropy associated with random genetic drift and mutation. The reason for quasi-equilibrium is a slow rate of adaptation associated with a traveling wave whose speed limited by the addition of new mutations [for review, see Rouzine (2020b)].

The limitations of UFE method are as follows:

(i) As with the other measures of linkage disequilibrium, one needs sufficiently diverse pairs of sites (Rouzine and Coffin 1999b; Batorsky et al. 2011).

(ii) On very long times, when the mutation-selection balance is established, UFE is wiped out (Fig. 2.2). The reason is that deleterious mutation (not important in adaptation) compensates the effect of selection and mixes haplotypes.

(iii) Above, E and s are assumed equal for all epistatic pairs, and haplotype frequency averaging is done over different pairs. In real life, E and s vary among pairs, and averaging has to be done, for each separate pair, over sequence sets isolated from independent populations, as is done in Section 2.3.

(iv) In the same vein, correlation between pairs can be caused not only by epistasis but also by linkage effects, and the two effects have to be separated (Section 2.3).

(v) An asexual population is considered. Strong recombination can partly compensate for epistasis. In the limit of very large population, the opposing forces of recombination and epistasis create another state, quasi-linkage equilibrium (Kimura 1965).

Despite these limitations, the development of the direct measure of epistatic strength that can take into account specific topology is very helpful for biomedical applications. Two of these limitations are overcome in Section 2.3.

2.1.7 Mathematical derivations

This section serves as a mathematical appendix for Sections 2.1.3 and 2.1.5. In eqs. (2.1) and (2.2), all selection coefficients are assumed to be equal, as are the coefficients of epistatic strength

$$s_i = -s_0, \quad s_0 > 0, \quad i = 1, \ldots, L$$

$$E_{ij} = E, \quad i = 1, \ldots, L; \quad j < i$$

Under these conditions, here and below we consider the following topologies (Fig. 2.6):
- Isolated pairs (a)
- Triple arches (b)
- Double arches (c)
- Connected long chain (d)
- Large binary tree (e)
- Double arches with unequal epistasis strength (c)

The value of W is fixed, and additional restrictions are imposed

$$0 < E < 1 \tag{2.14}$$

$$1 \ll k_i \ll L; \quad i = 1, \cdots, L \tag{2.15}$$

where k_i is a number of clusters of i mutated sites. The condition in eq. (2.14) is for clarity only, because many results given below can be extended from small positive to negative E. The condition in eq. (2.15) implies $f_{00} \approx 1, \ f \ll 1$.

2.1.7.1 Isolated pairs (Section for 2.1.3)

Consider a population of genomes where all sites interact only as isolated pairs (Fig. 2.6A). As discussed in Section 2.1.3, entropy is maximal under the restriction of total fitness W given by eq. (2.1), which, in the case of isolated pairs, can be written in the form

$$f_0 L \equiv -W/s_0 = k_1 + 2k_2(1 - E) \tag{2.16}$$

where normalized fitness $f_0 \ll 1$ is constant and considered given, and k_1, k_2 are the numbers of epistatic pairs with one and two mutated sites, respectively. Given k_1 and k_2, entropy S is the logarithm of the configuration number

$$e^S = N_{conf}$$

In the case of isolated pairs, one has

$$N_{conf}(k_1, k_2) = C_{L/2}^{k_2} C_{L-2k_2}^{k_1}$$

The task is to derive the condition $dS = 0$ subject to fixed fitness, eq. (2.16). Although $\log(N_{conf})$ is a function of integer values k_1 and k_2, if these values are large, the function changes slowly with its arguments and can be approximated with a smooth function. By differentiating S in k_1 and k_2, one gets

$$0 = dS = \log\left(\frac{N_{conf}(k_1+1,k_2)}{N_{conf}(k_1,k_2)}\right)dk_1 + \log\left(\frac{N_{conf}(k_1,k_2+1)}{N_{conf}(k_1,k_2)}\right)dk_2 \tag{2.17}$$

where function N_{conf} changes with the increments in k_1 and k_2 as

$$N_{conf}(k_1+1,k_2) \approx \frac{N_{conf}(k_1,k_2)}{f_{01}}$$

$$N_{conf}(k_1,k_2+1) \approx \frac{N_{conf}(k_1,k_2)}{f_{11}} \tag{2.18}$$

Here the relative error $O(k_i/L)$ is neglected due to the strong inequality, eq. (2.15). In eq. (2.18), the haplotype frequencies are calculated as

$$f_{01} = f_{10} = \frac{k_1}{L}$$

$$f_{11} = \frac{2k_2}{L} \tag{2.19}$$

Thus, eq. (2.17) simplifies to

$$0 = \log\left(\frac{1}{f_{10}}\right)dk_1 + \log\left(\frac{1}{f_{11}}\right)dk_2 \tag{2.20}$$

Differentiating eq. (2.16), which is the fixed fitness relation for k_1 and k_2, yields an additional condition

$$dk_1 = -2(1-E)dk_2$$

which is substituted into eq. (2.20). After setting the resulting factor at dk_2 to zero, one gets

$$f_{11} = f_{10}^{2(1-E)} \tag{2.21}$$

Equation (2.21) is the same as eq. (2.6), because $f_{00} \approx 1$ due to inequality in eq. (2.15), and $f_{01} = f_{01}$ from symmetry. Thus, the basic expression for UFE is confirmed.

Correlation coefficients and the average frequency. Traditional measures of allelic association are correlation coefficients

$$D_{11} = \frac{f_{11}}{f^2}, \quad D_{10} = \frac{f_{10}}{f(1-f)} \approx \frac{f_{10}}{f} \tag{2.22}$$

where

$$f = f_{10} + f_{11} \tag{2.23}$$

is the average allelic frequency. The lack of correlation corresponds to $D_{11} = D_{10} = 1$. The meaning of correlator D_{11} is the probability of two mutated interacting sites divided by the same quantity if they were independent. Because f also depends on the epistatic strength (below), it is desirable to express the correlation coefficients in terms of normalized fitness f_0 in eq. (2.16), which is a fixed input parameter. Dividing both sides of eq. (2.16) by L, one gets

$$f_0 = f_{10} + f_{11}(1 - E) \tag{2.24}$$

The haplotype frequencies f_{10} and f_{11} are found from solving eqs. (2.21) and (2.24) together. Although the system cannot be solved in the general form, it can be solved with an asymptotic accuracy in various intervals of E, using a small parameter $f_0 \ll 1$.

At the point of half-compensation, $E = 1/2$, eq. (2.21) becomes $f_{10} = f_{11}$ and eq. (2.24) takes the form $f_0 = (3/2)f_{10}$. From eqs. (2.23) and (2.24), one can express the correlation coefficients and f in terms of f_0, as given by

$$D_{11} = \frac{3}{8f_0}, \quad D_{10} = \frac{1}{2}, \quad f = \frac{4}{3}f_0 \quad E = \frac{1}{2} \tag{2.25}$$

At weak epistasis such that

$$0 \le E < \frac{1}{2} - \Delta E, \quad \Delta E \sim \frac{1}{-\log f}$$

from eq. (2.21) one obtains

$$f \approx f_{10} \approx f_0$$
$$f_{11} \approx f_0^{2(1-E)} \ll f_{10}$$

Hence, the correlation coefficients, eq. (2.22), and average frequency can be approximated with

$$D_{11} = \frac{1}{f_0^{2E}}, \quad D_{10} = 1, \quad f = f_0 \quad 0 < E < \frac{1}{2} - \Delta E \tag{2.26}$$

In the absence of epistasis, $E = 0$, one gets $D_{11} = 1$ implying the lack of correlations.

For large compensation, $1/2 + \Delta E < E < 1$, from eq. (2.21), one obtains $f_{10} \ll f_{11}$. Using eq. (2.24) yields

$$f \approx f_{11} \approx \frac{f_0}{1 - E}, \quad 1 - E \gg f_0 \tag{2.27}$$

From eqs. (2.21) and (2.27), one has

$$f_{10} = \left(\frac{f_0}{1 - E}\right)^{\frac{1}{2(1-E)}} \tag{2.28}$$

Substituting eqs. (2.27) and (2.28) into (2.22) yields

$$D_{11} = \frac{1-E}{f_0}, \; D_{10} = \left(\frac{f_0}{1-E}\right)^{\frac{E-\frac{1}{2}}{1-E}}, \; f = \frac{f_0}{1-E}, \; \frac{1}{2} + \Delta E < E < 1 \tag{2.29}$$

The plots of D_{11}, D_{10} and f as functions of E at $f_0 = 1/100$ are shown (Fig. 2.5).

2.1.7.2 Double arches

In a more complex topology, each group of three loci are connected by two bonds (Fig. 2.6C). The number of triple-mutated double arches is denoted k_3. Double arches where the noninteracting first and the third site have mutated are included into number k_1, as two single alleles. Fitness given by eq. (2.1) can be rewritten in the form

$$f_0 L \equiv -\frac{W}{s_0} = k_1 + 2k_2(1-E) + k_3(3-4E) \tag{2.30}$$

Compared to eqs. (2.16), eq. (2.30) has an additional term due to the triplets.

The number of combinations has the form

$$N_{conf} = C_{L/3}^{k_3} C_{L/3-k_3}^{k_2} C_{L/3-k_3-k_2}^{k_1} 2^{k_2} 3^{k_1} \tag{2.31}$$

To find the maximum of entropy S in variables k_i, $i = 1, 2, 3$, the change of $\log(N_{conf})$ in k_i needs to be evaluated. The arguments of N_{conf} are incremented by unit

$$N_{conf}(k_1+1, k_2, k_3) = N_{conf}\frac{3(L/3 - k_1 - k_2 - k_3)}{k_1+1} \approx \frac{N_{conf}}{f_1} \tag{2.32}$$

$$N_{conf}(k_1, k_2+1, k_3) \approx \frac{2N_{conf}}{3f_2} \tag{2.33}$$

$$N_{conf}(k_1, k_2, k_3+1) \approx \frac{N_{conf}}{3f_3} \tag{2.34}$$

where $f_i \equiv k_i/L$, and the strong inequality $1 \ll k_i \ll L$ is taken into account. From eqs. (2.32)–(2.34), the condition of maximal entropy has the form

$$dS = 0 = \log\left(\frac{1}{f_1}\right)dk_1 + \log\left(\frac{2}{3f_2}\right)dk_2 + \log\left(\frac{1}{3f_3}\right)dk_3 \tag{2.35}$$

Differentiating the fitness relation, eq. (2.30), we get

$$dk_1 = -2(1-E)dk_2 - (3-4E)dk_3 \tag{2.36}$$

Substituting eq. (2.36) into (2.35) and demanding that the factors multiplying dk_2 and dk_3 are both zero, one obtains

$$f_2 = \frac{2}{3} f_1^{2(1-E)}$$

$$f_3 = \frac{1}{3} f_1^{3-4E} \tag{2.37}$$

Correlation coefficients and the average mutant frequency. Dividing both sides of eq. (2.30) by L one gets

$$f_0 = f_1 + 2(1-E)f_2 + (3-4E)f_3 \tag{2.38}$$

The average mutant frequency is given by

$$f = f_1 + 2f_2 + 3f_3 \tag{2.39}$$

To find f_{11}, note that there are $k_2 + 2k_3$ epistatic pairs with two mutated sites, which must be divided by the total number of epistatic pairs, $2L/3$. Thus

$$f_{11} = \frac{3(k_2 + 2k_3)}{2L} = 3\left(\frac{1}{2}f_2 + f_3\right) \tag{2.40}$$

Frequency f_{10} is given by the difference

$$f_{10} = f - f_{11}$$

With the use of eq. (2.37), the last equation and eqs. (2.38)–(2.40) can be written in terms of f_1 as

$$\begin{aligned} f_0 &= f_1 + \tfrac{4}{3}(1-E)f_1^{2(1-E)} + \tfrac{1}{3}(3-4E)f_1^{3-4E} \\ f &= f_1 + \frac{4}{3}f_1^{2(1-E)} + f_1^{3-4E} \\ f_{11} &= f_1^{2(1-E)} + f_1^{3-4E} \\ f_{10} &= f_1 + \frac{1}{3}f_1^{2(1-E)} \end{aligned} \tag{2.41}$$

The next task is to express the correlation coefficients, eq. (2.22), in terms of input parameters E and $f_0 = k_0/L$, in different asymptotic intervals of E, as follows.

At special point $E = 1/2$, eqs. (2.41) give

$$f_1 = f_0/2, \quad f_{11} = f_0, \quad f_{10} = \frac{2}{3}f_0$$

Substituting these equations into eq. (2.22), the correlation coefficients and the average mutant frequency are

$$D_{11} = \frac{9}{25f_0}, \; D_{10} = \frac{2}{5}, \; f = \frac{5f_0}{3}, \quad E = \frac{1}{2} \tag{2.42}$$

Next, consider the interval of small compensation $[0,\; 1/2 - \Delta E)$, where $\Delta E \sim 1/(2|\log f|)$. In this interval, a double inequality applies

$$f_1 \gg f_1^{2-2E} \gg f_1^{3-4E}$$

whence

$$f_{10} \approx f \approx f_1 \approx f_0, \quad f_{11} \approx f_0^{2-2E}$$

The results for D_{11}, D_{10} and f have the form

$$D_{11} = \frac{1}{f_0^{2E}}, \quad D_{10} = 1, \quad f = f_0, \quad 0 < E < \frac{1}{2} - \Delta E \tag{2.43}$$

At $E = 0$, one has $D_{11} = 1$, as expected.

In the intermediate interval $E \in \left[\frac{1}{2} + \Delta E, \frac{3}{4}\right]$, the powers of f_1 are ordered differently

$$f_1^{3-4E} \gg f_1^{2-2E} \gg f_1$$

Then, eqs. (2.41) yield

$$f_1 \approx \left(\frac{3f_0}{3-4E}\right)^{\frac{1}{3-4E}}$$

$$f \approx f_{11} \approx f_1^{3-4E} = \frac{3f_0}{3-4E}$$

$$f_{10} \approx \tfrac{1}{3} f_1^{2(1-E)} = \tfrac{1}{3}\left(\frac{3f_0}{3-4E}\right)^{\frac{2-2E}{3-4E}}$$

and the correlation coefficients, eq. (2.22), and the average mutant frequency f are

$$D_{11} = \frac{3-4E}{3f_0}, \quad D_{10} = \frac{1}{3}\left(\frac{3-4E}{3f_0}\right)^{\frac{2E-1}{4E-3}}, \quad f = \frac{3f_0}{3-4E}, \quad \frac{1}{2} + \Delta E < E < 1 \tag{2.44}$$

The values of D_{11}, D_{10} and f are plotted as functions of E at $f_0 = 1/100$ (Fig. 2.5).

2.1.7.3 Triple arches

Consider now double arches with one additional bond (Fig. 2.6B). Now all three sites in one cluster interact epistatically. Notations for k_i and f_i, $i = 1, 2, 3$, are as in the previous section. The expression for fitness in eq. (2.1), in the case of triple arches, has a form

$$f_0 L \equiv -\frac{W}{s_0} = k_1 + 2(1-E)k_2 + 3k_3(1-2E) \tag{2.45}$$

The derivation is very similar to the previous case. The number of combinations at given k_1, k_2 and k_3 is

$$N_{conf} = C^{k_3}_{L/3} C^{k_2}_{L/3-k_3} C^{k_1}_{L/3-k_3-k_2} 3^{k_2} 3^{k_1}$$

To calculate the discrete derivative of $S = \log(N_{conf})$, one increments the arguments of N_{conf} as

$$\begin{aligned} N_{conf}(k_1+1, k_2, k_3) &\approx \frac{N_{conf}}{f_1} \\ N_{conf}(k_1, k_2+1, k_3) &\approx \frac{N_{conf}}{f_2} \\ N_{conf}(k_1, k_2, k_3+1) &\approx \frac{N_{conf}}{3f_3} \end{aligned} \tag{2.46}$$

With eqs. (2.46), the differential of entropy dS can be written as

$$dS = 0 = \log\left(\frac{1}{f_1}\right) dk_1 + \log\left(\frac{2}{f_2}\right) dk_2 + \log\left(\frac{1}{3f_3}\right) dk_3 \tag{2.47}$$

The differential of eq. (2.45) yields the relation

$$dk_1 = -2(1-E)dk_2 - 3(1-2E)dk_3 \tag{2.48}$$

Substituting eq. (2.48) into eq. (2.47), the latter is separated into two equations that yield

$$\begin{aligned} f_2 &= f_1^{2(1-E)} \\ f_3 &= \frac{1}{3} f_1^{3(1-2E)} \end{aligned} \tag{2.49}$$

Correlation coefficients and the average mutant frequency. The total number of epistatic pairs is L. Hence, the haplotype frequencies can be written as

$$\begin{aligned} f_{11} &= \frac{k_2 + 3k_3}{L} = f_2 + 3f_3 \\ f_{10} &= f_{01} = \frac{k_1 + k_2}{L} = f_1 + f_2 \end{aligned}$$

The average mutant frequency f and fitness frequency f_0 are

$$\begin{aligned} f &= f_{11} + f_{10} = f_1 + 2f_2 + 3f_3 \\ f_0 &= f_1 + 2(1-E)f_2 + 3(1-2E)f_3 \end{aligned}$$

where eq. (2.45) was used. Using eq. (2.49), the last four expressions can be written in terms of f_1

$$\begin{aligned} f_0 &= f_1 + 2(1-E)f_1^{2(1-E)} + (1-2E)f_1^{3(1-2E)} \\ f &= f_1 + 2f_1^{2(1-E)} + f_1^{3(1-2E)} \end{aligned}$$

$$f_{11} = f_1^{2(1-E)} + f_1^{3(1-2E)} \tag{2.50}$$

$$f_{10} = f_1 + f_1^{2(1-E)}$$

The approximate values for D_{11} and D_{10} for different values of E in interval $[0, 1/2)$ are derived from eqs. (2.50) and (2.22), as follows.

At special point $E = 1/4$, where the powers $3-6E$ and $2-2E$ are equal to each other, one has $f_{11} = 2f_1^{3/2}$. Under the condition $f_1 \ll 1$, eqs. (2.50) yield

$$f_0 = f \approx f_1 \approx f_{10} \tag{2.51}$$

and eqs. (2.22) simplify to

$$D_{11} \approx \frac{2}{\sqrt{f_0}}, \; D_{10} \approx 1, \; f = f_0, \; E = \frac{1}{4} \tag{2.52}$$

At point $E = 1/3$, the power $3(1-2E)$ is equal to 1. From eqs. (2.50), one obtains

$$f_{11} \approx f_{10} \approx f_1, \; f \approx 2f_1, \; f_1 = \tfrac{3}{4} f_0$$

and then

$$D_{11} \approx \frac{1}{3f_0}, \quad D_{10} \approx \frac{1}{2}, \quad f = \frac{3}{2} f_0, \quad E = \frac{1}{3}$$

In the interval of weak epistasis, $0 < E < 1/4 - \Delta E$, where $\Delta E = 1/(-2\log f)$, eqs. (2.50) result in eqs. (2.51) and $f_{11} \approx f_1^{2(1-E)}$. From eqs. (2.22), (2.51), one has

$$f = f_0, \; D_{11} \approx \frac{1}{f_0^{2E}}, \; D_{10} \approx 1, \; 0 < E < 1/4 - \Delta E$$

In the intermediate interval $E \in \left(\tfrac{1}{4} + \Delta E, \; \tfrac{1}{3} - \Delta E'\right)$ where $\Delta E' = 1/(-6\log f)$, the strong inequality applies

$$f_1^{2-2E} \ll f_1^{3-6E} \ll f_1$$

so that the relations in eq. (2.51) are still valid, and the frequency of double mutants is $f_{11} \approx f_1^{3-6E}$. Using eqs. (2.22) results in

$$f = f_0, \; D_{11} \approx \frac{1}{f_0^{6E-1}}, \; D_{10} \approx 1, \; \tfrac{1}{4} + \Delta E < E < \tfrac{1}{3} - \Delta E'$$

Finally, in interval of strongest epistasis $E \in \left(\tfrac{1}{3} + \Delta E', \; \tfrac{1}{2}\right)$, the order of powers is

$$f_1^{2-2E} \ll f_1 \ll f_1^{3-6E}$$

Using this double inequality, eqs. (2.50) produce

$$f_{10} \approx f_1,\ f \approx f_{11},\ f_0 = (1-2E)f$$

and eqs. (2.22) result in

$$D_{11} = \frac{1-2E}{f_0},\ D_{10} = \left(\frac{1-2E}{f_0}\right)^{\frac{2-6E}{3-6E}},\ f = \frac{f_0}{1-2E},\ \frac{1}{3} + \Delta E' < E < \frac{1}{2} \tag{2.53}$$

The values of D_{11}, D_{10} and f are plotted as functions of E at $f_0 = 1/100$ (Fig. 2.5).

2.1.7.4 Long chain

The next topology is a chain with interacting neighbors (Fig. 2.6D). Any subchains with length i, where $i = 1, \ldots, I$, and $1 \ll I \ll L$, are allowed. Due to the central assumption $f \ll 1$, subclusters must remain sparse and without much overlap, so that an upper limit of the cluster size, I, has to be introduced. As shown below, the results do not to depend on this parameter.

The fitness restriction from eq. (2.1), in the case of a long chain, has the form

$$f_0 L = k_1 + \sum_{i=2}^{I} (i - 2(i-1)E)k_i \tag{2.54}$$

The number of possible combinations, given the set of cluster numbers of each size $\{k_i\}$, has the form

$$N_{conf} = \prod_{i=1}^{I} C_L^{k_i} \tag{2.55}$$

provided that the sparsity condition is met. To find the maximum of S, the entropy differential found from eq. (2.55) must be equal to zero

$$0 = dS = \log\left(\frac{1}{f_1}\right) dk_1 + \sum_{i=2}^{I} \log\left(\frac{1}{f_i}\right) dk_i \tag{2.56}$$

where $f_i = k_i/L$. The differential of the fitness restriction, eq. (2.54), yields an additional condition

$$dk_1 = -\sum_{i=2}^{I} (i - 2(i-1)E)dk_i \tag{2.57}$$

Substituting this formula into eq. (2.56), one obtains

$$-\log\left(\frac{1}{f_1}\right) \sum_{i=2}^{I} (i - 2(i-1)E)dk_i + \sum_{k=2}^{I} \log\left(\frac{1}{f_i}\right) dk_i = 0$$

The expression for frequency f_i in terms of f_1 is obtained by setting the factor above multiplying each dk_i to zero, which yields

$$f_i = f_1^{i-2(i-1)E} = f_1^{2E}\left(f_1^{1-2E}\right)^i \tag{2.58}$$

Correlation coefficients and the average frequency. Dividing eq. (2.54) by L, one obtains

$$f_0 = \sum_{i=1}^{I}(i-2(i-1)E)f_i \tag{2.59}$$

The average mutant frequency and the haplotype frequencies are

$$f = f_1 + 2f_2 + \ldots = \sum_{i=1}^{I} if_i \tag{2.60}$$

$$f_{11} = f_2 + 2f_3 + \ldots = \sum_{i=2}^{I}(i-1)f_i$$

$$f_{10} = f - f_{11} = \sum_{i=1}^{I} f_i \tag{2.61}$$

Substituting eq. (2.58) into eqs. (2.59)–(2.61), all frequencies are expressed in terms of f_1 as

$$\begin{aligned} f_0 &= \sum_{i=1}^{I}(i-2(i-1)E)f_1^{i-2(i-1)E} \\ f &= f_1^{2E}\sum_{i=1}^{I} if_1^{i(1-2E)} \\ f_{11} &= f_1^{2E}\sum_{i=1}^{I}(i-1)f_1^{i(1-2E)} \\ f_{10} &= f_1^{2E}\sum_{i=1}^{I} f_1^{i(1-2E)} \end{aligned} \tag{2.62}$$

The last sum in eq. (2.62) represents the sum of geometric progression equal to

$$f_{10} = f_1^{2E}\left[f_1^{1-2E}\frac{1-f_1^{(1-2E)I}}{1-f_1^{1-2E}}\right] \approx \frac{f_1}{1-f_1^{1-2E}} \tag{2.63}$$

provided that it converges, which occurs if $E < 1/2$. The sums in eq. (2.62) for f_0 and f can be evaluated using an identity

$$\sum_{i=1}^{I} ia^i = \frac{d}{d\log a}\sum_{i=1}^{I} a^i$$

As a result, one gets

$$f_0 = \frac{d}{d(\log f_1)}\left(\frac{f_1}{1-f_1^{1-2E}}\right)$$

$$f = \frac{f_1^{2E}}{1-2E}\frac{d}{d(\log f_1)}\left(\frac{f_1^{1-2E}}{1-f_1^{1-2E}}\right)$$

It is convenient to introduce the notation

$$x = (1-2E)\log\left(\frac{1}{f_1}\right) \tag{2.64}$$

Then, $e^{-x} = f_1^{1-2E}$ and $d/d(\log f_1) = -(1-2E)d/dx$. In terms of x, all the frequencies take the form

$$\begin{aligned} f_0 &= \frac{e^{-\frac{x}{1-2E}}(1-2Ee^{-x})}{(1-e^{-x})^2} \\ f &= \frac{e^{-\frac{x}{1-2E}}}{(1-e^{-x})^2} \\ f_{10} &= \frac{e^{-\frac{x}{1-2E}}}{1-e^{-x}} \\ f_{11} = f - f_{10} &= \frac{e^{-\frac{2(x-E)}{1-2E}}}{(1-e^{-x})^2} \end{aligned} \tag{2.65}$$

As in the case of other topologies, eqs. (2.65) can be simplified in asymptotic cases. The first case is when epistatic strength is far from the full compensation point, $E < 1/2,\ 1/2 - E \gg 1/\ln(1/f_1)$, which is equivalent to $x \gg 1$. Neglecting small terms in eq. (2.65), one obtains

$$f_0 \approx f \approx f_{10} \approx f_1,\ f_{11} = f_0^{2-2E}$$

The correlation coefficients, eq. (2.22), and the average mutant frequency are

$$D_{11} \approx \frac{1}{f_0^{2E}},\ D_{10} \approx 1,\ f \approx f_0,\ \frac{1}{2} - E \ll \frac{1}{\log\left(\frac{1}{f_1}\right)}$$

The opposite case $x \ll 1$ corresponds to E close to the full compensation point $\frac{1}{2}$. Taking advantage of inequality $(1-2E) \ll 2Ex$, the expression for f_0 in eqs. (2.65) can be approximated as

$$f_0 \approx \frac{f_1}{x}$$

where eq. (2.64) is used. By substituting x from eq. (2.64) into the last equation, one obtains an equation for f_1

$$f_1 = f_0(1-2E)\log\frac{1}{f_1}$$

Because the logarithm is very large, a good approximate solution of the last equation is obtained by a single iteration

$$f_1 \approx f_0(1-2E)\log\frac{1}{f_0(1-2E)}$$

or in terms of x

$$x \approx (1-2E)\log\frac{1}{f_0(1-2E)}$$

From this expression and eqs. (2.65), the results for f, f_{10} and f_{11} have the form

$$f_{11} \approx f \approx \frac{f_0}{(1-2E)\log\frac{1}{f_0(1-2E)}}, \quad f_{10} \approx f_0$$

Finally, the correlation coefficients, eq. (2.22), and the average frequency are

$$D_{11} = \frac{1-2E}{f_0}\log\frac{1}{f_0(1-2E)}$$

$$D_{10} = (1-2E)\log\frac{1}{f_0(1-2E)}$$

The values D_{11}, D_{10} and f are plotted as functions of E at $f_0 = 1/100$ (Fig. 2.5).

2.1.7.5 Large binary tree

In a more complicated configuration, each site interacts with three other loci, and the network forms a binary tree. There are no loops, and any two sites are connected by a single path (Fig. 2.6E). Again, connected mutated clusters of any size i are possible, except now they represent subtrees of various shapes rather than finite chains, as in the previous section. The expressions for fitness, eq. (2.1), and for average and haplotype frequencies f, f_{11} and f_{10} in terms of the cluster frequencies f_i have the same form as for a long chain, eqs. (2.59)–(2.61). This is because, for infinite binary tree, as for infinite chain, any cluster of i connected mutated sites has $i-1$ bonds. Hence, the critical point of full compensation in E is also the same, $E_c = 1/2$.

The total number of configurations is expressed in terms of subtree numbers k_i, as follows

$$N_{conf} = C_L^{k_1} C_{L_2}^{k_1} (A_2)^{k_2} \ldots C_{L_i}^{k_i} (A_i)^{k_i} \ldots = \prod_{i=1}^{I} C_{L_i}^{k_i} (A_i)^{k_i}$$

Because $f \ll 1$, subtrees are sparse, and their overlap is neglected. Here $C_{L_i}^{k_i}$ is the number of ways to place k_i subtree roots on L_i available nodes, where L_i decreases with the number of nodes i from L for $i=1$ to 1 for $i=L$. This factor is multiplied by the number of configurations $(A_i)^{k_i}$, where A_i is the number of all possible subtrees with i nodes and the same root

$$A_i = \frac{(2i)!}{i!(i+1)!}$$

$$A_i \approx \frac{4^i}{\sqrt{\pi}\, i^{\frac{3}{2}}}, \quad i \gg 1$$

It will be convenient in what follows to use new notation

$$\beta_i \equiv \frac{L_i}{L}, \quad A_i = 4^i \alpha_i, \quad \alpha_i \equiv \frac{(2i)!}{4^i i!\ (i+1)!}$$

The maximum entropy condition has the form

$$0 = dS = \sum_{i=1}^{I} \log\left(\frac{N_{conf}(\ldots, k_i+1, \ldots)}{N_{conf}(\ldots, k_i, \ldots)}\right) dk_i = \sum_{i=1}^{I} \log\left(A_i \frac{\beta_i}{f_i}\right) dk_i \tag{2.66}$$

where $f_i = k_i/L$. Substituting dk_1 from eq. (2.57), which is valid for the binary tree topology as well, one obtains

$$\sum_{i=1}^{I} \log\left(A_i \frac{\beta_i}{f_i}\right) dk_i = \log\left(\frac{1}{f_1}\right) \sum_{i=2}^{I} (i - 2(i-1)E) dk_i$$

Differentials dk_i are independent variables, which results in a system of equations

$$f_i = A_i \beta_i\, f_1^{i-2(i-1)E} = \alpha_i\, \beta_i\, f_1^{2E} \left(4 f_1^{1-2E}\right)^i \tag{2.67}$$

where all frequencies f_i must be much less than 1, and $L_1 = L$ is used.

The role of large clusters versus small clusters is determined by the asymptotic behavior of f_i in eq. (2.67) at large $i \gg 1$. Assuming the most branched subtree, where every internal mutated node is connected to three mutated nodes, one has $\alpha_i \sim i^{-\frac{3}{2}}$ and $\beta_i \approx 2L/(i+1)$. An average node in typical tree is connected to two or three nodes, it is less compact than the most branched tree, leaving less place for possible root locations. One can approximate

$$\beta_i \approx \frac{2L}{(i+1)^{\gamma}}, \quad 1 < \gamma < 2$$

where coefficient γ is bounded between 1 and 2.

For the large clusters $i \gg 1$, the RHS of eq. (2.67) is dominated by the factor $f_1^{2E}\left(4 f_1^{1-2E}\right)^i$, because the remaining factor is a slow function of i. For the cluster frequencies f_i to decrease with the node number i, the inequality $4 f_1^{1-2E} < 1$ must be satisfied, which is the same as the condition

$$E < \frac{1}{2}\left[1 - \frac{\log 4}{\log(1/f_1)}\right] \tag{2.68}$$

Numeric results show that the curves of correlation coefficients and mutant frequency could be continued a bit further, until point $E = E_c = 1/2$ (Fig. 2.5). This is because coefficients $\alpha_i\beta_i$ neglected in eq. (2.68) decrease with i, which improves the convergence and slightly postpones the point of full compensation.

Correlation coefficients and the average mutant frequency. As already mentioned, f_0, f, f_{11} and f_{10} have the same general form as in eqs. (2.59)–(2.61) obtained for the chain topology of epistatic network. Using eq. (2.67) for the cluster frequencies specific for the tree, and new notation $x = -\log\left(4f_1^{1-2E}\right)$, one arrives at

$$f_0 = f_1^{2E}\sum_{i=1}^{I}(i - 2(i-1)E)\alpha_i\beta_i e^{-ix} \tag{2.69}$$

$$f = f_1^{2E}\sum_{i=1}^{I} i\alpha_i\beta_i e^{-ix} \tag{2.70}$$

$$f_{11} = f_1^{2E}\sum_{i=1}^{I}(i-1)\alpha_i\beta_i e^{-ix} \tag{2.71}$$

$$f_{10} = f_1^{2E}\sum_{i=1}^{I}\alpha_i\beta_i e^{-ix} \tag{2.72}$$

At large i, the common term in the sums in eqs. (2.69)–(2.72) has asymptotics

$$\alpha_i\beta_i e^{-ix} \approx \frac{e^{-ix}}{i^{\frac{3}{2}+\gamma}}, \quad i \gg 1 \tag{2.73}$$

If epistatic strength is not too close to the full compensation point, $1/2 - E \ll 1/\log(1/f_1)$, condition $x \gg 1$ holds, and the right-hand side of eq. (2.73) decreases very fast with i. Hence, only the first nonzero elements of sums in eqs. (2.69)–(2.72) matter. The final results are the same, as for the long chain topology

$$f_0 \approx f \approx f_{10} \approx f_1, \; f_{11} = f_0^{2-2E}$$

$$D_{11} \approx \frac{1}{f_0^{2E}}, \; D_{10} \approx 1, \; f \approx f_0, \; \frac{1}{2} - E \gg \frac{1}{\log\left(\frac{1}{f_1}\right)}$$

In the opposite case $E \approx \frac{1}{2}$ or $x \ll 1$, large clusters matter a lot, and the sum in eq. (2.72) simplifies to

$$\sum_{i=1}^{I}\frac{e^{-ix}}{i^{\frac{3}{2}+\gamma}} \approx \sum_{i=1}^{I}\frac{1}{i^{\frac{3}{2}+\gamma}}$$

which converges at large $i > 1$ if $I \gg 1$. One can approximate this sum by the first 50 terms with $\gamma \approx 1.5$, and take

$$\beta_1 = 1,\ \beta_2 = \beta_3 = \frac{1}{2},\ \beta_i \approx \frac{2}{2^{\log_{1.5}(i+1)}}\ \text{if}\ i > 3$$

As a result, from eqs. (2.69) to (2.72), one obtains

$$f_0 \approx f_{10} \approx 1.48 f_1,\quad f \approx 2.5 f_1,\quad f_{11} \approx 1.0 f_1$$

Then, the correlation coefficients and the average mutant frequency are

$$D_{11} \approx \frac{1}{4f_0},\quad D_{10} \approx 0.6,\quad f \approx 1.68\, f_0,\quad \frac{1}{2} - E \ll \frac{1}{\log\left(\frac{1}{f_1}\right)}$$

The values of D_{11}, D_{10}, and f as a function of E are plotted at $f_0 = 1/100$ (Fig. 2.5).

2.1.7.6 Double arches with unequal interactions

Consider, again, the topology of "double arches" (Fig. 2.6C), where the left arch has the epistatic strength E and the right arch has strength $E/2$. As before, notation k_1 and k_3 number singlets and triplets, respectively. Now we need to differentiate the numbers of doubles at the left and the right arch, denoted k_2 and k_2', respectively. Fitness, eq. (2.1), is expressed in terms of k_1, k_2, k_2', and k_3 as

$$f_0 L \equiv -\frac{W}{s_0} = k_1 + 2k_2(1-E) + k_2'(2-E) + 3k_3(1-E) \tag{2.74}$$

instead of eq. (2.30) for symmetric double arches. Next, the number of possible combinations is given by

$$N_{\text{conf}} \equiv e^S = C_{L/3}^{k_3} C_{L/3-k_3}^{k_2} C_{L/3-k_3-k_2}^{k_2'} C_{L/3-k_3-k_2-k_2'}^{k_1} 3^{k_1}$$

The increments of $N_{\text{conf}} = N_{\text{conf}}(k_1, k_2, k_2', k_3)$ are given by

$$\begin{aligned}
N_{\text{conf}}(k_1+1, k_2, k_2', k_3) &\approx \frac{N_{\text{conf}}}{f_1}\\
N_{\text{conf}}(k_1, k_2+1, k_2', k_3) &\approx \frac{N_{\text{conf}}}{3f_2}\\
N_{\text{conf}}(k_1, k_2, k_2'+1, k_3) &\approx \frac{N_{\text{conf}}}{3f_2'}\\
N_{\text{conf}}(k_1, k_2, k_2', k_3+1) &\approx \frac{N_{\text{conf}}}{3f_3}
\end{aligned} \tag{2.75}$$

where $f_i \equiv k_i/L \ll 1$.

The condition of the maximum of entropy is $dS = 0$. Using derivatives of $S = \log(N_{\text{conf}})$ from eq. (2.75), one gets

$$0 = dS = \log\left(\frac{1}{f_1}\right)dk_1 + \log\left(\frac{2}{3f_2}\right)dk_2 + \log\left(\frac{2}{3f'_2}\right)dk'_2 + \log\left(\frac{1}{3f_3}\right)dk_3 \tag{2.76}$$

Additional restriction from eq. (2.74) has a form

$$dk_1 = -2(1-E)dk_2 - (2-E)dk'_2 - 3(1-E)dk_3 \tag{2.77}$$

Substituting eq. (2.77) into (2.76), eq. (2.76) results in three equations

$$f_2 = f'_2 = \frac{1}{3}f_1^{2(1-E)}$$
$$f_3 = \frac{1}{3}f_1^{3-4E} \tag{2.78}$$

Correlation coefficients and the average mutant frequency. The mutant frequency is expressed in terms of cluster numbers as

$$f = f_1 + 2f_2 + 2f'_2 + 3f_3$$

Minding that the total number of epistatic bonds is $2L/3$, the haplotype frequencies are

$$f_{11} = \tfrac{3}{2}(f_2 + f'_2) + 3f_3$$
$$f_{10} = f - f_{11} = f_1 + \tfrac{1}{2}(f_2 + f'_2)$$

Fitness frequency $f_0 = \frac{k_0}{L}$ is found from eq. (2.74) and has the form

$$f_0 = f_1 + 2(1-E)f_2 + (2-E)f'_2 + 3(1-E)f_3$$

With the use of eq. (2.78), the four desired frequencies can be expressed in terms of f_1

$$\begin{aligned} f_0 &= f_1 + \frac{2}{3}(1-E)f_1^{2(1-E)} + \frac{1}{3}(2-E)f_1^{2-E} + (1-E)f_1^{3(1-E)} \\ f &= f_1 + \frac{2}{3}f_1^{2(1-E)} + \frac{2}{3}f_1^{2-E} + f_1^{3(1-E)} \\ f_{11} &= \frac{1}{2}f_1^{2(1-E)} + \frac{1}{2}f_1^{2-E} + f_1^{3(1-E)} \\ f_{10} &= f_1 + \frac{1}{6}f_1^{2(1-E)} + \frac{1}{6}f_1^{2-E} \end{aligned} \tag{2.79}$$

First, consider a special point $E = 1/2$. At this point, the various powers of f_1 entering eqs. (2.79) are in relations, as follows:

$$f_1 = f_1^{2(1-E)}, \quad f_1^{2-E} = f_1^{3(1-E)} = f_1^{3/2}, \quad f_1^{2(1-E)} \gg f_1^{2-E}$$

The last inequality indicates that singlets and the doubles with epistatic strength E, outnumber triplets and the doubles with epistatic strength $E/2$. As a result, eq. (2.79) approximates to

$$f \approx \frac{5}{3} f_1,\ f_{11} \approx \frac{1}{2} f_1,\ f_{10} \approx \frac{7}{6} f_1,\ f_0 \approx \frac{4}{3} f_1$$

Excluding f_1 and expressing all the frequencies in terms of mutant frequency f, one gets

$$f_{11} \approx \frac{3}{10} f,\ f_{10} \approx \frac{7}{10} f,\ f_0 \approx \frac{4}{5} f$$

The correlation coefficients, eq. (2.22), are expressed in terms of f_0

$$D_{11} = \frac{6}{25 f_0},\ D_{10} = \frac{7}{10},\ f = \frac{5}{4} f_0,\ E = \frac{1}{2}$$

Another important point of eq. (2.79) is $E = \frac{2}{3}$, where the following relations take place

$$f_1^{2(1-E)} = f_1^{\frac{2}{3}} \gg f_1,\ f_1^{2-E} = f_1^{\frac{4}{3}},\ f_1 = f_1^{3(1-E)}$$

implying that the doubles with strength E outnumber the singles and the triplets. The rarest clusters are the under-compensated doubles with $E/2$. The frequencies of interest are

$$f \approx \frac{2}{3} f_1^{\frac{2}{3}},\ f_{11} \approx \frac{3}{4} f,\ f_{10} \approx \frac{1}{4} f,\ f_0 \approx \frac{1}{3} f$$

and the correlation coefficients and the mutant frequency take the form

$$D_{11} = \frac{3}{4f},\ D_{10} = \frac{1}{4},\ f = 3 f_0,\ E = \frac{2}{3}$$

In the interval of weak compensation, $0 < E < \frac{1}{2} - \Delta E$, singles dominate

$$f_1 \gg f_1^{2(1-E)} \gg f_1^{2-E} \gg f_1^{3(1-E)}$$

whence

$$f \approx f_1 \approx f_{10} \approx f_0,\quad f_{11} \approx \frac{1}{2} f_1^{2(1-E)}$$

$$D_{11} = \frac{1}{2 f_0^{2E}},\quad D_{10} = 1,\ f = f_0,\ 0 < E < \frac{1}{2} - \Delta E$$

For the intermediate strength of epistasis, as given by $\frac{1}{2} + \Delta E < E < \frac{2}{3} - \Delta E$, the terms in (2.79) are ordered differently, with the double mutants of strength E on the top, as given by

$$f_1^{2(1-E)} \gg f_1 \gg f_1^{3(1-E)} \gg f_1^{2-E}$$

The four frequencies of interests take the form

$$f \approx \frac{2}{3} f_1^{2(1-E)}, \; f_0 \approx (1-E)f, \; f_{11} \approx \frac{3}{4}f, \; f_{10} \approx \frac{1}{4}f$$

and the correlation coefficients are

$$D_{11} = \frac{3(1-E)}{4f_0}, \quad D_{10} = \frac{1}{4}, \; f = \frac{f_0}{1-E}, \; \frac{1}{2} + \Delta E < E < \frac{2}{3} - \Delta E$$

Finally, for the largest epistatic strengths, such that $\frac{2}{3} + \Delta E < E < 1$, the term order in eqs. (2.79) is, as follows:

$$f_1^{3(1-E)} \gg f_1^{2(1-E)} \gg f_1 \gg f_1^{2-E}$$

Equation (2.79) yields

$$f \approx f_1^{3(1-E)}, \; f_0 \approx (1-E)f, \; f_{11} \approx f, \; f_{10} \approx \frac{1}{6}f^{2/3}$$

$$D_{11} = \frac{1-E}{f_0}, \; D_{10} = \frac{1}{6}\left(\frac{1-E}{f_0}\right)^{1/3}, \; f = \frac{f_0}{1-E}, \; \frac{2}{3} + \Delta E < E < 1$$

The results are compared between the cases of symmetric and asymmetric double arches (Fig. 2.8).

2.2 Detection of epistatic pairs in a single population: mission impossible

Interdependent evolution of different genomic regions is caused not only by epistasis. Common inheritance of alleles at closely situated loci or in the absence of recombination also causes allelic correlations even if epistasis is absent. Among such linkage effects are Fisher-Muller clonal interference (Muller 1932; Fisher 1958), genetic background and genetic hitchhiking effect (Rice 2004), Muller's ratchet (Felsenstein 1974) and the general interference between selection at different loci (Hill and Robertson 1968; Barton 1995; McVean and Charlesworth 2000). Sometimes, these related effects are all lumped together as "Hill-Robertson interference" (Felsenstein 1974); below, they will be referred to simply as "linkage effects." Linkage effects change completely all predictable quantities of asexual evolution (Gerrish and Lenski 1998; Rouzine et al. 2003; Desai and Fisher 2007; Brunet et al. 2008; Rouzine et al. 2008; Hallatschek 2011; Good et al. 2012; Good et al. 2014). The mathematical models demonstrate that linkage slows down adaptation a lot, helps deleterious mutations to accumulate, and modifies the form of phylogenetic tree (Walczak et al. 2012; Neher and Hallatschek 2013). The amplitude of linkage effects increases with the locus number, L. Recombination partly compensates them (Rouzine and Coffin 2005; Gheorghiu-Svirschevski, et al. 2007; Neher and Leitner 2010; Neher et al. 2010; Batorsky et al. 2011; Neher et al. 2013; Weissman and Hallatschek 2014; Xiao et al. 2016). In the setting of a large population size, recombination also partly compensates the effects of epistasis leading to quasi-linkage equilibrium (Kimura 1965; Neher and Shraiman 2009; Neher and Shraiman 2011). In the simplest two-locus models, epistasis was demonstrated to be important for the evolution of recombination (Barton 1995; Turelli and Barton 2006).

Thus, the consequence of linkage is the strong interaction between the evolutionary dynamics of different sites. In particular, the allelic association resulting from linkage is easy to confuse with the effects of epistasis. This linkage effect is probabilistic due to random sampling of genomes (genetic drift) and random mutations. Linkage effects become negligible become small only in giant populations increasing exponentially with the locus number L (Rouzine et al. 2003). In real DNA sequence data sets, it is often unclear how to tell apart the effect of shared ancestry from epistatic interaction, and which of the two dominates in a given DNA sample. Therefore, detecting epistasis from genomic data remains very difficult (Cordell 2002; Cordell 2009; Wei et al. 2014).

In order to solve a problem, one has to find its cause. This section based on Pedruzzi and Rouzine (2019) offers a detailed explanation for the repeated failure to detect epistasis in DNA samples. Artificial sequence data are generated from Monte-Carlo simulation of an evolving population. Several allelic correlation measures (and their combinations) are used to compare the form of their distributions between epistatic and epistatic pairs, in a broad range of model parameters. The time interval and the model parameter region where the epistatic pairs can be detected is mapped. The dominance of linkage effects is then explained from the random nature of phylogenetic tree and co-inheritance.

2.2.1 Epistatic pairs have a distinct signature in a narrow time window

The model of population with binary sequences is the same as in Section 2.1. To characterize the statistical correlation between alleles at different loci, termed linkage disequilibrium (LD), various measures are used. Let $f^{ij}_{\alpha\beta}$ be the average frequency of a chosen binary haplotype of the four possible, $\alpha\beta \in [00,\ 01,\ 11,\ 10]$ for loci i and j, and f^{i}_{α} be the frequency of allele of the two possible, $\alpha \in [0,\ 1]$ at site i. What all the measures of LD given below have in common is that they all vanish if $f^{ij}_{\alpha\beta} = f^{i}_{\alpha} f^{j}_{\beta}$, which implies the absence of allelic correlations between the two loci.

Lewontin measure has a form (Lewontin 1964)

$$\tilde{D}^{ij}_{\alpha\beta} = \frac{D^{ij}_{\alpha\beta}}{D^{ij}_{\max}}, \quad D^{ij}_{\alpha\beta} = f^{ij}_{\alpha\beta} - f^{i}_{\alpha} f^{j}_{\beta} \tag{2.80}$$

$$D_{\max}^{ij} \equiv \begin{cases} \max_{\alpha\beta}\left\{-f_\alpha^i f_{\beta'}^j, -(1-f_\alpha^i)\left(1-f_\beta^j\right)\right\} & \text{if } D_{\alpha\beta}^{ij} < 0 \\ \min\left\{f_\alpha^i\left(1-f_\beta^j\right), (1-f_\alpha^i)f_\beta^j\right\} & \text{if } D_{\alpha\beta}^{ij} > 0 \end{cases}$$

where $D_{\max}^{ij}$ is a normalization coefficient, such that $\tilde{D}_{\alpha\beta}^{ij} \in [0, 1]$. A related measure of LD, also defined for a specific haplotype, is Pearson correlation coefficient between pairs of loci, expressed as (Balding 2006)

$$r_{\alpha\beta}^{ij} = D_{\alpha\beta}^{ij} / \sqrt{f_\alpha^i(1-f_\alpha^i)\, f_\beta^j\left(1-f_\beta^j\right)} \tag{2.81}$$

Unlike Lewontin measure, which is always positive, Pearson coefficient varies between -1 and 1, showing either negative or positive correlation. A symmetric marker of LD was proposed by Wu et al. (2010)

$$WU = \log\frac{f_{11}f_{00}}{f_{01}f_{10}} \tag{2.82}$$

which represents the logarithm of an earlier measure proposed by Kimura(1965). Indices of loci i, j are dropped in eq. (2.82).

Universal Footprint of Epistasis introduced in Section 2.1 (Pedruzzi et al. 2018) has a form

$$\text{UFE} = 1 - \frac{\log(f_{11}/f_{00})}{\log(f_{01}f_{10}/f_{00}^2)} \tag{2.83}$$

As shown in Section 2.1, UFE is equal to the compensation degree E for an isolated interacting pair of deleterious alleles if the four haplotype frequencies are ensemble-averaged. The value $E = 0$ denotes the absence of compensation, and $E = 1$ means full mutual compensation. The singularity in eq. (2.83) at $f_{01}\, f_{10} = f_{00}^2$ does not affect results discussed below.

Of course, these are not the only possible pairwise measures of LD. In Chapter 1, the frequency of the least-represented haplotype $\min(f_{00}, f_{11}, f_{10}, f_{01})$ was used as an LD marker. Also, pairwise correlation is not the only method to detect epistasis. Techniques based on phylogenetic tree (Kryazhimskiy et al. 2011) and the quasi-linkage equilibrium approximation (Kimura 1965; Neher and Shraiman 2011) have been developed. The full summary is given in Wei et al. (2014). However, as it will become completely clear later on, the final conclusions do not depend on the detection method, and the four LD measures defined in eqs. (2.80)–(2.83) are more than enough to demonstrate the final results below.

In the original work by Pedruzzi and Rouzine (2019), all four measures were analyzed, and very similar conclusions were obtained. Hence, it is sufficient to show a couple of them, for example, Pearson coefficient and UFE. The histograms of three LD measures over all pairs of sites were calculated at different time points of Monte-Carlo simulation. The first two are Pearson coefficients for two haplotypes r_{11}, r_{01}, eq. (2.81), and the third is UFE, eq. (2.83). The distribution for the known epistatic pairs (black color) and for all the pairs (gray shade) is shown in Fig. 2.9. The area under each distribution is the total number of the pairs in the set.

At $t = 0$, LD has sharp maximum near zero, for both all pairs and its epistatic subset (Fig. 2.9, row 1).

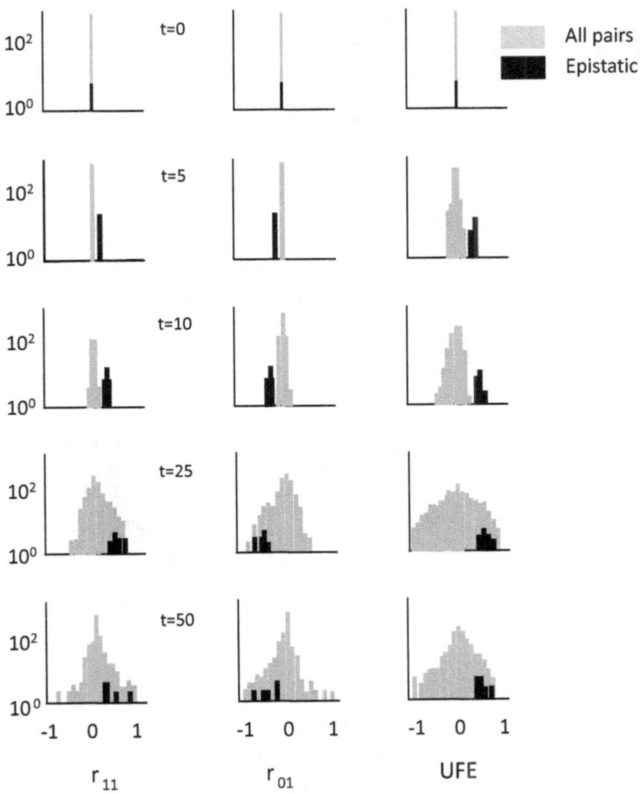

Fig. 2.9: Haplotype-based measures of epistasis identify a narrow time window of epistasis detection. Each column shows the profile of the distribution of a measure of epistasis: r_{11}, r_{01} from eq. (2.81) and UFE from eq. (2.83). Different rows correspond to different time points: $t = 1$, $t = 5$, $t = 10$, $t = 25$ and $t = 50$. The shaded regions correspond to the density distributions for all possible pairwise interactions (grey color) and the known epistatic pairs (black). Each shaded area is proportional to the pair number of each subset. Parameters: $N = 2 \cdot 10^4$, $s_0 = 0.1$, $L = 50$, E in the range $[0, 1]$, $\mu L = 7 \cdot 10^{-2}$. Network: each odd site interacts with its neighbor on the right $(1-2,\ 3-4, 5-6, \ldots)$ with epistatic strength $E = 0.75$. Initially, sequences are random with the average frequency of deleterious alleles $f = 0.4$. In the negative control with $E = 0$, the dark and light color always overlap (Pedruzzi and Rouzine 2019, fig S1). Based on Pedruzzi and Rouzine (2019).

During early times (Fig. 2.9, rows 2 and 3), the separation of the two distributions takes place, with epistatic subset moving outside of the distribution for the rest of sites. Later, the distribution of non-epistatic pairs expands and hides the smaller distribution of interacting pairs (Fig. 2.9, rows 4 and 5). Nonepistatic pairs of sites, due to the stochastic nature of mutation and random genetic drift, possess LD of random sign. Thus, except in a narrow time window, telling apart epistatic pairs from other pairs, in single population, is impossible.

2.2.2 Results are robust to the combinations of LD measures

One could hope to improve the chances of detection by combining different LD measures together. Several combinations of two and three LD measures were used in an attempt at detecting interacting pairs for $E = 0.75$, at two time points, without any improvement in detection. An example is shown in Fig. 2.10 where r_{11} is plotted against *UFE*. The use of WU and Lewontin measure or combining three measures at a time did not give any advantage for detection (Pedruzzi and Rouzine 2019).

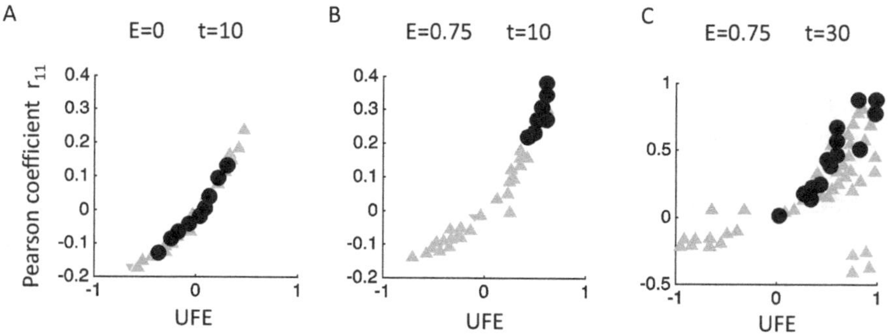

Fig. 2.10: Combination of two different measures of LD fails to detect epistasis.
(A) No epistasis. Epistatic pairs (black dots) and random pairs (gray triangles) overlap. (B) At intermediate time $t = 10$, epistatic pairs (black dots) cluster separately and, hence, are detectable. (C) At long times, substantial overlap with noninteracting pairs prevents detection. To optimize detection, a detection threshold for each statistics is used that minimizes the following quantity "DET + a FPOS", where a is a fitting parameter, DET represent the detection percentage, and FPOS is the percentage of false-positive, based on the a priori knowledge of the identity of epistatic pairs. Parameters are as in Fig. 2.9.

To obtain the best theoretically possible conditions for epistasis detection, an optimal threshold separating interacting from noninteracting pairs was calculated based on the a priori knowledge of epistatic pairs' identity (Fig. 2.10). The method, employing the principle of machine learning, did not give any substantial improvement on detection (Pedruzzi and Rouzine 2019, table S1). For a real data set, such a priori knowledge is unavailable, so that detection could be only worse. The other combinations of two and three measures give similar results (Pedruzzi and Rouzine 2019, table S1).

2.2.3 Parameter sensitivity analysis confirms the results

Detection is thus possible in a narrow time interval. The variation of the narrow time interval of detection with model parameters was analyzed, as follows.

Selection coefficient. The percent of detected epistatic pairs and the percent of false-positive pairs were calculated from simulations at different values of selection coefficient, s_0, using UFE (the use of the other measures produced similar results) (Fig. 2.11). An inverse scaling of the detection time window with s_0 was observed. At very small s_0, where evolution is almost selectively neutral, epistasis is never detectable.

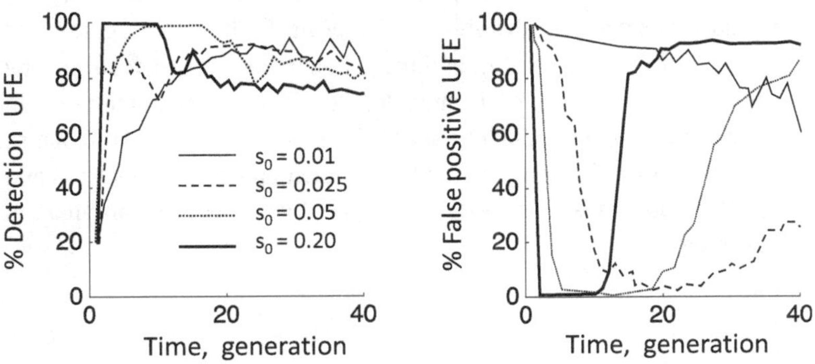

Fig. 2.11: Detection of epistasis is confined to the time interval controlled by the mean selection coefficient. Percentile of detection and false-positive fraction is shown as a function of time. The averaging is done over 25 random simulation runs per value of s_0 (shown). The measure of LD is UFE from eq. (2.83); the use of other measures produced similar results. A detection protocol that maximizes the difference between the detection percentile and the false-positive fractions by tuning the detection threshold shows the same trend. Time point $t = 1.5/s_0$ is the beginning of a transition which completely blurs detection. The initial allelic frequency $f_0 = 0.45$, s_0 is shown, the other parameters are as in Fig. 2.9. Based on Pedruzzi and Rouzine (2019).

Distributed selection coefficient. Next, the variation of s among sites was allowed according to a half-Gaussian distribution. The results are similar to the case with constant s, although false-positive detection rate is higher (Pedruzzi and Rouzine 2019, fig. S5).

Length of the genome and population size. In agreement with the previous results (Cordell 2009; Ueki and Cordell 2012; Wei et al. 2014), an increase in the sequence length L impairs the detectability of epistasis (Pedruzzi and Rouzine 2019, fig. S5). This famous result is known as "the curse of dimensionality": the number of possible locus pairs increases with L as L^2, while the number of epistatic pairs increases only as L, so that finding "a diamond in the rough" becomes more difficult.

Population size. Detection time window expands logarithmically with the increase in the population size N (Pedruzzi and Rouzine 2019, Fig. S5). This result is expected from the multi-locus evolutionary models, which predict logarithmic dependences on

N for all the observables (Rouzine 2020b). Only in populations with N exponentially large in L, linkage effects become negligible (Rouzine et al. 2003). If populations were infinite, epistasis would be easily detectable.

Initial standing variation. A detection window in time exists only when initial population is sufficiently diverse, as given by $f(0) > 10\%$ for the default parameter set (Pedruzzi and Rouzine 2019, fig. S5).

Recombination slightly improves detection. So far, evolution was asexual. The next task is to study the role of recombination in detection of epistasis. New model parameters are the average number of random crossovers per genome, M, and the outcrossing probability, r. Results show that a moderate rate of recombination helps to detect epistasis by disrupting linkage to a larger extent than the epistatic footprint. At the default parameter set (legend to Fig. 2.9), a significant reduction of linkage fluctuations is observed starting from $r = 20\%$ and $M = 5$ (Fig. 2.12). Thus, LD is much less sensitive to recombination than the adaptation rate, which increases at much smaller values of r (Neher et al. 2010; Batorsky et al. 2011). Very large rates of recombination break down epistatic correlations and render epistasis undetectable. Thus, there exists an optimal window of recombination rates for epistatic detection.

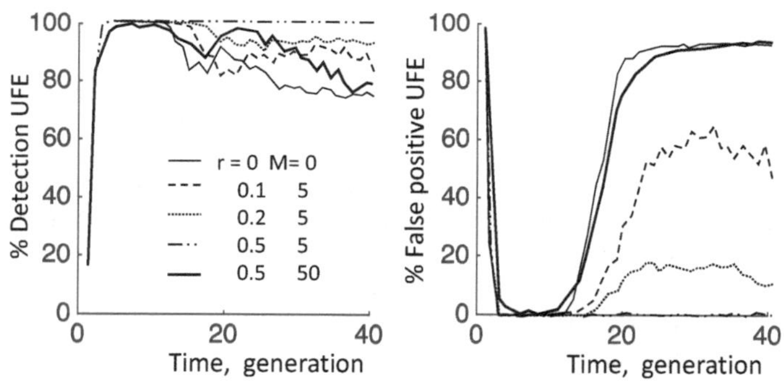

Fig. 2.12: Variation of the time window of detection with recombination.
Percentile of detection and false-positive fraction as a function of time are averaged over 25 random simulation runs at different parameters values. The values of s are randomly drawn from a half-Gaussian distribution. Similar results are observed for all measures of LD defined by eqs. (2.80)–(2.83) (Pedruzzi and Rouzine 2019, figs. S6–S10). $E = 0.75$, the other parameters as in Fig. 2.9. Based on Pedruzzi and Rouzine (2019).

2.2.4 Population divergence creates strong linkage effects

To understand the cause of strong linkage effects hiding epistatic signature, the phylogenetic tree of an asexual population was constructed using a hierarchical clustering algorithm (Fig. 2.13). The initial population has random location of alleles and, hence, a star-shaped phylogeny, with a similar distance between all sequences and the consensus

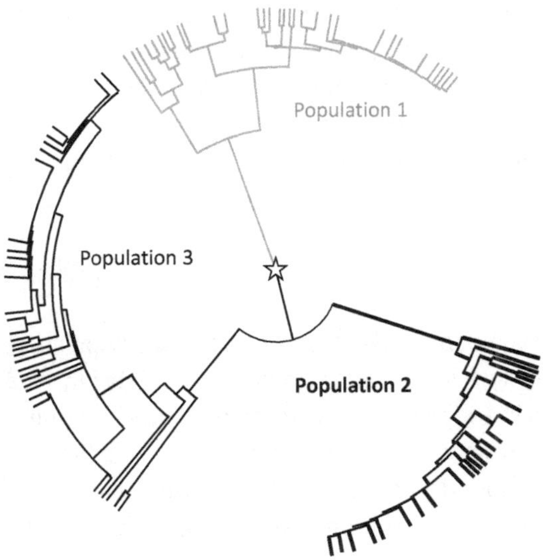

Fig. 2.13: Independently evolving populations have divergent random phylogeny of Bolthausen-Sznitman type. Phylogenetic tree was reconstructed from three populations, independently evolved from the same initial random seed for 30 generations. $N = 20,000$ genomes, initial average allelic frequency $f_0 = 0.40$, the other parameters as in Fig. 2.9. Based on Pedruzzi and Rouzine (2019).

sequence. As time elapses, sequences become increasingly closer genetically, and the tree becomes lopsided with a decreasing number of ancestors. Eventually, the tree evolves into Bolthausen-Sznitman (BS) coalescent, with a single common ancestor, as predicted for the traveling wave regime (Brunet et al. 2007; Walczak et al. 2012; Neher and Hallatschek 2013).

Emergence of BS phylogeny occurs roughly at the same time when LD of noninteracting pairs increase (Fig. 2.9). The reason is the divergence of the population from the initial state in random direction, as illustrated by the trees of three independently evolved populations in separate simulation runs (Fig. 2.13). The genetic distance between their most recent common ancestors increases linearly in time due to fixation of beneficial mutations at random loci. Because haplotype configurations of the common ancestors are inherited by all members of the populations, the mutual divergence of populations results in strong LD with a random sign. This phylogenetic behavior makes it obvious that any epistatic signature is irretrievably lost against the background of random co-inheritance, regardless of the detection method used.

2.2.5 The use of multiple populations rescues epistatic signature

Because linkage fluctuations emerge due to stochastic divergence of populations, the natural idea is to average out haplotype frequencies over multiple populations. To test it, multiple populations were evolved in simulation at the same parameter set. The haplotype frequencies in eqs. (2.80)–(2.83) were averaged over populations, for each locus pair separately. As expected, using a sufficient number of populations resulted in a substantial reduction of the noise, and the detection interval was expanded indefinitely (Pedruzzi and Rouzine 2019). In Section 2.3, we consider this in more detail.

2.2.6 Discussion

Using a Monte-Carlo simulation of the evolutionary dynamics of an asexual population, the distribution density of several pair-wise measures of linkage disequilibrium and their combinations was calculated, for epistatic and nonepistatic pairs of sites separately. In a single population, the signature of epistatic pairs can be detected only under very special conditions: when the initial population has standing variation with randomized alleles, and in the early transitional period inversely proportional to the average selection coefficient. During later adaptation, linkage measures of noninteracting pairs cannot be told apart from that of epistatic pairs. Random linkage effects dominate over the effects of epistasis in a broad parameter region and for all the measures of pairwise linkage disequilibrium.

To understand the reason for linkage fluctuations, the phylogenetic tree of a population was investigated. The increase in fluctuations is coincident with the establishment of BS coalescent predicted for the traveling wave regime (Brunet et al. 2007; Walczak et al. 2012; Neher and Hallatschek 2013). Once BS genealogy is established, individual sequences become similar to their common ancestor sequence, which diverges in time from common ancestors in other independent populations (Fig. 2.13). This divergence is the cause of the strong fluctuations of LD.

Further, averaging the haplotype frequencies over a sufficient number of simulation runs representing independent populations rescues the detection of epistatic pairs, in a broad parameter range. However, one should keep in mind that the multiple-population averaging was conducted under the ideal conditions, in which every population evolved with the same parameter set and represented the same fraction of the total ensemble. Unequal sampling in real data sets may create underperformance. Most importantly, the number of populations required for averaging in a long genome may be extremely large, and additional technique of clearing residual noise is required. Such a technique is considered in Section 2.3.

Several simplifying assumptions were made, as follows. (i) Selection coefficient is constant in time. (ii) Epistatic strength is fixed for all pairs and constant as well. (iii) The simplest topology of epistatic network is considered. While these assumptions are justified given the present focus on linkage, more complex network topology might pose additional challenges (Section 2.3).

Examples of independent replicate populations include influenza virus sampled simultaneously in different countries (before it spread between them), virus passage in replicate tissue cultures, and cancer cell evolution in different organs. In human genetics, independent isolates can also be obtained from subpopulations that split long time ago, such as different countries of common origin. For example, European nations, despite interbreeding, remain genetically distinct after a split from a common root about 6,000 years ago (Behar et al. 2010). Therefore, they can be approximately viewed as independent populations with weak gene flow. The above results imply that comparing genetic data from distinct ethnicities (e.g., on predisposition to various diseases) allows to study epistatic interaction much more reliably than for a single ethnic group.

2.3 Detection of epistasis by the three-way correlation method

As discussed in the previous Section, random co-variation due to linkage completely hides epistatic signature from genomic samples (Pedruzzi and Rouzine 2019). This limitation cannot be cured by any detection method including the popular QLE approach (Kimura 1965; Neher and Shraiman 2011). The most recent common ancestor of a population moves away from the origin along a stochastic trajectory. As a result, any measure of correlation, or even the use of the entire tree, produces only strong noise with random sign. The only way to help detection is to average out haplotype frequencies over many independent populations evolving under similar conditions. Without multiple populations, it is not possible to infer epistasis in principle. But how many populations one needs for reliable averaging? Even in a relatively short genome of 40 loci, 200 populations are not enough to filter out the false-positive interactions (see below). For longer genomes of thousands of loci, the number of populations required for effective averaging increases due to the "curse of dimensionality."

This section follows the work by Pedruzzi and Rouzine (2021) that offers a method to eliminate this residual noise at a fairly modest number of independent populations. The intuitive idea is, when calculating the LD measure, to select only the sequences with a consensus allele at a site connected to the site pair of interest, do the same at other neighbor sites, and thus find the most sensitive site, which interrupts a path that creates a false-positive interaction. In other words, a conditional filter of sequences based on the sensitivity of the pairwise correlation to neighbor alleles splits the genome into independent blocks.

Two parallel methods are used to demonstrate the high fidelity of this detection technique: analytic derivations for two simplest network topologies and Monte-Carlo simulation. The simulation considers the case of positive (compensatory) epistasis, $0 < E < 1$, but the method works at any value of E, including negative epistasis. As in Section 2.1, the system is assumed to be in the process of adaption and still far away from the steady state caused by mutation-selection balance. The selection pressure must be either constant or changing slowly compared to the inverse average selection coefficient (Rouzine 2020b).

After testing, the method is applied to neurominidase protein sequences of influenza A virus adapting to human population. Using $8,000+$ influenza sequences obtained from diverse geographic locations before and after the pandemic of 2009, epistatic interactions are separated them from linkage effects. The specific protein has changed by 20% when it was replaced by a new strain in 2009. At some point in the past, the two strains had a common ancestor. The primary and secondary mutations that allowed the new strain to outcompete the old strain are inferred.

It is worth noting that the widely used technique of elimination of indirect interaction based on quasi-linkage-equilibrium approximation (Kimura 1965) is not suitable for eliminating linkage correlations. They are applied to sequences from different animal species that diverged millions of years ago and assume the absence of linkage effects a priori (Weigt et al. 2009; Cocco et al. 2018).

Below a method of noise-free epistatic detection is derived analytically, tested on simulated sequence data, applied to a real sequence set, and applied to make a testable prediction for a protein network of a virus.

2.3.1 Simulation model to generate sequences for testing the method

The evolution of a haploid asexual population is simulated using a Wright–Fisher process including the factors of random mutation, random genetic drift, and constant directional selection (Fig. 2.14A). The Wright–Fisher model is the same as in Section 2.1, with the uniform selection coefficient s. The case of variable s and its measurement at separate loci is addressed in Section 2.4 (Barlukova and Rouzine 2021).

Some pairs of sites are chosen to interact, with the relative positive amplitude $E = 0.75$. A simple epistatic network of double arches is chosen for the first demonstration (Fig. 2.14B). The network produces three types of allelic correlations due to (i) real epistatic interactions, (ii) indirect interactions through a third site, and (iii) linkage correlations due to having a common ancestor (Section 2.2). A more complex topology with loops is analyzed below as well.

2.3.2 First step: averaging over populations

The first task is to detect all potential epistatic interactions using correlation analysis. For each pair of sites (i,j), the correlation measure introduced in Section 2.1 has a form

$$\mathrm{UFE}_{ij} = 1 - \frac{\log(f_{11}/f_{00})}{\log(f_{01}f_{10}/f_{00}^{2})} \qquad (2.83)$$

where $f_{00}, f_{10}, f_{01}, f_{11}$ are the haplotype frequencies averaged over the simulation runs. Other correlation measures, such as Pearson coefficient r^2, are found to generate similar

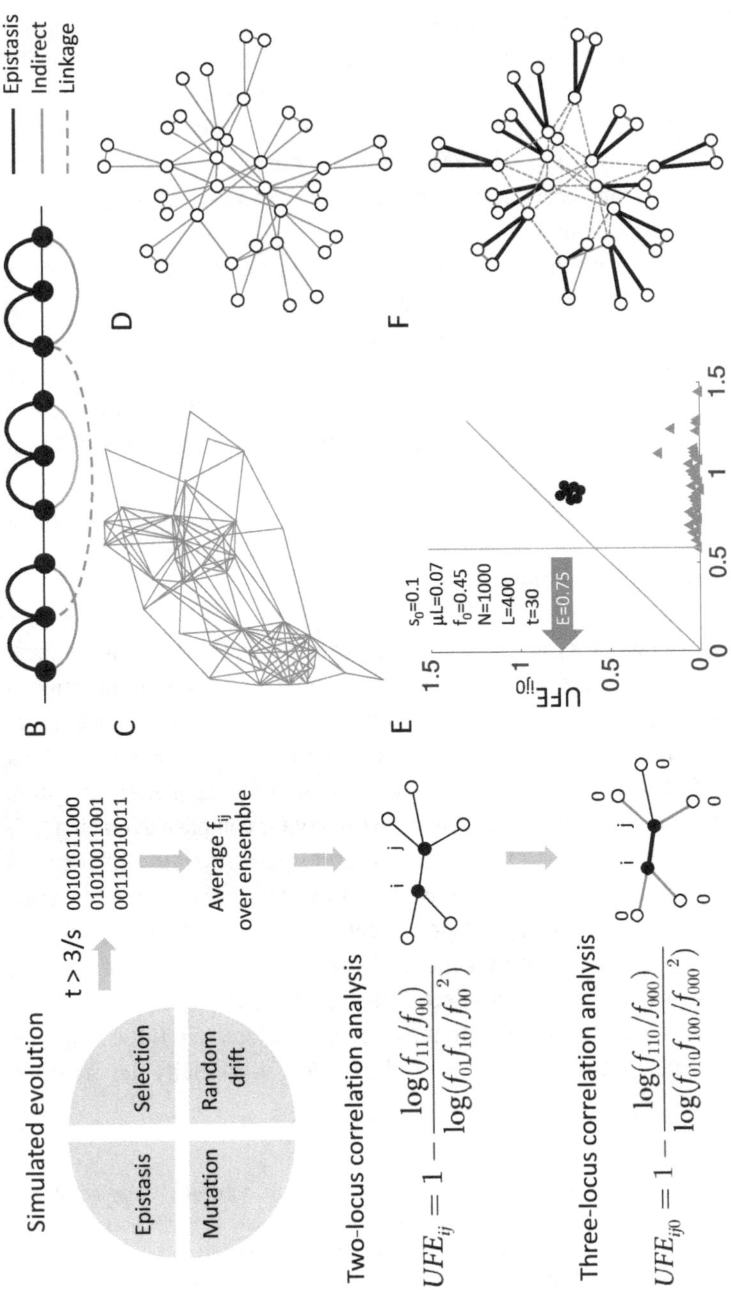

Fig. 2.14: Schematic diagram of the method and its testing on simulated sequence data.
(A) The computer model of asexual evolution includes the factors of random mutation, selection, epistasis, and random genetic drift. Pairwise haplotype frequencies f_{ij} are averaged over simulation runs (independent populations). The pairwise correlation measure UFE_{ij} is calculated from eq. (2.83). The indirect links and the

stochastic noise (Fig. 2.9). As compared to these statistics, UFE_{ij} has the advantage of measuring the degree of mutual compensation of two alleles, $\text{UFE} = E$, if the interacting pair does not interact with the other sites (Fig. 2.5D). Because the RHS of eq. (2.83) has a singular point when one of the four haplotype frequencies is zero, only site pairs such that $f_{ij} > f_{\text{cut}}$ are considered, where f_{cut} is a low cutoff set here at $f_{\text{cut}} = 0.05$. In addition, below only the pairs with a high correlation strength will be considered, $\text{UFE}_{ij} > 0.5$. The actual value of E is set in simulation at 0.75.

Without averaging f_{ij} over simulation runs, the graph of inferred bonds has many of false-positive bonds, which completely mask the 24 true interactions (Fig 2.14C). For a genome longer than $L = 40$, the number of false-positive bonds would be even larger proportionally to L^2. By averaging f_{ij} over 200 simulation runs (independent populations), the number of false-positive interactions decreases by a lot (Fig. 2.14D). However, about a half of remaining links are still false-positive, because their total number is still larger than 24 (Fig. 2.14B).

2.3.3 Second step: three-way correlation

The residual false-positive links are caused by averaging over a limited ensemble. To purge them, one can go in two directions, as follows. Either genomic data have to be sampled from thousands of independent populations, which are unlikely to be found, or one has to find a workaround. The first approach is quite unpractical and not even possible in most cases. Such data are often unavailable. Linkage noise increases with the genome length L, and so does the number of independent populations necessary for complete averaging.

The proposed purging protocol cuts "detours" around pairs of sites which create false-positive bonds. Detours come in two kinds: (i) chains of indirect epistatic interactions and (ii) allelic combinations with common ancestors (Fig. 2.14B). To cut a potential detour of a site pair of interest, only genomes where a neighbor site of the pair has allele 0 (better-fit or consensus) are selected for analysis. The other genomes in the sample are ignored. Then, the correlation measure is re-calculated. If the neighbor site is located at an important detour, this procedure will substantially decrease the

Fig. 2.14 (continued)
residual linkage are detected and filtered out by using the tri-way correlation measure, UFE_{ij0}, from eq. (2.84). **(B)** Preset epistatic network for 40 sites. Black thick curves: real epistatic interactions. Gray thin lines: indirect interactions. Gray dashed line: an example of linkage. **(C) and (D)** The network of strong ($\text{UFE}_{ij} > 0.5$) candidate epistatic interactions predicted **(C)** from a single population and **(D)** after averaging over 200 populations. **(E)** Scatter plot of the three-way haplotype, $\min(\text{UFE}_{ij0})$ shown against UFE_{ij} for the pairs identified in **(D)**. The dashed sector corresponds to the direct interactions. The upper dashed line is the diagonal, $\text{UFE}_{ij} = \text{UFE}_{ij0}$, and the lower dashed line separates direct (black dots) from indirect interactions and linkage (gray triangles). **(F)** Predicted network is exactly the same as the network set in simulation. Parameters: initial allele frequency $f_0 = 0.45$, mutation rate per genome $\mu L = 0.07$, fixed selection coefficient $s = 0.1$, $N = 1{,}000$, $L = 40$, epistatic strength $E = 0.75$. Based on Pedruzzi and Rouzine (2021).

indirect correlation. Correlations due to direct epistatic interactions are, of course, not sensitive to detours and their cutting.

For each candidate bond i, j in Fig. 2.14D, one finds the three-locus measure

$$\text{UFE}_{ij0} = 1 - \frac{\log(f_{110}/f_{000})}{\log(f_{010}f_{100}/f_{000}{}^2)} \tag{2.84}$$

where 0 in the third position denotes consensus allele at a site linked to site i or j. All linked sites are considered as the 0-node, one by one. Then, the minimum value over all 0-nodes, $\min(\text{UFE}_{ij0})$, is calculated to detect the most important detour if there is more than one. The false-positive links are those that have a low ratio $\min(\text{UFE}_{ij0})/\text{UFE}_{ij}$. The procedure is repeated for every potential link detected in Fig. 2.14D.

The scatter plot confirms that, for the false-positive pairs, $\min(\text{UFE}_{ij0})$ is several-fold smaller than UFE_{ij} (Fig. 2.14E). For direct epistatic interactions, as expected, the two LD measures are nearly the same. The exact value of the threshold in $\min(\text{UFE}_{ij0})/\text{UFE}_{ij}$ is not important, as long as the haplotype frequencies are averaged over, at least, ~ 20 populations (for the parameter set chosen). The inferred network is exactly the network set in simulation: 100% detection rate and 0% false-positive rate (Fig. 2.14B). In addition to perfect detection, the use of UFE yields quite accurate estimates of epistastic strength: $\text{UFE} \approx E$ within 15% accuracy (Fig. 2.14F, numbers at links). Thus, the present method provides 100% fidelity at $L = 40$ sites or less, for the number of replicate populations 200, and epistatic strength $E = 0.5 - 0.75$. Performance is less impressive at smaller $E < 0.25$.

2.3.4 Analytic test of the method

The next task is to validate this method analytically. For the epistatic network in Fig. 2.14B, for indirect pairs of sites, one can show that

$$\text{UFE}_{ij} \approx \frac{1}{4(1-E)}$$

$$\text{UFE}_{ij0} \approx 0$$

which explains the separation of direct and indirect interactions (Fig 2.17E). The analytic derivation given below uses the method of quasi-equilibrium (Pedruzzi et al. 2018) described in detail in Section 2.1 [not to be confused with the quasi-linkage equilibrium of (Kimura 1965)].

2.3.4.1 Derivation for topology without loops

Consider the network of double arches, as in the simulation above (Fig. 2.14B). The general fitness expression is given by eqs. (2.1) and (2.2). As in simulation, all selection coefficients are assumed to be equal, $s_i = s_0$, and epistatic strength to be the same for all epistatic pairs, $E_{ij} = E$. In this case, each genome can be fully described by a set of numbers of interacting clusters of different size (Section 2.1). Let k_i denote the number of clusters with i alleles and b_i bonds. Then, from eqs. (2.1) and (2.2), log fitness W is given by a sum over clusters, eq. (2.9). The bond numbers b_i at $i \geq 3$ depend on a network topology; $b_1 = 0$, $b_2 = 1$. For the topology of double arches, there are single, double, and triple clusters of mutated alleles, and $b_3 = 2$ (Fig. 2.6B).

As in Section 2.1, quasi-equilibrium is assumed due to the slow movement of traveling wave (Pedruzzi et al. 2018). The numbers k_i are obtained from the condition that the entropy of genome is maximal given fitness determined by eq. (2.9). The evolving population reaches the maximum-entropy state with respect to the polymorphic sites restricted by the condition that fitness is given. Entropy S is defined as the log number of possible sequence configurations, eq. (2.10).

From eq. (2.37) in Section 2.1.7.2, the frequencies of clusters of size $i = 1, 2$ and 3, denoted $f_i = k_i / L$, are related as

$$f_2 = \frac{2}{3} f_1^{2-2E}, \; f_3 = \frac{1}{3} f_1^{3-4E} \; E < \frac{3}{4} \tag{2.85}$$

Indirect pair. The 1st and 3d site in each triplet in Fig. 2.6B interact only indirectly through site 2. For these two sites, the haplotype frequencies are

$$f_{11} = 3f_3 + f_1^2, \; f_{10} = \frac{3}{2} f_2 + f_1 \tag{2.86}$$

When epistatic interaction is sufficiently strong, as given by the condition $E > 1/2$, triplets dominate numerically over single alleles and doubles, as given by $f_1 \ll f_2 \ll f_3$, eq. (2.85). From eqs. (2.85) and (2.86), using these strong inequalities, we can approximate the haplotype frequencies as

$$f_{11} \approx f_1^{3-4E}, \; f_{10} \approx f_1^{2-2E}, \; \frac{1}{2} < E < \frac{3}{4} \tag{2.87}$$

Using covariance measure UFE_{ij} defined in eq. (2.83) with $f_{00} \approx 1$, $f_{01} = f_{10}$, one gets

$$\mathrm{UFE}_{ij} = \frac{1}{4(1-E)} \tag{2.88}$$

Direct pair. For directly interacting sites 1 and 2 (Fig. 2.6B), the two equations before eqs. (2.44) in Section 2.1.7.2 yield

$$f_{11}^{\mathrm{dir}} \approx f_1^{3-4E}, \; f_{10}^{\mathrm{dir}} \approx \frac{1}{3} f_1^{2-2E}$$

which also results in eq. (2.88). Thus, correlation between indirectly interacting sites 1 and 3 is equal to the correlation for directly interacting sites 1 and 2.

However, if one calculates, instead of UFE_{ij}, the three-way measure UFE_{ij0} given by eq. (2.84) by including only haplotypes with consensus allele 0 at site 2, instead of eqs. (2.87) and (2.88), one gets a completely different result

$$f_{101} \approx f_1^2, \quad f_{100} \approx f_1 \qquad (2.89)$$
$$\mathrm{UFE}_{ij0} \approx 0$$

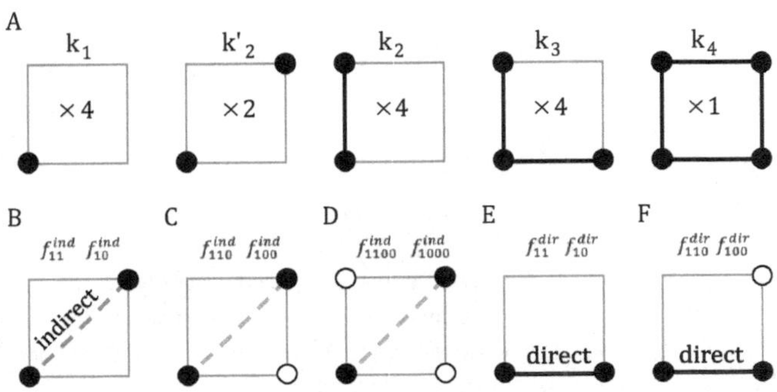

Fig. 2.15: UFE derivation for closed square topology of epistatic network.
(A) Possible configurations and their symmetry. (B) Indirect interaction pairwise. (C) Indirect interaction with a fixed zero at a node. (D) Indirect interaction with two fixed zeros. (E) Direct interaction pairwise. (F) Direct interaction with two fixed zeros. Based on Pedruzzi and Rouzine (2021).

Thus, indirect covariance is destroyed by selecting the sequences with a consensus allele inserted between the two sites. Indeed, by the definition of fitness, eq. (2.1), only minority alleles interact. Once a consensus allele is demanded in the middle of a detour imitating an interaction, the detour is cut, and the false interaction is gone. Although the derivation was presented only for indirect interactions, the same method happens to be equally effective for eliminating linkage effects, the main foe of detection (compare Fig. 2.14B with Fig. 2.14E). Although the exact mechanism remains to be established, probably, the method works by disrupting the chains of minority alleles with a common ancestor in the same way it interrupts detours.

2.3.4.2 Derivation for topology with closed squares

Using the same methodology, one can also consider a slightly more complex network, which consists from L isolated closed squares (Fig. 2.15A). In this case, indirect interaction takes place between the opposite corners, and the sites of a side interact directly with strength E. The main difference from the previous topology is that the square forms a loop, so that two paths can connect the opposite corners (connectivity 2). The results below show that, at $E > 1/3$, direct and indirect correlations have the same amplitude. The 3-way correlation method, again, decreases indirect correlation to a larger degree than direct correlation (Fig. 2.15B, Tab. 2.3). The main difference from the loop-less network

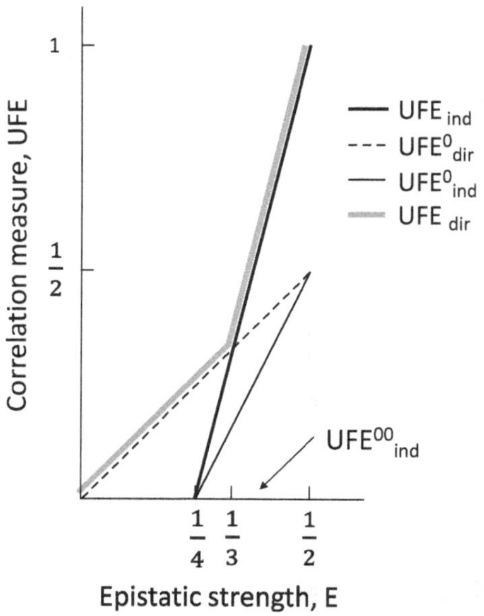

Fig. 2.16: Dependence of 5 different types of linkage measure (Fig. 2.15B–F) on epistatic strength predicted for square topology. Based on Pedruzzi and Rouzine (2021).

above is that indirect correlation does not disappear and preserves the same order of magnitude as the direct interaction. This result is intuitively expected, because there are two equal detours connecting the corners, and both must be cut to obtain a strong effect. The natural way to address this issue is to add another 0 and measure a 4-locus correlation to interrupt both detours. This generalized method, indeed, removes indirect interactions completely in the entire interval of E, as given by $\mathrm{UFE}_{\mathrm{ind}}^{00} \equiv 0$. The results are shown in Fig. 2.16 and Tab. 2.3, and the derivation is as follows.

Tab. 2.3: Direct and indirect UFE values. Index 0 indicates a 3-site measure and 00 a 4-site measure.

Interval	$E < \frac{1}{4}$	$\frac{1}{4} < E < \frac{1}{3}$	$\frac{1}{3} < E < \frac{1}{2}$
$\mathrm{UFE}_{\mathrm{ind}}$	0	$4E-1$	$4E-1$
$\mathrm{UFE}_{\mathrm{ind}}^{0}$	0	$\frac{4E-1}{2}$	$\frac{4E-1}{2}$
$\mathrm{UFE}_{\mathrm{ind}}^{00}$	0	0	0
$\mathrm{UFE}_{\mathrm{dir}}$	E	E	$4E-1$
$\mathrm{UFE}_{\mathrm{dir}}^{0}$	E	E	E

Fitness defined by eq. (2.9) takes the form

$$-\frac{W}{s_0} = k_1 + 2k_2' + 2k_2(1-E) + k_3(3-4E) + k_4 4(1-2E) \tag{2.90}$$

where k_i are the numbers of the five possible types of configurations: single, double, opposing double, triplet, full square (Fig. 2.15A). Each configuration of size i has $n_i = 1, 2,$ or 4 symmetric variants. Hence, the entropy given by eq. (2.10) assumes the form

$$S = \log\left[C_L^{k_4} C_L^{k_3} 4^{k_3} C_L^{k_2} 4^{k_2} C_L^{k_1} 4^{k_1} C_L^{k_2'} 2^{k_2'}\right] \tag{2.91}$$

Setting derivative of S in k_i to zero, one obtains

$$0 = df_1 \log\frac{4}{f_1} + df_2' \log\frac{2}{f_2'} + df_2 \log\frac{4}{f_2} + df_3 \log\frac{4}{f_3} + df_4 \log\frac{1}{f_4} \tag{2.92}$$

The additional condition is that fitness, eq. (2.90), is fixed, whence

$$-df_1 = 2\,df_2' + 2(1-E)df_2 + (3-4E)df_3 + 4(1-2E)df_4 \tag{2.93}$$

Substituting eq. (2.93) into (2.92) and demanding that a coefficient at each independent differential is equal to zero, one gets

$$\frac{f_2'}{4} = \left(\frac{f_1}{4}\right)^2, \ \frac{f_2}{4} = \left(\frac{f_1}{4}\right)^{2(1-E)}, \ \frac{f_3}{4} = \left(\frac{f_1}{4}\right)^{2-4E}, \ \frac{f_4}{4} = \left(\frac{f_1}{4}\right)^{2(1-2E)} \tag{2.94}$$

Without the loss of generality, consider an interval $0 < E < 1/2$. Below 0, indirect interaction does not exist (see below), and the point $E = 1/2$ is the point of full compensation, where the system loses stability (Section 2.1). At higher E, the system cannot be stable, because mutations accumulate without limit. As usual, deleterious alleles are assumed to be sparse, $f_1 \ll 1$. Then, the frequencies of clusters in eqs. (2.94) are ordered, as given by

$$f_2' \ll f_2, \ f_1 \gg f_2 \gg f_3 \tag{2.95}$$

How the probability of a 4-site cluster, f_4, is ordered with respect to the others, depends on the further subdivision of this interval of E, as given by

$$\begin{array}{cc} E < \frac{1}{4} & f_4 \ll f_3 \\ \frac{1}{4} < E < \frac{1}{3} & f_3 \ll f_4 \ll f_2 \\ \frac{1}{3} < E < \frac{3}{8} & f_2 \ll f_4 \ll f_1 \\ E > \frac{3}{8} & f_4 \gg f_1 \end{array} \tag{2.96}$$

Consider now the opposite corners of a square, which correspond to indirect interaction (Fig 2.15B). Combining all possible configurations that produce haplotype 11 at the corners yields

$$f_{11}^{\text{ind}} = \frac{f_2'}{2} + \frac{f_3}{2} + f_4 \tag{2.97}$$

$$f_{10}^{\text{ind}} = \frac{f_1}{2} + \frac{f_2}{2} + \frac{f_3}{4} \tag{2.98}$$

The 3-locus haplotype frequencies are obtained by adding 0 at a third site (Fig 2.15C). Because some configurations are no longer allowed, one gets smaller values than above in eqs. (2.97) and (2.98)

$$f_{110}^{\text{ind}} = \frac{f_2'}{2} + \frac{f_3}{4} \tag{2.99}$$

$$f_{100}^{\text{ind}} = \frac{f_1}{2} + \frac{f_2}{4} \tag{2.100}$$

The 4-locus haplotype frequency, when two zero neighbors are required (Fig. 2.15D), is given by

$$f_{100}^{\text{ind}} = \frac{f_2'}{2} + f_{1000}^{ind} \frac{f_1}{4} \tag{2.101}$$

Now the same procedure is repeated for the side of a square, which corresponds to direct interaction (Fig. 2.15E and F), which yields

$$f_{11}^{\text{dir}} = \frac{f_2}{4} + \frac{f_3}{2} + f_4 \tag{2.102}$$

$$f_{10}^{\text{dir}} = \frac{f_1}{4} + \frac{f_2'}{4} + \frac{f_2}{4} + \frac{f_3}{4} \tag{2.103}$$

$$f_{110}^{\text{dir}} = \frac{f_2}{4} + \frac{f_3}{4} \tag{2.104}$$

$$f_{100}^{\text{dir}} = \frac{f_1}{4} + \frac{f_2}{4} \tag{2.105}$$

Using assumption $f_1 \ll 1$, one gets $f_{00} \approx 1$, and the definition of UFE, eq. (2.83), simplifies to

$$\text{UFE}_{ij} = 1 - \frac{\log(f_{11})}{\log(f_{10}^2)} \tag{2.106}$$

where we used $f_{10} = f_{01}$ due to the symmetry of the topology. Using strong inequalities in eqs. (2.95) and (2.96) and substituting eqs. (2.99)–(2.105) into (2.106), one obtains every type of UFE as a function of E (Tab. 2.3 and Fig. 2.16).

Several conclusions can be drawn:

i) When epistasis is weak or negative epistasis, $E < 1/4$, indirect interactions are absent. For direct interaction, UFE = E, as for isolated pairs (Section 2.1).
ii) At large epistasis $1/3 < E < 1/2$, direct and indirect correlation have exactly the same magnitude, and direct UFE > E. (Because direct and indirect UFE is determined by the same 4-allele clusters.)
iii) Adding 0 at a third site makes indirect correlation smaller than direct correlation, and adding another 0 at the remaining site eliminates indirect correlation completely (by excluding 4-allele clusters).

In the general case of an epistatic network with a higher connectivity than two, the number of the additional zeros required to eliminate an indirect interaction is equal to the number of independent detours, i.e., the connectivity of network. In practice, the connectivity of the network is a priori unknown, and one needs to add extra zeros iteratively until the inferred network stops changing. This procedure is not used for virus protein data below, because the inferred network is almost a tree (connectivity close to unit) already after the three-locus test.

2.3.5 Application to influenza A virus

Now that the detection method is tested both analytically and in simulation, it can be applied to infer an epistatic network from genomic samples of an evolving population. One or two surface proteins of influenza A H1N1, neuraminidase (NA) is a determinant of virus infectivity and a target of drug therapy. Together with hemagglutinin protein, it enables the virus entry into a host cell. The task is to identify the specific mutations and their interactions that caused the emergence of pandemic strain of 2009. For this goal, the sequences of the new strain sampled worldwide are compared to the sequences of the old strain, which have, on the average, 80% of mutual homology. The main assumption here is that both strains evolved nearly independently and in parallel in different geographic locations, with very slow virus migration between them. In fact, virus in different places represents parts of one evolving metapopulation. However, it is the best approximation one can get when working with data on the world pandemics.

Approximately 8000 influenza A H1N1 sequences were downloaded from database https://www.fludb.org for the period of 2000 – 2010. The focus was on the difference between the strains before and after pandemic of 2009 and the epistatic network related to that difference. Similar numbers of sequences were re-sampled from the first and second strains several hundred times.

The old and the new strains were found to be diverse genetically, the new strain gradually replacing the old strain over the period of several years. In terms of travelling wave theory, this can be described as two traveling waves with overlapping fitness distribution, one gradually waning over several years and another replacing it (Rouzine 2020b).

The aligned sequences were binarized by denoting a consensus allele at each amino acid site by 0 and a nonconsensus allele by 1. This simplification is adequate unless more than two amino acid variants per site occur at similar frequencies, which was found to be rare. Only strongly polymorphic sites were considered (genetic distance more than 5%). The mutant allele frequency per genome displayed a bimodal distribution, with two separate maxima of different height, at $f = 0.05$ and 0.2, the first peak being taller. The bimodal distribution represents the mixture of the two strains, with 80% homology between them. The pre-2009 and the 2009 pandemic strain differed in NA protein at about 100 sites.

The robustness of the results to the sampling balance between strain was investigated. In order to compensate for unequal sampling from the two strains, the more abundant sequences with smaller allelic frequency, $f < d_v$, were randomly sampled with the sample number decreased by a coefficient, D_w, ranging from 5% to 50%. The average pairwise haplotype frequencies $\langle f_{ij} \rangle$ were obtained by resampling 200 times.

To infer the intraprotein network of interactions, the procedure tested on simulated sequences (Fig. 2.14C–F) was repeated with the real NA sequences. The two-site and three-site UFE was calculated as defined in eqs. (2.83) and (2.84). To avoid divergence in these equations, only genetically diverse pairs when all four haplotypes were present in excess of cutoff $\langle f_{ij} \rangle > 0.05$ were included. Depending on sampling parameters (d_v, D_w), up to 22 epistatic links were inferred. To ensure robustness of that number to the presentation of local subpopulations in the database, the flattest region of the dependence of the number of links on (d_v, D_w) was chosen (Pedruzzi and Rouzine 2021, fig. 2F). The resulting network with 16 links is shown in Fig. 2.17.

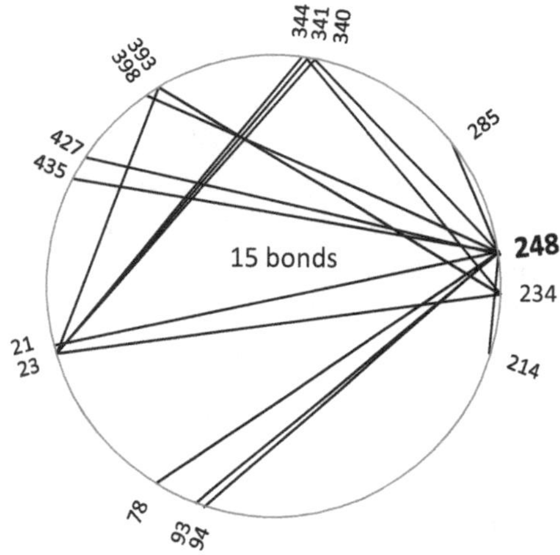

Fig. 2.17: Epistatic network predicted from sequence data on surface protein sequences of Influenza A H1N1 obtained between years 2005 and 2010.
The circular diagram shows the network of interactions between variable amino acid sites in the neuraminidase protein. Based on Pedruzzi and Rouzine (2021).

2.3.5.1 A primary mutation and compensatory sites

Site 248 in NA protein appears to be the primary site connected to multiple compensatory sites (Fig. 2.17). The same typical appearance of a *fitness valley* network is frequently observed in HIV for various drug-resistant mutations.

One possible concern is that the inferred pairs mutations in Fig. 2.17 are not epistatic interactions, but rather represent a driving mutation with hitchhikers riding the invading strain. However, as easy to show, the method would not produce any network all in this case. Indeed, consider a limiting case, in which the virus population represents a mix of the uniform pre-2009 strain and the uniform post-2009 strain. Consider the 20% of the genome where the two strains differ, and suppose their sequences are 101100... and 010011...., which differences occurred due to either driving or hitchhiking mutations. As is easy to see, for any site pair, only two haplotypes f_{ij} out of the four are present. For example, for the first and second sites, haplotypes 10 and 01 are observed, and haplotypes 11 and 00 are absent. When calculating correlation UFE_{ij} and UFE_{ij0} in eqs. (2.83) and (2.84), this pair would be eliminated, because the protocol requires that, for all four haplotypes, $f_{ij} > 0.05$. Thus, the proposed protocol ignores the mere difference between the old and new strains and detects a more subtle association between allelic frequency fluctuations.

2.3.5.2 Structural interpretation

To understand the biochemical nature of inferred compensatory sites in NA protein (Fig. 2.17), they were marked on a three-dimensional protein structure (Fig. 2.18). The active pocket of NA has the function of binding the virus receptor, sialic acid, on target cell surface. Note that the inferred primary mutation 248 is located at its edge and was previously shown to enhance the low-pH stability of NA (Takahashi et al. 2013). Mutation 248 is ubiquitous in all influenza A H1N1 variants isolated after the 2009 pandemic, regardless of a geographic location (Seibert et al. 2012; Byarugaba et al. 2016; Otte et al. 2016). The inferred compensatory mutations are all located on the protein surface inside of α-helixes connecting and guiding the spatial orientation of β-sheet. None of them are located inside of the protein or in a β-sheet. One can assume that their biochemical role is to restore the proper location of β-sheets upset by the primary mutation at residue 248. This is a clear case of allosteric interaction not related to direct binding.

In contrast to the epistatic links (Figs. 2.17 and 2.18), the false-positive links eliminated by the present method are not tied to any specific biochemical structure. A simple consequence of stochastic phylogeny, linkage is indiscriminatory in the biochemical sense. Virtually any genetically diverse pair in the database can form a false-positive bond (Fig 2.14C).

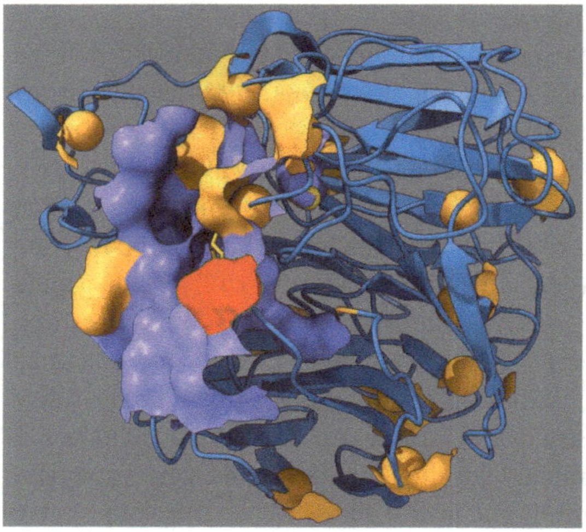

Fig. 2.18: Structural location of the predicted epistasis network for the neuraminidase of influenza virus. The figure shows the three-dimensional structure of Influenza A H1N1 neuraminidase (PDB ID code 4QVZ). Colored spheres represent predicted epistatic residues from Fig. 2.17. Red sphere: Predicted primary mutation (residue 248). Orange spheres: Compensatory residues from Fig. 2.17. Based on Pedruzzi and Rouzine (2021).

2.3.6 Discussion

Thus, Pedruzzi and Rouzine (2021) developed a method to detect epistasis against the background of stochastic linkage effects and indirect interactions. After the observed haplotype frequencies are averaged over a modest number of independent populations, correlators are calculated, the site pairs with high correlation are selected for analysis, and a 3-site (or 4-site, in more complex cases) haplotype test is applied to remove the false-positive interactions.

The entire method is rooted in the phenomenon of quasi-equilibrium predicted and tested by Pedruzzi et al. (2018) (Section 2.1). The concept assumes that adaptation is slow, and that the distribution of alleles has enough time to attain the most probable (most disordered) state given the average fitness of the population. The high fidelity of the method is demonstrated analytically and by computer simulation, using a sequence set evolved under a known epistatic network. For a short genome of 40 loci sampled from 200 independent populations, the method eliminates every false-positive bond and detects all true interactions. Against the background of pessimistic reviews on the prospects of epistatic detection (Wei et al. 2014), these results offer a gleam of hope.

The technique is applied to identify primary and secondary mutations responsible for the post-2009 strain, without addressing the origins or the history of the strain. The results predict a single primary mutation and 15 – 20 strong compensatory mutations, which number is on the same order as observed, for example, in drug-resistant or immune-resistant strains of HIV (Nijhuis et al. 1999; Rouzine and Coffin 1999c; Kelleher et al. 2001) (also Section 1.1). The inferred primary mutation, 248, has been observed after 2009 in various geographic locations (Seibert et al. 2012; Byarugaba et al. 2016; Otte et al. 2016) and is known to enhance virus infectivity (Takahashi et al. 2013). Although empiric evidence for or against the inferred network of compensatory mutations was not found, the technique of deep mutational scanning (Lee et al. 2018; Hom et al. 2019) would be suitable for testing this prediction.

It is important to stress that the other existing techniques are not suitable for eliminating linkage correlations. For example, a popular technique designed to filter out indirect interaction is based on the quasi-linkage equilibrium approximation (QLE) assuming that recombination and epistasis are the only opposing forces and linkage effects are absent, which approximation is valid only for very large population sizes (Kimura 1965; Neher and Shraiman 2011). Hence, the technique can be applied only to data from different animal species that diverged millions of years ago (Weigt et al. 2009; Cocco et al. 2018). The key limitation of QLE method is ignoring linkage effects, which exist in real life even for different species, because any two species have a common ancestor, and a part of common proteins are always co-inherited. The 3-site method does not have this shortcoming, because it does not assume the absence of linkage effects, but eliminates them together with indirect interactions. Hence, the 3-site method works on genomes and proteins sampled from the same species, as long as 20 – 100 independently evolving populations are available.

In some cases, both QLE method and the 3-site method can be applied. However, the second method is more accurate and faster computationally. Indeed, the 3-site method evaluates one pair of loci at a time, without the need of the simultaneous optimization of $L^2/2$ parameters of the full interaction matrix (in the binary case; with 20 amino acids, there is more) (Weigt et al. 2009; Cocco et al. 2018). Still another potential problem of the QLE method that the system is over-defined, and some of these optimal values are too loose and have to be tweaked to get a good agreement with the available experiments. This issue does not exist in the 3-site method; the need to run the 4-site test if the inferred topology has some loops is the only extra element required.

The principal limitation of all above methods is that both assume constant selection, and neither will work under rapidly changing selection conditions (Schiffels et al. 2011). While the evolution of influenza in a population is under the time-dependent selection pressure of accumulating immune memory B cells, the change takes decades, so it can be mapped to the case with constant selection (Rouzine and Rozhnova 2018; Yan et al. 2019; Marchi et al. 2021). The case of virus evolution in epitopes under the CD8 T-cell response requires more careful treatment (Batorsky et al. 2014).

Another limitation is the assumption that migration between independent subpopulations (demes of virus metapopulation) used for averaging the haplotype frequencies is sufficiently slow, so that their evolution is mostly independent. For that, the migration rate into a deme from neighboring demes must be much smaller than mutation rate. In the opposite limit, when migration from any deme to any deme is faster than mutation, the entire metapopulation is well mixed, and the method ceases to apply. Between the two extremes, there is a broad interval of migration rate, where the adjacent demes form well-mixed clusters, and the population has a large number of such clusters. In this case, the method will work, and the linkage noise will decrease with the number of well-mixed clusters.

To summarize, the 3-site haplotype technique is useful to infer the epistatic pairs and tease them apart from stochastic linkage effects. Hopefully, further development of this technique will prove useful for any researcher interested in evolutionary genetics.

2.4 Estimation of selection coefficients from DNA data

Sections 2.1–2.3 focused on detection and measurement of epistatic interactions at polymorphous loci. The use of UFE statistics allowed to exclude selection coefficients and measure directly the degree of mutual compensation of two deleterious alleles. The complete knowledge of fitness landscape includes the values of selection coefficients as well. They are essential for predicting the evolutionary trajectory of a population, such as the development of drug resistance or immune escape (Chapter 4). There is no surprise that enormous effort has been put into the development of various techniques of inference of selection coefficients.

Direct laboratory measurements of selection coefficients in viruses and bacteria require specially designed experiments (Imhof and Schlotterer 2001; Kassen and Bataillon 2006; Acevedo et al. 2014; Stern et al. 2014; Wrenbeck et al. 2017). Simplified analytic models and Monte-Carlo simulation were used to estimate the average selection coefficients over genome based on genomic samples from HIV-infected individuals (Rouzine and Coffin 1999c; Frost et al. 2000; Neher and Leitner 2010; Batorsky et al. 2011) (Sections 1.1 and 1.3). Estimating selection coefficients for individual sites is much more challenging. For example, Illingworth and Mustonen (2012) inferred selection coefficients for individual sites of the hemagglutinin gene of Influenza A/H3N2 by fitting the deterministic one-locus model and its approximate extension for two-loci to the time series data on the allelic frequencies. (Keightley and Eyre-Walker 2007) proposed a method of estimation of the parameters of the distribution of fitness effects (DFE) in mutation-selection-drift equilibrium based on the assumption that it represents a gamma-distribution. Until the work of Barlukova and Rouzine (2021) reviewed in this section, a more general method of inference independent on artificial assumptions was absent.

Such a general method is necessary due to the presence of linkage effects making the dynamics of individual genomic sites intertwined even in the absence of epistasis. Accurate and general expressions for the average speed of adaptation, genetic diversity, the probability of allele fixation, and phylogenetic properties were obtained for various multi-locus models (see introductions to Sections 2.2 and 2.3). As discussed in the previous sections, the evolution of a multi-site genome can be described as a narrow distribution of genomes in fitness moving toward a higher or lower fitness depending on population parameters (Tsimring et al. 1996; Rouzine et al. 2003; Rouzine and Coffin 2005; Desai et al. 2007; Brunet et al. 2008; Hallatschek 2011; Good et al. 2012), which was also observed experimentally (Nguyen Ba et al. 2019). DFE serves as an important input parameter in any model, in addition to the population size, mutation rate, and recombination rate. This section presents a rather general approach to measure the relative value of selection coefficients for each locus, which applies when the system is still adapting and far from equilibrium.

The hint at this method was provided by the experimental observation that DFE for beneficial mutations, as a function of selection coefficient, has often an exponential form (Fig. 2.19) (Imhof and Schlotterer 2001; Kassen and Bataillon 2006; Acevedo et al. 2014; Stern et al. 2014; Wrenbeck et al. 2017). The first biologically consistent explanation of this effect was offered by Barlukova and Rouzine (2021) who showed that an exponential DFE emerges naturally, as a consequence of the adaptation process. The results presented below emphasize the difference between DFE observed in tissue culture experiments on virus evolution and the "inherent" distribution obtained in a directed-mutagenesis experiment. While the first has often the universal exponential form, the second can have any shape depending on a protein and a species. Taking advantage of this theoretical result, a new method to measure the fitness effect of individual mutations not sensitive to linkage effects was developed (Barlukova and Rouzine 2021; Likhachev and Rouzine 2023).

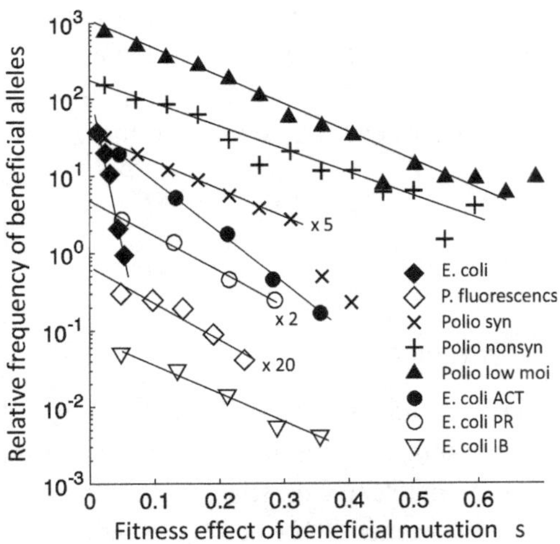

Fig. 2.19: Different studies on distribution of fitness effects of beneficial mutations demonstrate an exponential form.*Y*-axis: Frequency of beneficial alleles (arbitrary units), $DFE(s,t)\varphi(s)$ in eq. (2.107). *X*-axis: Mutation gain in fitness due to a beneficial mutation (selection coefficient). Symbols represent results obtained for different sites of the genome in experiments on *Escherichia coli* (Imhof and Schlotterer 2001), *Pseudomonas fluorescens* (Kassen and Bataillon 2006), poliovirus synonymous mutations, poliovirus nonsynonymous mutations (Acevedo et al. 2014), poliovirus low MOI (Stern et al. 2014)}, *E. coli* acetamide (ACT), propionamide (PR), and isobutyramide (IB) (Wrenbeck et al. 2017). Based on Barlukova and Rouzine (2021).

2.4.1 Experimental distribution of selection coefficients

The key to interpreting the exponential shape of DFE is that beneficial mutations can emerge only at the sites occupied by less-fit alleles. (As everywhere in the book, two possible alleles per site are assumed. Although each site can have four nucleotides A, C, T, G, on moderate time scales of several thousand generations, most genetically diverse sites have no more than two alleles.) A well-adapted population has best-fit alleles at most sites. Therefore, the observed DFE is determined by the few remaining less-fit sites, whose number decreases in time, t, the sites with larger s disappearing first. The average frequency of less-fit alleles at a site with fitness effect s denoted $f(s,t)$ is the frequency of sites available for beneficial mutations. If the system is fully adapted, we have $f=0$, and no further beneficial mutation is possible.

In experiments, researchers passage bacteria or virus for a short time in culture and observe the rise of new, naturally occurring mutations that confer a fitness benefit. (Acevedo et al. 2014) used a deep-sequencing technique CirSeq to monitor the frequency of minority alleles in polio virus as a function of time and fit it with a simple one-site model to estimate the values of s for each site. (Imhof and Schlotterer 2001) did a similar procedure with *E. Coli.* Then, the researchers counted the number of selection sweeps belonging to an interval of s (X-axis in Fig. 2.19). The first task is to understand, with mathematical precision, what exactly these experiments observed.

Suppose, a random mutation with the fitness benefit s has occurred at a less-fit site. Due to random genetic drift and linkage effects, a new allele can be either established in a population with the probability $\varphi(s)$ or lost (Good et al. 2012). Experiments, which results are shown in Fig. 2.19, detected established mutations with the observable density $\mathrm{DFE}_{\mathrm{obs}}(s,t)$ given by

$$\mathrm{DFE}_{\mathrm{obs}}(s,t)=\mathrm{DFE}(s,t)\varphi(s) \tag{2.107}$$

where $\mathrm{DFE}(s,t)$ is defined as the product

$$\mathrm{DFE}(s,t)\equiv f(s,t)g(s) \tag{2.108}$$

and $g(s)$ denotes the intrinsic density of sites with the selection coefficient s. Function $g(s)$ can be measured in experiment by performing site mutagenesis for each site, one by one, and evaluating fitness differences between the wild type and mutant strains (for example, in a growth competition experiment) (Lee et al. 2018; Hom et al. 2019). Thus, the observable and the intrinsic density of sites in s differ qualitatively (Schiffels et al. 2011).

In eqs. (2.107) and (2.108), the average frequency of deleterious alleles $f(s,t)$ depends on the state of population and time. Intrinsic distribution $g(s)$ is a property of the virus and the cell culture and does not depend on time or the state of population. It is not universal and varies broadly between viruses and proteins. In this section, ensemble-average mutant frequency $f(s,\ t)$ is assumed to have preevolved before the experiment, for a sufficient time and under similar conditions. The population is assumed to be far from steady state (mutation-selection equilibrium), that is, it is not well-adapted yet.

As shown below in the long term, $f(s,t)$depends sharply and exponentially on s, with the derivative increasing in time. At the same time, the scale of $g(s)$ in s stays, of course, constant. Therefore, sooner or later, $f(s,t)$will become a function that changes much faster with s than $g(s)$ and dominates the observable DFE, which explains the experimental results (Fig. 2.19).

2.4.2 Model

The population model is the same asexual model with preexisting genetic variation that was used in Sections 2.1–2.3, except the epistatic term in eq. (2.1) is now dropped assuming additive contributions from sites to fitness. The approximation is appropriate as long as evolution is only moderately long. In this case, the many epistatic interactions only renormalize the current values of s_i. If protein evolves for a very long time, a large part of its sequence will change, affecting the values of s_i (Rouzine and Coffin 1999c; Shah et al. 2015; McCandlish et al. 2016; Lee et al. 2018; Hom et al. 2019). Pairwise epistatic effects were considered in Sections 2.1–2.3; for a global epistasis model, see Good and Desai (2015) and the references therein. On a short-term scale, few sites are polymorphic and even fewer interact, so that not including epistasis is a fair approximation for most sites.

Inherent distribution density $g(s)$ is an input parameter of the model. As already mentioned, the form of $g(s)$ varies broadly between organisms and proteins. In the simulation below, the values of s_i are randomly sampled from a half-normal intrinsic distribution

$$g(s) = \frac{2}{s_{av}\pi} \exp\left(-\frac{s^2}{\pi {s_{av}}^2}\right), \quad s > 0 \tag{2.109}$$

where s_{av} is the average. In the analytic derivation below (Section 2.4.4), no assumptions regarding the form of $g(s)$ are made.

2.4.3 Monte-Carlo simulation of genetic evolution

Evolution of an asexual Wright–Fisher population is simulated by Monte-Carlo algorithm, as described in Sections 2.1–2.3. The first 100 sequences of a larger population are shown at different time points (Fig. 2.20A–E). In X-axis, genomic sites i are ranked in the ascending order of the selection coefficient s_i, and the rank is shown. In the process of adaptation, the density of deleterious alleles is being redistributed according to the rank of s_i. Deleterious alleles with a high mutation cost become extinct first, which offers a qualitative interpretation of the experiments in Fig 2.19. One can observe two distinct intervals of time, as follows.

(i) *Selection of preexisting alleles*: In early evolution, $t \ll 1/s_{av}$, where s_{av} is the average of $g(s)$, new beneficial mutations are not important, and the natural selection of preexisting alleles is the only acting factor of evolution (Fig 2.20A and B). This regime has been observed experimentally for vesicular stomatitis virus evolving in culture (Dutta et al. 2008). In this time interval, the distribution of alleles over genomes has random appearance.

(ii) *Traveling wave*: In the late phase, $t > 1/s_{av}$, new beneficial mutations are critically important, because they create new genomes that are more fit than the fittest of preexisting sequences (Fig 2.20C–E).

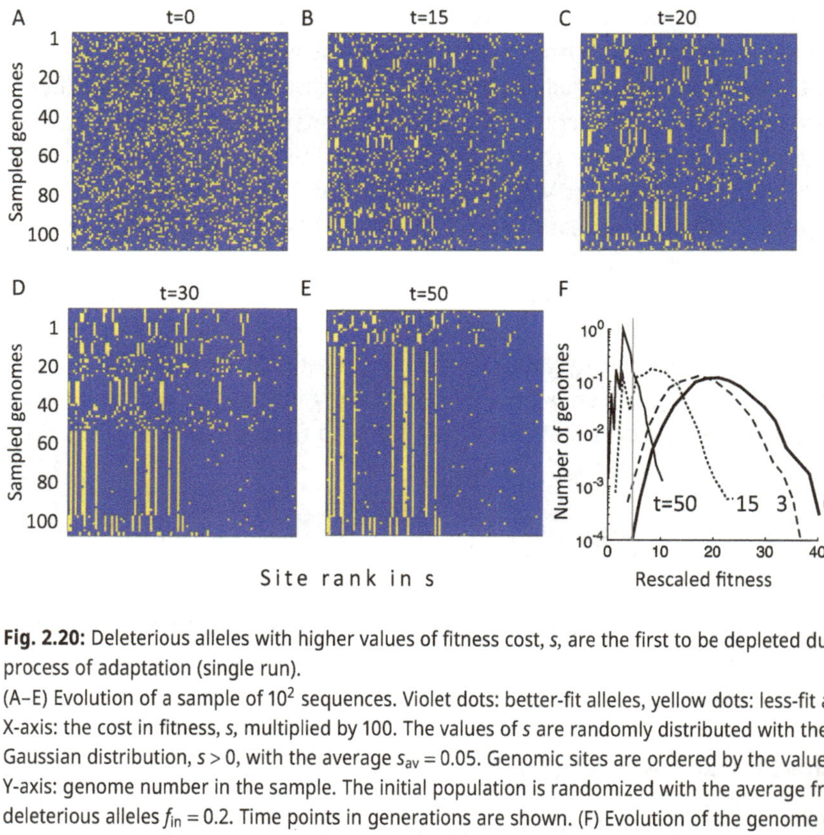

Fig. 2.20: Deleterious alleles with higher values of fitness cost, s, are the first to be depleted during the process of adaptation (single run).
(A–E) Evolution of a sample of 10^2 sequences. Violet dots: better-fit alleles, yellow dots: less-fit alleles. X-axis: the cost in fitness, s, multiplied by 100. The values of s are randomly distributed with the half-Gaussian distribution, $s > 0$, with the average $s_{av} = 0.05$. Genomic sites are ordered by the value of s. Y-axis: genome number in the sample. The initial population is randomized with the average frequency of deleterious alleles $f_{in} = 0.2$. Time points in generations are shown. (F) Evolution of the genome distribution in fitness. X-axis: the effective number of deleterious alleles, defined as $k = -W/s_{av}$. where W is genome fitness. Four discrete time points are shown. Vertical grey line shows the most-fit class of genomes at $t = 0$. The emergence of clonal structure in (A–E) reveals the transition from the selection of preexisting sequences to the traveling wave regime. Parameters: population size $N = 10^4$, number of sites $L = 100$, genomic mutation rate $\mu L = 0.05$. Based on Barlukova and Rouzine (2021).

In the second interval, groups of identical sequences (clones) grow at the high-fitness edge of the traveling wave (Fig. 2.20F). The clones can be seen as the vertical columns of linked deleterious alleles. The fitness distribution moves, as whole, toward high fitness. As the wave moves on, the clonal structure becomes more pronounced (Fig 2.20C–E). Eventually, all genomes have a single common ancestor, which is one of the best-fit genomes in the initial population.

2.4.4 Analytic derivation of universal DFE

Consider the ensemble-averaged mutant frequency $\langle f(s,t)\rangle$ and assume that populations are not close to the steady state yet, so that deleterious mutation events can be neglected. Below, the exponential form of $\langle f(s,t)\rangle$ is obtained by three independent derivations.

2.4.4.1 Early evolution

At early times $t < s_{av}$, natural selection of preexisting alleles is the only factor and mutation can be neglected. Consider a genomic site with selection coefficient s. The other selection coefficients in the genome are unknown and obey the distribution density $g(s)$, which is assumed to be independent between sites. In the initial population, less-fit alleles are distributed randomly between alleles and individuals with average frequency f_{in}.

Let $I_0(s,t)$ and $I_1(s,t)$ be the frequencies of all possible sequences that have alleles 0 and 1, respectively, at the site of interest at time t. The average mutant frequency is given by

$$\langle f(s,t)\rangle = \frac{I_1(s,t)}{I_0(s,t)+I_1(s,t)} \tag{2.110}$$

The copy number of any binary sequence, $\{K_i\}$, decays in time proportionally to $e^{-t\sum_{i=1}^{L-1} s_i K_i}$. Averaging $I_1(s,t)$ and $I_0(s,t)$ in eq. (2.210) over configurations yields

$$\begin{aligned} I_0(s,t) &= (1-f_{in})Z \\ I_1(s,t) &= f_{in}\, e^{-ts} Z, \end{aligned} \tag{2.111}$$

$$Z = \int_0^\infty ds_1 \int_0^\infty ds_2 \ldots \int_0^\infty ds_{L-1} \sum_{K_1=0}^{1}\sum_{K_2=0}^{1}\cdots\sum_{K_{L-1}=0}^{1}\prod_{i=1}^{L-1} p(K_i)g(s_i)\, e^{-t\sum_{j=1}^{L-1} s_j K_j} \tag{2.112}$$

The probabilities of having less-fit and better-fit alleles in the initial population are $p(1) = f_{in}$ and $p(0) = 1 - f_{in}$, respectively, where f_{in} is the initial mutant allele frequency.

Taking out the product symbol outside of the integrals in eq. (2.112), summation and integration is performed for different i independently, and one obtains

$$\langle f(s,t)\rangle = \frac{f_{in}}{(1-f_{in})\ e^{\ ts} + f_{in}} \tag{2.113}$$

Note that eq. (2.113) does not depend on the form of the intrinsic distribution $g(s)$. At small f_{in}, it can be approximated with an exponential (Fig. 2.21A)

$$f(s,t) \approx f_{\text{in}} \exp(-ts),\ f_{\text{in}} \ll 1$$

The exponential slope of the distribution of deleterious alleles defined as

$$\beta = -\frac{\partial \log(f)}{\partial s} \tag{2.114}$$

is, in this case, approximately equal to time, $\beta = t$ (Fig. 2.21B). This result explains the experiments (Fig. 2.19) if the virus preevolved for a short time. In the case of a longer period of evolution, one needs another approach.

2.4.4.2 Traveling wave regime

In the regime of traveling wave, which starts around $t_0 \sim 1/s_{\text{av}}$, the derivation is more complex. Firstly, beneficial mutations must be included into consideration, because they allow the high-fitness edge of the wave to move forward. Secondly, different realizations of a population differ strongly, and genetically diverse sites can be neglected when calculating $\langle f(s,t)\rangle$. Most deleterious alleles are found at rare uniformly deleterious sites, which differ between individual populations (vertical columns of dots in Fig 2.20). Their ensemble-averaged density determines $\langle f(s,t)\rangle$, which decrease occurs due to the fixation of new beneficial alleles at these uniformly deleterious sites. The dynamic balance equation describing this process has a form of ODE

$$\frac{\partial \langle f(s,t)\rangle}{\partial t} = -\mu N \varphi(s) \langle f(s,t)\rangle, t > t_0 \sim 1/s_{\text{av}} \tag{2.115}$$

The initial condition can be obtained from the early time phase, eq. (2.113), where selection of preexisting genomes is the dominant process

$$\langle f(s,t_0)\rangle \approx f_{\text{in}} e^{-t_0 s} \tag{2.116}$$

The solution of eq. (2.115) with that initial condition has the form

$$\langle f(s,t)\rangle = f_{\text{in}}\ e^{-t_0 s - \mu N \varphi(s)(t-t_0)} \approx f_{\text{in}} e^{-\beta(t)s} \tag{2.117}$$

$$\beta(t) = t_0 + \mu N \varphi'(0)(t - t_0) \tag{2.118}$$

Here the argument of exponential in eq. (2.117) is approximated with its linear expansion. This approximation applies in the long-term, when $\beta(t)$ is large and hence, the characteristic s is small.

Fixation probability $\varphi(s)$ was found in Good et al. (2012)

$$\varphi(s) = A\left[e^{-\frac{s^2}{2v}}\left(\frac{e^{s\frac{x_c}{v}}-1}{s}\right) + \frac{e^{\frac{x_c^2}{2v}}}{vx_c}\int_{x_c}^{\infty} dx\, xe^{-\frac{(x-s)^2}{2v}}\right],$$

$$A = \frac{1}{N}\left[\frac{x_c}{v} + \frac{1}{x_c}\right]^{-1} \tag{2.119}$$

Here $\varphi(0) \approx 1/N$ corresponds to the selectively neutral limit, v is the average adaptation rate (fitness increase per generation), x_c is the lead of traveling wave (i.e., relative fitness corresponding to the boundary of clonal interference), and N is the population size.

For sufficiently large population sizes, which correspond to the multiple-mutation regime, one has $x_c^2 \gg v$ (Good et al. 2012), and the second term in eq. (2.119) can be neglected, which yields

$$\varphi(s) = \frac{1}{N}\frac{v}{x_c}e^{-\frac{s^2}{2v}}\left(\frac{e^{s\frac{x_c}{v}}-1}{s}\right) \tag{2.120}$$

Assuming, again, small $s \ll v/x_c$ (but still comparable to $1/\beta$), one obtains

$$\varphi(s) \approx 1/N + \varphi'(0)s$$

$$\varphi'(0) = \frac{x_c}{2Nv} \tag{2.121}$$

Parameters v and x_c are expressed in terms of N, the beneficial mutation rate per genome per generation, U_b, and the initial distribution density of s before evolution started (Good et al. 2012). The last parameter is the same as the intrinsic density $g(s)$.

As a specific example of $g(s)$, consider a half-Gaussian form, eq. (2.109). At large population sizes, where the traveling regime applies, such that $v \gg s_{av}^2$, or $\log(Ns_{av}) \gg \log^2(s_{av}/U_b)$, we get

$$x_c = \sqrt{2V\log(Ns_{av})} \tag{2.122}$$

$$x_c/v = \frac{1}{\sqrt{\pi}\, s_{av}}\log\left(\frac{s_{av}}{U_b}\right) \tag{2.123}$$

(Good et al. 2012, eqs. (2.19), (2.20), and (2.22), with $\sigma \equiv \sqrt{\pi}s_{av}$.) Substituting eq. (2.123) into (2.121), one obtains

$$\varphi'(0) = \frac{1}{2\sqrt{\pi}\, Ns_{av}}\log\left(\frac{s_{av}}{U_b}\right) \tag{2.124}$$

which is then substituted into eq. (2.118).

Thus, in the long-term adaptation (but before steady state), the model predicts, again, the exponential form for the ensemble-averaged density of less-fit alleles $\langle f(s,t)\rangle$, with the slope increasing in time, as given by eqs. (2.116) and (2.118). This prediction interprets DFE data in Fig. 2.19, assuming these pathogens preevolved under the same conditions for a moderately long time.

2.4.4.3 Quasi-equilibrium argument

In the previous sections, an exponential DFE was derived for a specific population model. The same result can be obtained in the more general case using results from Section 2.1. In the traveling wave regime, $NU_b \gg 1$, when adaptation is very slow, the distribution of alleles is in the state of quasi-equilibrium, where entropy is maximal conditioned by the average fitness slowly changing in time (Section 2.1) (Pedruzzi et al. 2018).

What limits the speed of adaptation is the rareness of mutations or recombinants at the high-fitness edge of fitness distribution (Rouzine et al. 2003; Rouzine and Coffin 2005). As a result, when the traveling wave is inching up in fitness, each fitness class in the bulk of the wave has a sufficient time to reach the most disordered state, in the ensemble-averaged sense. Hence, for each fitness class, the allelic distribution across sites and genomes, in a statistical ensemble, is governed by the condition that entropy is maximal. In the traveling wave regime, the fitness distribution is always narrow, $\Delta W \ll |W|$, hence, entropy is at the conditional maximum, $S(W)$, restricted by the average value of fitness, W. Maximal entropy known as the Second Law of Thermodynamics is a very general property of stochastic systems, which assume the most probable (most disordered) state in equilibrium (Landau and Lifshitz 1969).

From this principle, the probability to have a deleterious allele at a given site is given by exp(S) and eq. (2.4) in Section 2.1, which can be written as

$$\langle f(s,t)\rangle = [1 - \langle f(s,t)\rangle\,]e^{-\beta(t)s} \tag{2.125}$$

where $\beta = \partial S/\partial W$. Equation (2.125) generalizes the results in eqs (2.113), (2.114), (2.117) and (2.118), which contain the expressions for a specific population model and time interval.

Note that $\beta(t)$ in all these expressions depends on time (linearly) because the adapting population is assumed to be still far from the steady state, and deleterious mutation events are neglected. Hence, the *quasi* in quasi-equilibrium. The true steady state, which is eventually reached, after a sufficiently long time, is dominated by the balance between deleterious mutation, genetic drift, and natural selection (Rouzine et al. 2003; Goyal et al 2012). In the deterministic limit of very large populations (Section 1.1), the opposing factors are only deleterious mutation and natural selection.

Thus, the exponential dependence of the observed DFE on s is interpreted by the fact that the fastest-changing term in DFE is the frequency of deleterious alleles $\langle f(s,t)\rangle$ which quickly assumes that shape the course of evolution. The result is quite universal, because it is robust to the intrinsic density of s, initial conditions, and population model.

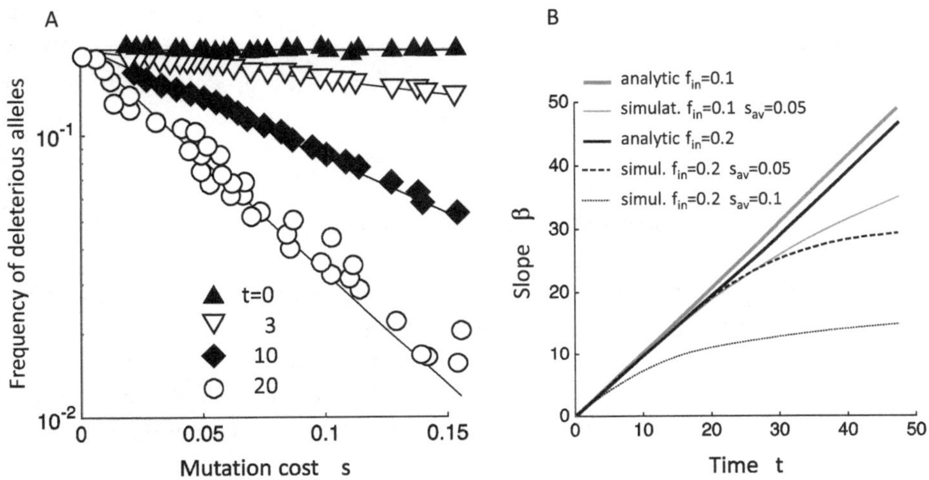

Fig. 2.21: The frequency of deleterious alleles decays exponentially with their fitness effect, with the slope increasing in time.
(A) Analytic prediction for the frequency of deleterious alleles from eq. (2.113) agrees with Monte-Carlo simulation. *X*-axis: Mutation cost of deleterious allele at a genomic site, s. *Y*-axis: Frequency of deleterious alleles at such a site, $f(s)$. The mutant frequency f is averaged over 20 random simulation runs, the straight lines are linear regression. Different symbols show simulation results at different time points. The lines are the analytic predictions. (B) The slope of the distribution of deleterious alleles β, as a function of time t: the short-term prediction of eq. (2.113) (thick grey or black lines) and the interpolated simulation result (thin, dashed, or dotted lines). The log slope for the simulated curves of mutant frequency in (A) is obtained by an exponential fit. The deviation of the simulated slope from eq. (2.113) coincides with the establishment of the traveling regime (Fig 2.20F). In the long term, the traveling wave prediction, eq. (2.118) applies. The diagonal shows $\beta(t) = t$. Parameters are as in Fig 2.20 unless shown otherwise (f_{in} and s_{av}). Based on Barlukova and Rouzine (2021).

2.4.4.4 Monte-Carlo test confirms results

The analytic result for $\langle f(s,t)\rangle$ in eq. (2.113) (early phase) was compared to the results of Monte-Carlo simulation at different time points (Fig 2.21A). Initially, all sites are set to have the allelic frequency $f = f_{in}$. Thus, the initial slope β is equal to 0 (blue line). As time goes, the slope increases linearly in time, as it should, and function $\langle f(s,t)\rangle$ depends exponentially on s, as it should. Thus, the analytical formula, eq. (2.113), agrees with simulation in the short term.

Next, the analytic prediction for the time dependence of slope β is compared with interpolated simulation results, at different values of s_{av} (Fig 2.21B). At early times, the simulation results match the short-term prediction for β from eq. (2.113). As expected, at longer times, $t > 1/s_{av}$, the results deviate from eq. (2.113) due to the transition to the traveling wave regime (Fig 2.20F). In this time interval, eqs. (2.118) and (2.124) determine the slope $\beta(t)$. The value of t_0 in eq. (2.118) is adjusted to make the simulated and analytical curves match each other (single-parameter fit). Despite the use of fitting, the comparison between simulation and analysis remains meaningful, because adjusting t_0 does change the time derivative of the slope β. The results show an excellent agreement in $d\beta/dt$ between simulation and the analytic prediction (Fig 2.21B). To summarize, computer simulation confirms the proposed interpretation of the experimentally observed exponential form of DFE (Fig. 2.19).

2.4.5 Estimating selection coefficients from a sequence set

These theoretical results have an important practical application: a method of the inference of selection coefficients in relative units, $s_i\beta$, that is not affected by omnipresent linkage effects. For the method to work, the system has to meet several conditions, which follow from the previous sections. The proposed method is rather model-insensitive, but these conditions must be met. Whether they are met, becomes clear in the end of the procedure, hence, the method is self-testing:

(i) Natural selection is directional and either constant or changing little on the time scale $t \sim 1/s_{av}$.
(ii) A sufficient number of independent replicate populations (passages, different geographic locations) is available to average out $\langle f(s,t)\rangle$.
(iii) The organism (virus) is still adapting and far from the full equilibrium, as given by $\langle f(s,t)\rangle \gg f_{eq}$.

The first step of the procedure is obtaining a database of aligned DNA, RNA, or protein sequences for the species of interest at three time points (the third is for testing validity). Consensus sequence is determined at each site, as the most abundant variant. The sequences are binarized by replacing consensus alleles with 0, and minority alleles with 1. The frequency of 1 for each site, $\langle f_i(t)\rangle$, is calculated for the entire database including all populations. Insertions and deletions are eliminated, because they cannot be described by a single selection coefficient per site, unless they have a fixed length and can be treated as a binary substitution. In principle, the method can be generalized to a nonbinary format, but here the binary version is described.

Based on eq. (2.125), the relative value of selection coefficient s_i at amino acid position i can be estimated from

$$\beta(t)s_i = -\log\left[\frac{\langle f_i(t)\rangle}{f_{norm}}\right] \tag{2.126}$$

where $\langle f_i(t)\rangle \ll 1$ is assumed. Below angular brackets showing ensemble averaging $\langle \ldots \rangle$ are dropped.

The presence of an additional factor f_{norm} in eq. (2.126) absent in eq. (2.125) is necessary, because of the change in the definition of $f_i(t)$. In the previous sections, f_i was defined as the less-fit allele frequency, and $s_i > 0$ was always positive and denoted the fitness gain from beneficial mutation. In real genomic samples, the best-fit sequence unknown, and one has instead to use the consensus sequence as the reference sequence. Therefore, anti-consensus allele can be better-fit, in which case $s_i < 0$. The sites with negative s_i are those with $f_i > f_{\text{norm}}$, hence the need for normalization factor $f_{\text{norm}} < 1$.

The left-hand side of eq. (2.126) represents a product of two factors: $\beta(t)$, which depends only on time, and s_i, which depends only on locus number, i. This remarkable fact points to the existence of a fixed point $s_i = 0$, which determines the normalization factor f_{norm}, as follows.

For each time t, genomic sites i are ranked in the descending order in s_i to produce a map $i \to m_i$, where m_i is the site number after ranking. The result is plotted as a monotonic curve $s_{\text{rank}}(m, t)$. Then, one finds the intersection point between the curves $s_{\text{rank}}(m, t)$ obtained at three different times t (Fig. 2.22). The intrinsic distribution density discussed in Section 2.4.1 is given by $g(s) = [ds/dm]^{-1}$.

To test the method, the three curves should have a clear separation and a single intersection point (Fig. 2.22A), otherwise, one of the three above conditions is not satisfied. If the curves have many intersections or collapse onto each other (Fig. 2.22B), this behavior implies either insufficient averaging over populations, or being in a steady state, or the time points are too close, or time-dependent selection. In any of these cases, the method will not work.

Next, the value of f_{norm} in eq. (2.126) is tuned to obtain $s_{\text{rank}} = 0$ at the intersection point. Finally, the ranked sites are re-ordered back, $i \leftarrow m_i$, and βs_i are plotted as bars of different height against their actual sites, i.

To conclude, the method estimates selection coefficients in relative units of $1/\beta$. The scaling factor β, the same for all sites in the genome, has to be found from an independent observation. The detailed code of the method is found in Likhachev and Rouzine (2023) and the site indicated therein.

2.4.6 Discussion

Many experiments observe an exponential DFE (Fig. 2.19). Consequently, population models often assume it as an input parameter (Gerrish and Lenski 1998; Good et al. 2012; Walczak et al. 2012). Above findings provide a simple evolutionary interpretation for this observation: faster depletion of deleterious alleles with higher fitness costs. The distribution is not constant but shrinks linearly with time. The result follows both for specific population models and in the general form, from the Second Law of Thermodynamics.

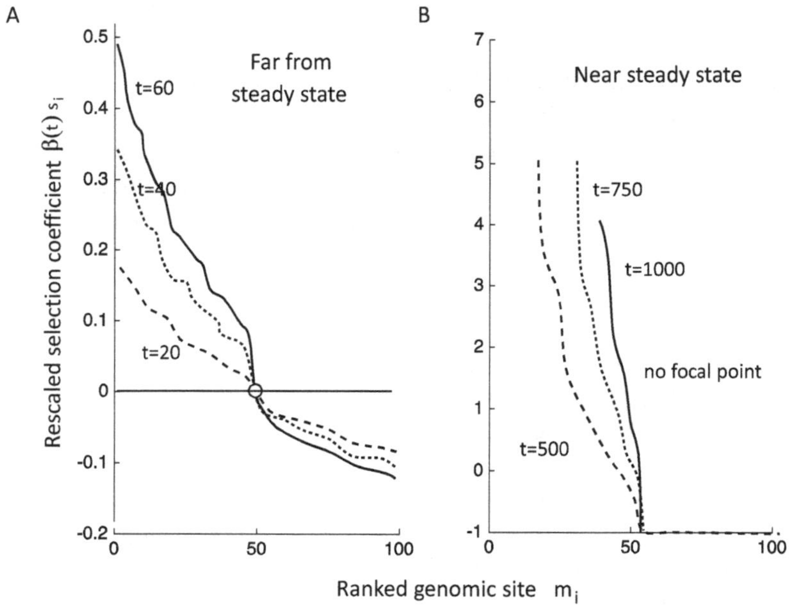

Fig. 2.22: Ranked curve $s_i(m_i)$ for three time points.
(A) Successful separation of curves, with a proper focal point, when far from a steady state. B) Failure of the method close to a steady state. Parameters: (A) $\langle s \rangle = 0.05$, $L = 100$, $N = 1{,}000$, $t = 60$, $f_0 = 0.1{,}100$ runs, $\mu L = 0.1$, $t_1 = 20$, $t_2 = 40$, $t_3 = 60$; (B) $t_1 = 500$, $t_2 = 750$, $t_3 = 1{,}000$, $t = 1{,}000$, and the other parameters as in (A). Based on Likhachev and Rouzine (2023).

At longer times, when mutations de novo are crucial for adaptation, the theory predicts that most deleterious alleles are found, mostly, in monomorphic sites, which location varies between random realizations. Hence, the exponential law is valid only after averaging over realizations (populations, passage replicas).

Based on these theoretical results, a method is developed to estimate the relative selection coefficients at each polymorphic site from the genomic sequence sets sampled from multiple adapting populations at three time points. This method allows to measure the "intrinsic" DFE, $g(s)$, which differs from the naturally occurring DFE(s) and can assume any shape depending on a virus and a protein.

The distribution of the form eq. (2.125) was first obtained for the energy distribution of fermions, the quantum particles that do not share quantum states with other identical particles. Using a similar thermodynamic argument, it was derived for various biological two-state systems in equilibrium (Berg et al. 2004; Sella and Hirsh 2005; Rice et al. 2015). The reason why the equilibrium statistics turns out to be valid in a nonequilibrium case is the slow speed of adaptation (Pedruzzi et al. 2018).

Strikingly, the results obtained for the adaptation regime do not apply in the true steady state, where deleterious mutation balances the effect of selection and the derivations in Section 2.4.4 become invalid. The case of steady-state was considered by Rice et al. (2015). As expected, the cited authors did not derive an exponential for DFE(s). Instead, they demonstrated an exponential for the ratio between the beneficial and deleterious branches DFE(s)/DFE($-s$). In contrast to the adaptation regime studied here, which has an exponential slope of DFE(s) increasing in time, $\beta(t)$, the exponential slope predicted by Rice et al. (2015) for DFE(s)/DFE($-s$) is constant and equal to the average time to the most recent common ancestor for a pair of sequences.

Early attempts to explain the ubiquitous exponential form of DFE did not employ any evolutionary arguments and were based on the extreme-value theory (Gillespie 1982; Orr 2003; Joyce et al. 2008). Their aim was to prove an exponential for the intrinsic distribution of selection coefficient, as measured in site-mutagenesis experiment, $g(s)$, and in the limit of large s. The present theory predicts that, DFE is exponential in the broad range of s due to the evolving allelic frequency $\langle f(s,t)\rangle$. Another difference is that the cited approach predicts that DFE(s) is constant in time, while the present analysis, in agreement with experiment, demonstrates that slope β is not constant and varies in time (cf. Figures 2.19 and 2.21). The third difference is that an exponential is predicted for all values of s, while Gillespie (1982), Orr (2003) and Joyce et al. (2008) predicted it only in the limit of large s.

The main result, eq. (2.125), is robust to the details of a population model. Its main limitation is the assumption of constant, directional selection. In real populations, selection can also be of balancing type due to diploid dominance. Also, the external conditions can change due to changing environment or migration. Another example is a virus population under the mounting immune response (Batorsky et al. 2014) or the interaction between a virus and its defective interference particles (Metzger et al. 2011; Rouzine and Weinberger 2013b; Rast et al. 2016) (Chapter 4). Unless the time changes are very slow compared to the selection time, $1/s_{av}$, such complex cases may require another approach.

In conclusion, the exponential form of DFE observed in many pathogens appears automatically during adaptation. Complementing the epistasis inference in Sections 2.1–2.3, the theory offers a general method to measure the entire set of selection coefficients from genomic data, which completes the chapter on the inference of fitness landscape.

Chapter 3
Evolutionary role of a trait

Nothing in biology makes sense except in the light of evolution.
Theodosius Dobzhansky

The problem with this famous statement is that *not everything* in biology makes sense even in the light of evolution. Many features of organisms emerged during a long process of evolution where natural selection acts against the background of stochastic factors, such as mutation, random genetic drift, linkage, and recombination (Rouzine 2020b). Hence, many features simply happened as accidents in a track of evolutionary history and may not give any selective advantage. There are also many instances where the purpose of a selected trait is not clear in advance, and serious research is needed to find out what use to an organism this trait could be. This chapter investigates three examples where an apparently accidental trait has a deep evolutionary purpose discovered after investigation.

3.1 Evolution of HIV toward AIDS

99.5% of untreated individuals infected with HIV develop the acquired immunodeficiency syndrome (AIDS) (Barre-Sinoussi et al. 1983; Gallo et al. 1984; Levy et al. 1984). The demographic groups at a high risk of exposure comprise men who have unprotected sex with men, intravenous drug users who reuse needles, and unprotected sex workers (Quagliarello 1982). AIDS symptoms span a broad range of opportunistic diseases in previously healthy individuals including Kaposi's sarcoma, pneumonia, and *Mycobacterium tuberculosis* (Gottlieb et al. 1981; Masur et al. 1981; Small et al. 1983; Vieira et al. 1983). The cause of the acquired immunodeficiency in HIV-infected individuals is the depletion of CD4 T cells by the virus accompanied by functional impairment, such as low lymphocyte proliferative responses, causing the onset of AIDS symptoms within 2–20 years (Gottlieb et al. 1981; Masur et al. 1981; Small et al. 1983). The depletion of CD4 T cells creating gaps in their antigenic repertoire is linked to the fact that HIV selectively infects professional antigen-presenting cells expressing CD4, including helper CD4 T cells, macrophages, and dendritic cells. The low level of CD4 T cells achieved in the course of infection correlates strongly with the onset of AIDS symptoms (Klatzmann et al. 1984; Masur et al. 1989).

During the last two decades, potent cocktails of anti-retroviral drugs have been developed to avert or postpone this outcome and restore HIV-infected individuals to normal life. Strong international efforts were put in place to make them widely available all over the globe, including developing countries. Antiretroviral therapy (ART) suppresses HIV replication, which increases $CD4^+$ T-cell counts in peripheral blood and restores normal immune function (Ho et al. 1995; Wei et al. 1995). These results confirm,

https://doi.org/10.1515/9783110697384-003

beyond reasonable doubt, that the immunodeficiency in untreated individuals is caused by the virus-induced decrease in the $CD4^+$ T-cell count and the loss of $CD4^+$ T-cell function. An extra confirmation comes from experimental infection of nonhuman primates where the infection with pathogenic strains of SIV, a simian virus related to HIV, or chimeric simian/human immunodeficiency viruses leads to rapid CD4 T-cell depletion and death within 1–2 years (Reimann et al. 1996; Nishimura et al. 2005; Letvin et al. 2006; Sun et al. 2006).

At the same time, as it is evident from several facts, infection of CD4 cells by the virus does not suffice to explain the progression to AIDS. The disease progression is not caused by the consumption of a fixed stock of preexisting CD4 cells. HIV infection represents a steady state with a rapid turnover and replacement of all relevant cell types. The death of infected cells in about a day is almost exactly compensated with the supply of newly infected cells (Ho et al. 1995; Wei et al. 1995; Mohri et al. 1998; Rouzine and Coffin 1999a; Ribeiro et al. 2002a, b). The main target cells of HIV, memory $CD4^+$ $CCR5^+$ T cells, are also continuously replenished, with the life span of several weeks (Mohri et al. 1998; Grossman et al. 1999; Rouzine and Coffin 1999d). The steady state is almost exact, but not quite, with a very slow decline in CD4 T-cell count. Before this quasi-steady state sets up, target cells are deeply depleted during the virus peak of the acute phase of infection and then partly restored to about a half of normal level (Chun, Carruth et al. 1997; Grivel et al. 2000; Brenchley et al. 2004; Picker et al. 2004). Across HIV-infected subjects, the time to the onset of AIDS symptoms negatively correlates with the virus load (Arnaout et al. 1999; Rodriguez et al. 2006) and the level of immune activation (Giorgi et al. 1999; Hazenberg et al. 2003; Deeks et al. 2004; Rodriguez et al. 2006).

The existence of a steady state was explained long ago using simple mathematical models of hunter-prey type (De Boer and Perelson 1998; Rouzine and Coffin 1999a; Nowak and May 2000). However, all these models predict a perfect steady state. The reasons for the slow decline of CD4 cell count on time scales much longer than the life spans of target and infected cells remain unclear. The existence of many SIV strains, which do not cause progression to AIDS in their natural hosts (Silvestri et al. 2003; Goldstein et al. 2006; Gordon et al. 2008; Pandrea et al. 2008; Chahroudi et al. 2012), demonstrates directly that the biological factors creating a chronic steady-state infection differ from the mechanisms causing AIDS. Despite the differences observed between the immune response to SIV in its natural host and HIV in humans (Norley et al. 1999; Milush et al. 2011; Chahroudi et al. 2012), the mechanism of progression to AIDS remains unknown.

One can reasonably hypothesize that viruses and their natural hosts are mutually adapted, while pathogenic infections occur after a recent cross-species transfer that launches the process of virus adaptation to the new host. This section is based on a model (Rouzine 2020a), which predicts a gradual CD4 T-cell depletion driven by virus adaption to a new host. The high diversity and fast evolution rate of HIV within and between individuals has been studied intensely (Rouzine and Weinberger 2013a). HIV has the fastest average evolution rate among RNA viruses, up to $3\ 10^{-3}$ substitutions per nucleotide position per year, and 5 times as fast in its envelope protein (Dapp et al. 2017;

Patino-Galindo and Gonzalez-Candelas 2017). The virus adaptation to a host was described by a family of related mathematical models, including asexual evolution at a very low multiplicity of infection Rouzine et al 2003, evolution with recombination at a moderate multiplicity of infection (Rouzine and Coffin 2005, 2007, 2010; Neher et al 2010, 2013) (Section 1.3), and evolutionary escape from the immune response (Sections 4.1 and 4.2).

The time of the onset of AIDS symptoms in untreated individuals varies broadly among individuals. As already mentioned, the time to AIDS correlates negatively with the virus load and the degree of immune activation (Arnaout et al. 1999; Giorgi et al. 1999; Hazenberg et al. 2003; Deeks et al. 2004; Rodriguez et al. 2006). Notably, the correlation exhibits strong individual fluctuations, revealing the existence of hidden parameters depending on an individual and a virus. Below, the existing models of virus evolution and immune response are combined with data from patients to identify the parameters controlling the decline of CD4 T cell and thus make prognosis for an individual patient who is either untreated or on a suboptimal therapy regimen.

Thus, the main hypothesis of this section is that the slow slide toward AIDS is caused by the virus adaptation to a host. An alternative hypothesis postulates that the damage to homeostatic mechanisms is the cause for the gradual CD4 cell depletion [see Okoye and Picker (2013) for a review]. Below, this alternative view is used to develop two additional models (Section 3.1.4), and their predictions are compared with the predictions of the main model, in order to find out which model fits the data better (Fig. 3.1). Experimental tests of the main model are proposed as well. If confirmed experimentally, this work will offer the first method of the accurate prognosis of progression to AIDS based on patient data and, in the long shot, a method of functional cure designed to prevent progression to AIDS.

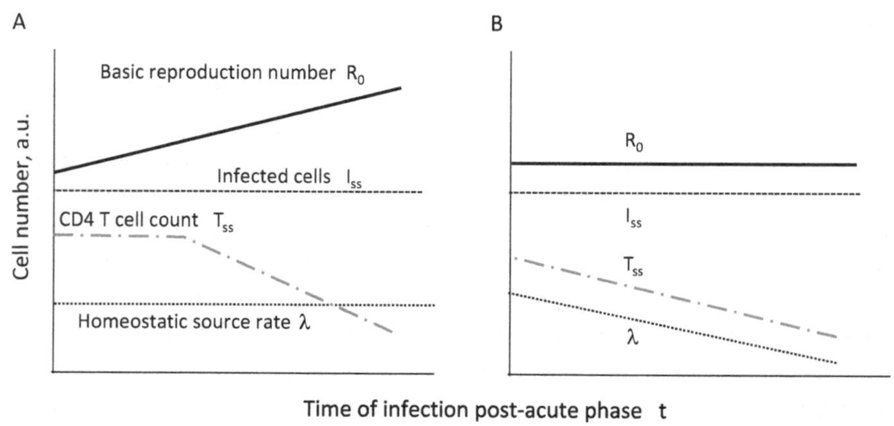

Fig. 3.1: Two alternative hypotheses of the gradual decline of CD4 T-cell count tested in the present article.
Dash-dotted line shows the quasi-steady-state level of CD4 T cells predicted by eq. (3.10) if (A) basic reproduction ratio R_0 changes due to gradual viral adaptation, (B) R_0 stays constant, but the homeostatic source of CD4 T cells becomes weaker in time. Based on Rouzine (2020a).

3.1.1 Model of virus dynamics

3.1.1.1 Model 1: no immune response

The starting point are two simplest models of viral dynamics predicting an exact steady state, with and without the immune response. Viral evolution and progression to AIDS will be considered in Section 3.1.2.

In infants of 1 year and younger, the T-cell response is immature and can be neglected. Antibodies are ineffective due to resistance mutations (Burns et al 1994; Simmonds et al 1990). The standard model of virus dynamics includes uninfected target cells and infected cells, as follows (Mohri et al. 1998; Nowak and May 2000):

$$\frac{dT}{dt} = \lambda - (d_T + kI)T \tag{3.1}$$

$$\frac{dI}{dt} = kIT - d_I I \tag{3.2}$$

where T denotes the number of uninfected $CD4^+$ $CCR5^+$ T cells infectable by virus, I is the infected cell number, λ is the constant rate of homeostatic replenishment of effector CD4 T cells from precursors, d_T is the inverse life span of target cells, d_I is the death rate of infected cells, and k is the infectivity parameter. The virion number is not included explicitly, because virions are very short-lived in vivo, about 6h (Perelson et al. 1997), and their number is proportional, at any time, to the number of infected cells.

The model predicts an exact steady state, which corresponds to chronic, clinically latent phase of HIV infection (Nowak and Bangham 1996; Mohri et al. 1998; Rouzine and Coffin 1999a; Ribeiro et al. 2002a, b). Putting the right-hand sides of eqs. (3.1) and (3.2) to zero, one obtains the steady-state levels for uninfected CD4 T cells, T_{ss}, and infected CD4 T cells, I_{ss}

$$T_{ss} = \frac{T_0}{R_0} \tag{3.3}$$

$$I_{ss} = \frac{d_T}{k}(R_0 - 1) \tag{3.4}$$

Here T_0 denotes the CD4 cell number in uninfected patients ($I = 0$), and R_0 is the basic reproduction number, as given by

$$T_0 \equiv \frac{\lambda}{d_T}, \; R_0 \equiv \frac{k\lambda}{d_I d_T} > 1 \tag{3.5}$$

For an infection to occur, R_0 must be larger than 1.

3.1.1.2 Model 2: CD8 T-cell immune response

The simplest model with the cytotoxic immune response has the form (De Boer and Perelson 1998; Nowak and May 2000)

$$\frac{dT}{dt} = \lambda - (d_T + kI)T \tag{3.6}$$

$$\frac{dI}{dt} = kIT - d_I I\left(1 + \frac{E}{E_0}\right) \tag{3.7}$$

$$\frac{dE}{dt} = aE\frac{I}{I + I_0} - d_E E \tag{3.8}$$

Equation (3.8) describes the dynamics of the CD8 T effector cells E, where a is the maximum proliferation rate, I_0 is the characteristic infected cell number at which half-maximum immune cell expansion is attained, and d_E is the natural death rate of effector cells. As compared to eq. (3.2), eq. (3.7) has an extra term describing the killing of infected cells by the effector cells, E. New model parameter E_0 represents effector cell number that increases the cell death rate two-fold. Setting the right-hand side of eqs. (3.6)–(3.8) to zero, the steady-state values of state variables are

$$\begin{aligned} T_{ss} &= \frac{T_0}{1 + \dfrac{R_0}{R^*}}, \quad R^* \equiv \frac{T_0 d_T}{I_{ss} d_I} \gg 1 \\ I_{ss} &= \frac{I_0 d_E}{a - d_E} \\ E_{ss} &= E_0 \left[\frac{R_0}{1 + \dfrac{k d_E I_0}{d_T(a - d_E)}} - 1\right] \end{aligned} \tag{3.9}$$

with T_0 and R_0 defined by eq. (3.5).

The model is designed for a typical untreated chronic infection and does not include latently infected cells. In the mucosal phase of infection, in HIV controllers, and under anti-retroviral therapy, latent cells are critically important (Perelson et al. 1997; Rouzine et al. 2015; Rouzine 2022). The model also does not include the helper T-cell response, which is very important in low-viremia patients and virus controllers (Lambotte et al. 2005; Letvin et al. 2006; Vingert et al. 2010). The arms race between helper CD4 T cells and virus may be responsible for HIV persistence and control (Rouzine et al. 2005; Rouzine et al. 2006; Sergeev, Batorsky, Coffin et al. 2010; Sergeev, Batorsky and Rouzine 2010; Rouzine 2022). Because the focus is on the progression to AIDS of patients with high to moderately low viremia, the helper cell response and latent cells can be safely neglected.

3.1.2 The rate of HIV adaptation to a host

The next task is to calculate the virus adaptation rate and then link it to the decline of CD4 T-cell count (Section 3.1.2). The adaptation rate is defined as the rate at which the logarithm of virus basic reproduction number R_0 changes per unit time, as given by $V = d\log R_0/dt$. The value of R_0, which is the average number of secondary infected cells per an infected cell, is expressed in terms of parameters of a model of virus dynamics (Section 3.1.1). The adaptation rate V can be expressed, in the general form, in terms of parameters of an evolutionary model and depends on its choice (Rouzine et al. 2003; Desai and Fisher 2007; Brunet et al. 2008; Rouzine et al. 2008; Neher et al. 2010; Rouzine and Coffin 2010; Hallatschek 2011; Good et al. 2012; Neher et al. 2013). In the simplest scenario of asexual multi-locus evolution, the virus is far from evolutionary equilibrium, deleterious mutation can be neglected, and recombination is absent. Under these assumptions, which apply to patients with low virus loads (<10^5 RNA/mL) (Batorsky et al. 2011), the adaptation rate is given by the expression (Rouzine et al. 2008)

$$V = 2s^2 \frac{\log\left(I_{ss}\sqrt{sU_b}\right)}{\log^2 \dfrac{s \log I_{ss}}{U_b}} \tag{3.10}$$

where U_b is the beneficial mutation rate per genome per generation of infected cells (1 day), s is the increase in $\log R_0$ per mutation, and I_{ss} is the number of infected cells in chronic infection (steady state). The value of I_{ss} is given by eq. (3.4) or (3.9), depending on the immunocompetence (age) of the patient.

The arguments of both logarithms in eq. (3.10) must be much larger than 1, which implies that I_{ss} cannot be too small. This condition may not hold in "controllers," rare patients with anomalously low virus loads, where beneficial mutations arise to fixation rarely and do not overlap in time, in which case $V = 2s^2 Iss\ Ub$. Still, even in the best controllers, the value of I_{ss} is rarely below 100–1000 cells, and the relevant $s \sim 0.01$, $U_b \sim 0.01 - 0.1$ (Rouzine and Coffin 1999c), so that the multi-locus fixation always takes place, and eq. (3.10) is valid.

So far, the value of selection coefficient s is assumed to be the same for all loci. In real life, it varies among loci according to a distribution density, $\rho(s)$. For example, exponential distribution $\rho(s) = (1/s_{av}) \exp(-s/s_{av})$ is often seen in experiment (Fig. 2.19, Section 2.4). At sufficiently large values $I_{ss}U_b$, this case can be reduced to the constant-s case given by eq. (3.10), using the effective parameters (Good et al. 2012)

$$s_{eff} = s^*, \quad U_{eff} \sim U_b\sqrt{2\pi V}\ \rho(s^*)$$

where the effective selection coefficient, s^*, was derived analytically by Good et al. (2012) in terms of s_{av}, population size, and mutation rate. The substitution rate V is found self-consistently by plugging these effective parameters into eq. (3.10). However, the specific expression for s_{eff} is not essential for our aim, because its value can easily be estimated from patient sequence data (Sections 1.1 and 1.3). The model can be used with the experimentally estimated value s_{eff} and mutation rate U_b replaced by U_{eff} in eq. (3.10). Because U_{eff} is inside of a large logarithm, the tolerance to its error is high. In contrast, s_{eff} must be estimated as accurately as possible, because the viral adaptation rate (and hence the inverse time until the onset of AIDS symptoms) is proportional to its square (Section 3.1.3).

Asexual evolution model applies when the virus load is sufficiently low, and cells are rarely co-infected with two virus variants (low MOI). In patients with a high viremia, $>10^5$ RNA copy/mL plasma, HIV recombination is frequent enough to accelerate adaptation several-fold (Rouzine and Coffin 2005; Batorsky et al. 2011). The effective recombination probability per genome (outcrossing rate), r, depends on MOI and can be estimated directly from patient sequence data (Neher and Leitner 2010; Batorsky et al. 2011). In an average patient with viremia 3×10^5 RNA copy/mL, recombination accelerates adaptation rate by the factor of 2 (Batorsky et al. 2011). In this case, adaptation occurs due to synergetic effect of mutation and recombination (Neher et al. 2010). An expression for the adaptation rate in the long-term evolution in the presence of both selection and mutation was derived analytically using a scaling argument (Neher et al. 2013; Good et al. 2014; Weissman and Hallatschek 2014).

3.1.3 Predicting the speed of progression to AIDS

The steady-state number of CD4 T cells T_{ss} given by either eq. (3.3) or eq. (3.9), depending on the immunocompetence of a person, decreases with the basic reproduction number R_0, also known as "virus fitness," which increases slowly in time in the course of adaptation to a patient (Rouzine and Coffin 1999c) (Section 1.1). An important feature is separation of time scales between adaptation, which takes years, and time scales associated with dynamics, which are days and weeks (Mohri et al. 1998). Therefore, when parameter R_0 increases very slowly, the system remains in a quasi-steady state at any one time, except T_{ss} slowly decreases over the course of HIV infection. This process explains the pathogenesis of HIV.

Let us estimate the time to the onset of AIDS. The average initial CD4 T-cell level in a chronic HIV infection is $T_{\mathrm{ss}}(0)=500$ cell/mkl blood. The early AIDS symptoms start when CD4 cell count falls below $T_{\mathrm{ss}}(t_{\mathrm{AIDS}})=200$ cell/mkl blood. The time to the onset of symptoms can be calculated, as follows:

$$t_{\mathrm{AIDS}}=\frac{\log R_0(t_{\mathrm{AIDS}})-\log R_0(0)}{\dfrac{d\log R_0}{dt}} \tag{3.11}$$

which assumes that the adaptation rate

$$V \equiv \frac{d \log R_0}{dt}$$

is constant in time. Using eq. (3.3) for T_{ss} in the absence of the immune response, for the time of progression to AIDS one obtains

$$t_{\text{AIDS}} = \psi / V, \psi \equiv \log \frac{T_{ss}(0)}{T_{ss}(t_{\text{AIDS}})} \tag{3.12}$$

Note that ψ is close to 1 on average but may vary between patients. In the initially immunocompetent individuals, from eqs. (3.9) for T_{ss} and (3.11), the time to AIDS is given by

$$t_{\text{AIDS}} = \frac{1}{V} \left\{ \psi + \log \left[1 + \frac{R^*}{R_0(0)} \left(1 - e^{-\psi} \right) \right] \right\} \tag{3.13}$$

which is several-fold longer than in the absence of the immune response, eq. (3.12). This is consistent with the observation that the average time to AIDS is much longer in untreated adults (10 years) than in infants (1 to 2 years). In eqs. (3.12) and (3.13), the adaptation rate of HIV, V, can expressed in terms of evolutionary parameters depending on the leading factors evolution. For the asexual evolution that takes place in low-to-average viremia patients, V is given by eq. (3.10).

3.1.3.1 Negative correlation between time to AIDS and virus load

Prediction in eq. (3.13) appears to be consistent with the negative correlation between the time to AIDS and the virus load observed between patients (Arnaout et al. 1999). To test the quantitative agreement, one can simplify the expression by using two composite parameters (Fig. 3.2). Please note that the measured virus load, v, is proportional to the total number of infected cells, as given by $I_{ss} = v\theta$. The predicted time to AIDS in adults as a function of viremia, eq. (3.13), can be rewritten in the form

$$t_{\text{AIDS}} = \frac{A}{\log(Bv)} \left[1 + \log \left(1 + \frac{C}{v} \right) \right] \tag{3.14}$$

Equation (3.14) is obtained by expressing V and R^* in terms of I_{ss} from eqs. (3.10) and (3.9), respectively, and then substituting them into eq. (3.13). Here new composite parameters A, B, C are defined as

$$A \equiv \frac{1}{s^2} \log^2 \frac{s}{U_b}, \ B \equiv \theta v \sqrt{s U_b}, \ C \equiv \frac{T_0 d_T}{\theta d_I R_0(0)} \left(1 - e^{-1} \right)$$

Parameters A and B are used as fitting parameters, that is, adjusted to fit data in Fig. 3.2, and C is fixed at any large value, $C > 10^6$ RNA copy/mL. Fitting is done using the least-squares method and the program fminsearch.m in MATLAB™. The experimental estimates of parameters from Tab. 3.1, which is designed for patient prognosis described below, are not used. The variation in C has a negligible effect on the fitting accuracy. A good fit of model predictions to data is observed (Fig. 3.2).

Moreover, the main hypothesis that AIDS progression is driven by HIV adaptation agrees with the positive correlation between AIDS progression and the level of the immune activation observed in humans and simians (Arnaout et al. 1999; Giorgi et al. 1999; Hazenberg et al. 2003; Deeks et al. 2004; Rodriguez et al. 2006). In Section 1.1, the adaptation of HIV was linked to the early immune escape in epitopes and subsequent compensatory mutations. The immune pressure measured by s and the number of highly diverse sites (and hence genomic mutation rate U_b) correlate with the intensity and the breadth of the cytotoxic immune response, of which the cytokine level measured in cited works is a marker. Note that the adaptation rate is proportional to the square of selection coefficient, eq. (3.10). If the immune response is the main cause of viral adaptation, correlations between its magnitude and the speed of progression AIDS are expected.

Tab. 3.1: Model parameters relevant in the quasi-steady state.

Notation	Name	Units	Range
ψ	Drop in log CD4 T-cell count	1	~1
I_{ss}	Infected cell number	Cell	$100-3\times10^9$
L	Total number of sites	1	10–1000
μ	Mutation rate	1/gener./nucl.	10^{-6}–10^{-4}
$U_b=\mu L$	Mutation rate per genome	1/gener./genome	$3\times10^{-4}-0.03$
b	Effective selection coefficient	1	0.1 – 0.005
r	Recombination probability	1	0–1
T_0	Target cell number in uninfected person	cell/mm^3	700–1400
d_T	Inverse life span of target cells	1/day	0.1 – 0.01
d_I	Inverse life span of infected cells	1/day	0.7 – 1.2
$R_0(0)$	Initial basic reproduction ratio	1	5–11
$R^*=\frac{T_0 d_T}{I_{ss} d_I}$	Composite parameter	1	

3.1.3.2 Parameter estimates for AIDS prognosis

In addition to explaining the correlation between viremia and progression to AIDS, this model can be used to predict the time to AIDS in individual patients. The model parameters in eq. (3.13) (Tab. 3.1) can be informed from the virological data from individual patients, which are measured directly or inferred, as follows.

Coefficient ψ is found from eq. (3.12), where $T_{ss}(0)$ is the CD4 cell count in a patient in the early chronic infection.

The number of infected cells, I_{ss}, can be estimated from viremia in RNA copy/mL blood measured by qPCR using a standard unit converter θ. To find θ, we use the fact

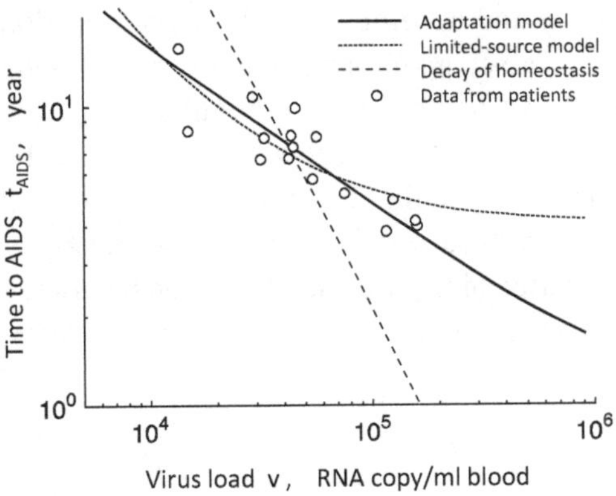

Fig. 3.2: Time to AIDS predicted by three models with two composite fitting parameters compared to data isolated from HIV-positive men seroconverted during study (Arnaout et al. 1999).
Out of 16 patients, 11 were untreated and 5 received some form of suboptimal anti-retroviral treatment (not triple cocktails) that did not affect the outcome. Circles: Survival time versus virus load in patient data. Solid thick curve: Best-fit prediction of the virus evolution model, eq. (3.14). Dotted curve: The model of a limited homeostatic source, eq. (3.16). Dashed line: The model of the virus-induced decay of homeostatic environment, eq. (3.18). Based on Rouzine (2020a).

that an average untreated patient with $v = 3 \times 10^4$ RNA/copy mL blood has 3×10^8 HIV RNA-producing infected cells in lymphoid tissue (Haase 1999). Hence $\theta \approx 10^4$ cell · mL/RNA copy.

Mutation rate $U_b = \mu L$, where the mutation rate per site, μ, was measured for each pair of substitutions with average $\mu \approx 3 \times 10^{-5}$ (Mansky and Temin 1995), and L is the number of highly variable sites (SNPs) in a virus strain in a patient, which can be counted directly.

Average selection coefficient s and outcrossing rate r required for patients with higher viremia can be estimated from patient genomic sequences by several methods (Rouzine and Coffin 1999c; Frost et al. 2000; Neher and Leitner 2010; Batorsky et al. 2011; Illingworth and Mustonen 2012). The first estimate of effective s in untreated patients was obtained by Rouzine and Coffin (1999c) using a one-locus deterministic evolution model (Section 1.1). Neher and Leitner (2010) used a similar method to estimate s and an additional approximation to estimate r. Frost et al. (2000) also used a similar technique to estimate s for drug-resistance mutations. Later Batorsky et al. (2011) performed direct Monte-Carlo simulation using the algorithm described in Section 2.1 without epistasis but with recombination, and compared two measures of linkage disequilibrium calculated for simulated sequences with real sequence data (Section 1.3); this last method is the most accurate, because it takes into account linkage effects, recombination, and any factors you wish to include.

Input parameters T_0, d_T, and d_I are less sensitive, because they vary little between patients, and the respective values are inside of a large logarithm in eq. (3.13). Hence, they can be replaced with their well-known averages over patients (Sergeev,

Batorsky, Coffin et al. 2010; Sergeev, Batorsky and Rouzine 2010). Basic reproduction ratio, $R_0(0)$, can be measured in the primary infection; for a patient observed already in chronic phase, it is unknown. Fortunately, its value is also inside of a large logarithm in eq. (3.13), hence it can be approximated by its average value, $\langle R_0(0)\rangle = 8$ (Ribeiro et al. 2010). Thus, all model parameters can be either determined accurately for an individual, or are not sensitive and can be replaced with a population average.

The relative sensitivity of the predicted value of t_{AIDS}, eq. (3.13), to model parameters measured in patients or estimated can be expressed in terms of their relative experimental errors, as follows:

$$\frac{\partial \log t_{\mathrm{AIDS}}}{\partial \log s} = -2$$

$$\frac{\partial \log t_{\mathrm{AIDS}}}{\partial \log \log I_{\mathrm{ss}}} = -1$$

$$\frac{\partial \log t_{\mathrm{AIDS}}}{\partial \log R_0} = -\frac{\partial \log t_{\mathrm{AIDS}}}{\partial \log R*} = -\frac{D}{(1+D)(\psi + \log(1+D))}$$

$$D \equiv \frac{R*}{R_0}\left(1 - e^{-\psi}\right) \tag{3.15}$$

The 95% confidence interval for log t_{AIDS}, denoted by ε, can be estimated as

$$\varepsilon = \sqrt{\sum_{i=1}^{4}\left[\frac{\partial \log t_{\mathrm{AIDS}}}{\partial \log X_i}\right]^2 \varepsilon_i^2}$$

Here X_i is any of the above four parameters: s, $\log I_{\mathrm{ss}}, R_0$, and $R*$, the double-log derivatives are given by eqs. (3.15), and ε_i denotes the 95% confidence interval for $\log X_i$. Variation of different parameters across patients is assumed to be statistically independent. The variance of $\log R* = \log \frac{T_0 d_T}{I_{\mathrm{ss}} d_I}$ is the sum of variances the logarithms of T_0, d_T, I_{ss}, d_I. For the details regarding patient classification and sampling, see Rouzine (2020a).

3.1.4 Alternative models of progression to AIDS: impaired homeostasis

A plausible alternative explanation of progression to AIDS, based on data, is the deregulation of homeostasis (Okoye and Picker 2013). Because the closed model based on that idea has not been developed, below two models based on this alternative hypothesis are analyzed, and their predictions are compared with patient data.

Above, a fixed linear replenishment rate λ was assumed. In fact, homeostatic mechanism shuts down replenishment when T increases to a threshold, for example, linearly

$$\lambda(T) = \lambda_0 \left(1 - \frac{T}{T_0}\right) \tag{3.16}$$

where T_0 and λ_0 are two independent constant parameters. The first way to introduce impaired homeostasis is to postulate that the maximal number of divisions is finite (i.e., due to the telomere length restriction and insufficient expression of telomerase). In this case, the total production of target cells has a finite limit, Λ, as given by

$$\lambda(T_{ss}) t_{AIDS} = \Lambda$$

Substituting eq. (3.16) into eqs. (3.5) and (3.9), from the above equation, we get

$$t_{AIDS} = \frac{\Lambda}{\lambda_0} \frac{(\lambda_0/T_0) + kI_{ss}}{kI_{ss}} \equiv a \frac{b + I_{ss}}{I_{ss}} \tag{3.17}$$

Here a, b are new composite parameters that do not depend on I_{ss} and are used for fitting data. In eq. (3.17), a strong inequality is assumed, $kI_{ss} \gg d_T$, which is equivalent to $R_0 \gg R^*$. The inequality holds for an average untreated infected person, because kI_{ss}/d_T is the turnover rate increase of CD4 T cells due to HIV infection, which value is between 3 and 6 (Mohri et al. 1998). Fitting eq. (3.17) to data in Arnaout et al. (1999) predicts a bent curve in the semi-log scale, comprising a steep decline crossing over to a plateau (Fig. 3.2). Such a curvature is not seen in data, so that this version of impaired homeostasis does not fit data well.

Instead of limited production of target cells, the progress to AIDS could also be caused by the gradual degradation of homeostatic environment, due to the accumulation of virus products (Okoye and Picker 2013). One could postulate that the viral products gradually erode the maximal homeostatic source intensity, λ_0, as given by

$$\frac{d\lambda_0}{dt} = -aI_{ss}\lambda_0(t) \tag{3.18}$$

Solving this simple equation, one obtains an exponential decay in time for source intensity, with the decay rate proportional to the number of infected cells

$$\lambda_0(t) = \lambda_0(0) e^{-aI_{ss}t}$$

which causes the decline in T_{ss}, per eq. (3.9). As already mentioned, the onset of AIDS occurs when the CD4 count drops to a characteristic value

$$T_{ss}(t_{AIDS}) = T_{AIDS}$$

Inserting $\lambda_0(t)$ instead of λ into eqs. (3.5) and (3.9), one gets

$$t_{AIDS} = \frac{1}{aI_{ss}} \log\left(\frac{\lambda_0(0)}{kI_{ss}T_{AIDS}}\right) \equiv \frac{1}{aI_{ss}} \log \frac{\beta}{I_{ss}} \tag{3.19}$$

where $kI_{ss} \gg d_T$ is assumed, as before, and composite parameters α and β do not depend on I_{ss} and are used for fitting data. The best-fit curve for t_{AIDS} decreases far too rapidly with the virus load compared to the actual correlation (Fig. 3.2). This drastic misfit excludes the "source degradation" model completely.

Thus, both basic models of homeostatic impairment, either with the division counter or with virus-induce degradation, fail to provide a fit of comparable quality with the main model based on viral adaptation using the same number of fitting parameters: two (Fig. 3.2). The reason for the good fit of the adaptation model is the slow, logarithmic dependence of the evolution rate on virus load. Importantly, this feature is a robust property of a large class of models of evolution with multiple evolving sites (Rouzine 2020b). This feature is also completely independent on the particular choice of the model of the immune response. Thus, the proposed explanation not only agrees with data, but is also remarkably robust to the factors of evolution and the details of immune response.

3.1.5 Discussion

As demonstrated above, progression to AIDS can be explained by the evolution of HIV which becomes more fit with time and brings down the steady-state target cell number. Models based on that assumption match the data on the negative correlation between the time to AIDS and virus load much better than the majority opinion based on the hypothesis of impaired homeostasis. Despite some evidence that homeostasis is, indeed, impaired, the models based on this fact do not explain progression to AIDS.

Of course, the comparison with alternative models does not end here. As shown by a large number of studies, an enhanced activation of T cells is also implicated in a faster disease progression (Giorgi et al. 1999; Hazenberg et al. 2003; Deeks et al. 2004). However, this effect alone does not lead to a predictive model of AIDS progression (Alizon and Magnus 2012). The model of "diversity threshold model" has the same problem and received serious critique (Alizon and Magnus 2012).

The viral adaptation model successfully explains this correlation by noting that the antigenic escape helped by epistasis is likely the principal cause of HIV adaptation in patients (Section 1.1, Model 4), and that the patients with stronger immune response impose a larger selection pressure on the virus to evolve which, paradoxically, accelerates the onset of immunodeficiency.

The alternative models could be compared better using the samples of CD4 T-cell counts obtained from patients at different time points. The predicted CD4 count decline slope, dT_{ss}/dt, could be compared with the observed decline averaged over patients with a similar virus load. Note that the limited-division homeostasis model predicts an abrupt drop in CD4 T-cell count before the onset of AIDS, instead of the gradual decline usually observed in patients. At the present moment, the viral adaptation model matches data better than the homeostasis-based models. The better performance is due to a very slow dependence of the adaptation rate on population size of infected cells universal for any regime of multi-locus evolution (Rouzine 2020b). In a homeostasis-based model, the slow dependence is much harder to predict without making many artificial assumptions.

For the exact prognosis of the time to AIDS, the details of the immune dynamics may be important. The models above did not include the helper-dependent CD8 T-cell immune response, generally, very important for virus control (Lambotte et al. 2005; Letvin et al. 2006; Vingert et al. 2010). The models of the immune dynamics may also include memory cells (Rouzine et al. 2005), latent cells (Rouzine et al. 2015), and more complex models of T-cell replenishment (Grossman et al. 1998; Grossman et al. 1999; Rouzine and Coffin 1999d).

So far, untreated patients were discussed. Most patients today receive anti-retroviral therapy (ART) which slows down progression to AIDS significantly. Not all of these cases, however, show completely successful treatment. The present analysis can easily be extended to the more general case of suboptimal anti-retroviral therapy (ART), which does not cause the full suppression of viral replication. In the study by (Arnaout et al. 1999) whose data are used for fitting (Fig. 3.2), 5 patients of 16 were in this class. In this situation, replication of virus is not suppressed, a quasi-steady state is preserved, and the virus can adapt, causing progression to AIDS, as described. To introduce such suboptimal ART into the model, the basic reproduction number, R_0, is replaced with $R_{ART} = (1-\varepsilon)R_0 > 1$, where ε, such that $0 < \varepsilon < 1$, is ART efficacy. To show that this replacement is valid, it suffices to note that any existing ART affects only the infectivity coefficient, k, which includes both the number of virions per cell and their average infectivity. As long as the ART dose is suboptimal, as given by the condition $\varepsilon < 1 - 1/R_0$, and the steady state is maintained, the above results apply, with R_{ART} replacing R_0.

Therefore, according to the present model, a suboptimal ART does not affect the outcome significantly. This prediction was actually observed in early patients including the five patients in Fig. 3.2. Monotherapy and low-dose treatment are associated not only with rapid progression to AIDS but also with resistant mutations. They are not in clinical use today. Yet, due to the low efficacy of ART in some individuals or their poor adherence to regimens, such cases still exist.

When high-dose ART is successfully applied, $\varepsilon > 1 - 1/R_0$, causing the virus load drop down to very small levels limited by latent infection (Section 3.2), the immune response turns off, and the steady state is destroyed. In this case, the present model is no longer appropriate. To describe this scenario, one needs to introduce latent cells into the picture (Rouzine et al. 2015) (Section 3.2) and consider stochastic virus evolution occurring mostly during short spikes of viremia. Because the spikes are short, and replication (and hence evolution) of virus is suppressed by ART, a highly efficient therapy is expected to delay the onset of AIDS a lot, which is what is seen in most patients.

To summarize, the model of AIDS progression based on viral adaptation matches the existing correlation data better than the models based on impaired homeostasis, including the negative correlation between the time to AIDS and viremia and the positive correlation between the time to AIDS and immune activation levels. The model also naturally explains the lack of progression to AIDS in SIV strains infecting their natural host. The present model can also be used clinically for making individual prognosis.

3.2 An evolutionary role for HIV latency

In infected individuals, HIV replicates in memory $CD4^+$ T lymphocytes. Most infected cells produce high amounts of virus and die in about a day after their infection (Ho et al, 1995; Wei et al 1995). However, the virus integrated in human DNA can also be in a quiescent state called "latency" (Chun, Stuyver et al. 1997; Finzi et al. 1997). The size of latent cell population is relatively small, about 1 in 10^6 $CD4^+$ T cells, and does not contribute much to the total viral RNA load (Pierson et al. 2000). However, the latent reservoir is important, because it makes the virus persist during prolonged antiretroviral therapy (ART). Due to HIV latency, patients have to remain on lifelong therapy to avoid viral rebound. Once a patient interrupts ART, latent virus reactivates, and HIV raises to the pretreatment level within a couple of weeks. Hence, latency is the main barrier to HIV cure (Richman et al. 2009).

While latency helps HIV to persist during drug therapy, its evolutionary benefit during the millions of years of natural lentiviral evolution in the absence of ART remains unclear. In fact, latency appears to be only deleterious to fitness, because latently infected cells generate no virus and thus decrease viral levels in individuals. Given that fact, the prevailing hypothesis is that latency is an accident of evolution, which happened some millions of years ago and was inherited along phylogenetic tree. It is thought that the mechanistic reason for latency is the infection of $CD4^+$ T cells transitioning between the activated and resting phases of cell cycle (Han et al. 2007; Eisele and Siliciano 2012; Coffin and Swanstrom 2013). Latency is therefore considered a rare side effect that can only occur after the start of the immune response activating host CD4 T cells.

This popular view has some gaps. If latency were a neutral or deleterious side-effect, it could have been lost due to either negative natural selection or random genetic drift, especially, given a rapid evolutionary rate of HIV. Next, a study in rhesus macaques infected with SIV, a relative of HIV infecting simians, demonstrated that latent cells reach high levels very early, within a few days of infection (Whitney et al. 2014), long before the formation of the adaptive immune response to the virus (Kuroda et al. 1999; Schmitz et al. 1999). Finally, an experimental study in cell culture (Razooky et al. 2015) demonstrated that latency is built into the regulatory circuit of HIV gene, in agreement with another set of data showing that a half of infections in cell culture result in a latent state (Calvanese et al. 2013; Dahabieh et al. 2013). Several other studies demonstrated that the auto-regulatory Tat circuit of HIV amplifies the stochastic fluctuations in the expression of HIV provirus that are sufficient to induce a switch to a latent state (Weinberger et al. 2005; Weinberger et al. 2008; Burnett et al. 2009). In general, the stochastic noise of gene expression is deleterious and filtered out from regulatory circuits (Fraser et al. 2004; Batada and Hurst 2007). Therefore, the persistence of the latent state built into the regulation of HIV provirus gene suggests an unknown fitness benefit that outweighs fitness costs.

A possible fitness advantage for latency is that it could enhance the chances of viral survival under harsh environmental conditions. (Cohen 1966) proposed similar explanation for lysogeny of bacteriophage-λ (Arkin et al. 1998) and persistence of bacteria (Balaban 2011). However, such advantage exists only under the risk of extinction due to fluctuations of environment. In a real host, HIV and other lentiviruses are rarely in danger of a population bottleneck, because they successfully evade immune clearance. Lentiviruses evolved for millions of years in the absence of antiretroviral drugs. In the absence of treatment, they maintain high viral loads of ~10^5 particles/mL of blood plasma and ~10^8 infected cells per host for years. A possible mechanism for the evasion of the immune response and viral persistence is the infection of helper $CD4^+$ cells (Rouzine et al. 2006; Rouzine 2022). Furthermore, HIV only infects a tiny percentage of available target cells, which makes their number not limiting during a chronic infection.

There is, however, one phase of infection when virus loads are always low: initial mucosal infection. Numerous studies show that the probability of an HIV infection per sex act is very low. In HIV-discordant couples, less than 1% of encounters result in a systemic HIV infection (Gray et al. 2001; Wawer et al. 2005; Fraser et al. 2007). When systemic infections are established, they usually expand from a single founder RNA sequence (Keele et al. 2008; Kearney et al. 2009) demonstrating that only one genomic sequence in the transmitted doze (inoculum) starts the systemic infection. The use of a simian model of SIV (simian immunodeficiency virus) reveals a ~6-day delay between the time of experimental inoculation and the extrapolated onset of a systemic infection (Zhang et al. 1999; Haase 2011). Thus, the first 6 days of HIV infection are not suitable for viral growth.

The unfavorable conditions early after infection are easy to explain. The replication of HIV passed by the usual sexual route starts in the mucosal tissue of vagina, mouth, or rectum (Haase 2011). SIV in nonhuman primates also usually spreads through mucosal transmission: either during sexual activity or after fighting, with subsequent licking of the inflicted wounds by other animals (Santiago et al. 2005). Experiments in primates provide evidence that the mucosa are initially unfavorable to SIV growth. Large inoculations of ~10^9 infectious units become extinct within 5 days (Miller et al. 2005). Each infected cell lives for about 1 day (Markowitz et al. 2003), so the actively infected cell number after 5 days is proportional to $(R_0^{\mathrm{muc}})^5$ where R_0^{muc} here denotes the initial basic reproductive ratio in the mucosa. Because the actively infected cell number falls below detection within 5 days (Miller et al. 2005), $(R_0^{\mathrm{muc}})^5$ is much less than unit.

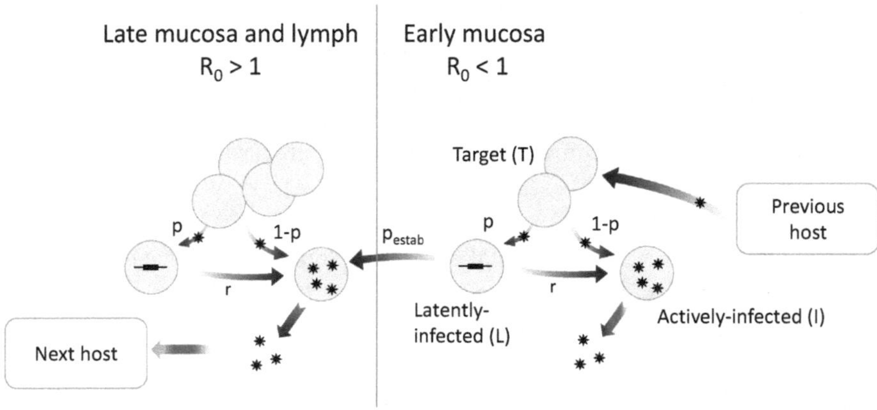

Fig. 3.3: HIV latency is a bet-hedging strategy for optimizing viral transmission and Trojan horse. Schematic of the two-compartment model. Right to left: HIV enters a host at a mucosal site but, due to the small number of permissive target cells in the early mucosa (before day 6), $R_0 < 1$. To successfully establish systemic infection, the virus must avoid extinction until $R_0 > 1$ by waiting in latently infected cells. Based on Rouzine et al. (2015).

The work discussed below (Rouzine et al. 2015) quantitatively tested the hypothesis that latency provides a fitness advantage at the stage of intrahost transmission (Fig. 3.3). The key idea is that the increased in the probability of latency upon infection of a cell, which we henceforth denote p_{lat}, increases the fraction of survived infected cells in mucosa. As already mentioned, the probability of an actively infected cell surviving to day 6 to initiate systemic infection is very small. In contrast, latently infected cells are long-lived and can reactivate once the conditions change to $R_0 > 1$ to initiate systemic virus expansion. At the same time, increasing p_{lat} has also a negative effect of decreasing the viral load in host in the phase of systemic infection. This effect reduces the amount of virus transmitted to new hosts and hence, the probability of transmission. Because the survival rate of initial infected cells will be higher, but their number will be lower, it is unclear whether the first outweighs the second. If it does, latency likely represents an evolutionarily beneficial trait that is maintained by natural selection. By entering long-lived latency in some infected cells, the strain would increase its potential of surviving the harsh conditions of initial infection to establish systemic infection.

Models and results are presented in Sections 3.2.1–3.2.5. Discussion of results is found in Section 3.2.6. Mathematical derivations and details are given in Sections 3.2.7–3.2.10.

3.2.1 Mathematical models of lentiviral transmission

To quantify the net impact of latency on lentiviral transmission, three mathematical models are analyzed below (Fig. 3.4). Each model includes latency and early mucosal infection with $R_0^{\text{muc}} < 1$.

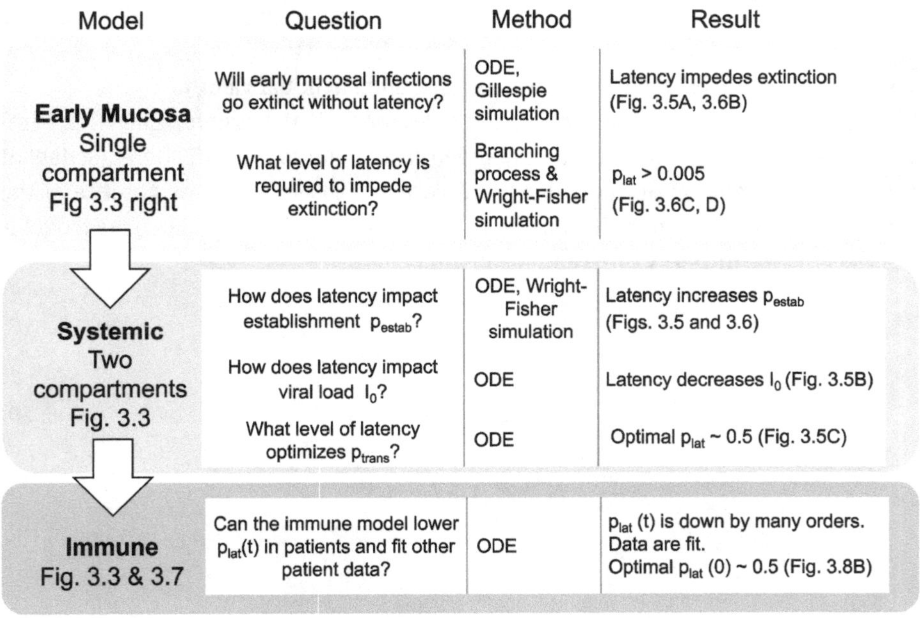

Fig. 3.4: Flowchart of models. Based on Rouzine et al. (2015).

The first model describes the infection in the mucosa alone (Section 3.2.7). Given a small number of cells infected by transmitted virus, mucosal infection is essentially stochastic (Pearson et al. 2011), and one has to use Monte-Carlo simulation and probability theory. Below both stochastic model and the deterministic approximation are analyzed, and experimentally estimated parameters are used to determine how the probability of viral survival in the mucosa depends on latency probability, p_{lat}.

The second model is a two-compartment model that describes both mucosal infection and the systemic infection in the lymphoid tissue (Fig. 3.3 and Section 3.2.8). To start a systemic infection, the virus must enter $CD4^+$ T-cell-rich lymphoid tissue, where > 98% of $CD4^+$ T cells are found (Murphy 2011). The mucosal and systemic infection models are two identical models, with the difference only in a single parameter value, R_0 (Fig. 3.3 and Tab. 3.2). As mentioned, R_0^{muc} is much smaller than 1 during initial infection (Miller et al. 2005), and R_0^{LT} is large and ~10 during systemic infection (Nowak et al. 1997). The dynamics in the two compartments can be solved separately. As shown below, the two-compartment model predicts an optimal level of latency probability, $p_{\text{lat}}^{\text{opt}} \approx 0.5$, which agrees with the results of experiments in cell culture (Calvanese et al. 2013; Dahabieh et al. 2013) and mucosal infection in simians (Whitney et al. 2014). However, this prediction is at odds with the much lower frequencies of latency observed in chronically infected patients (Chun, Carruth et al. 1997; Ho et al. 2013).

To account for this discrepancy, the host immune response is added to the systemic infection model (Nowak and May 2000). Indeed, the presence of the immune response is a key difference between a chronic infection in a patient from a cell culture experiment. All additional parameters are known from fitting data in patients (Tab. 3.3). Importantly, the extended model predicts the same optimal initial value of p_{lat}.

The initial model, without the immune response, is the simplest model of viral dynamics (Nowak and May 2000) expanded to include latent infections (Sedaghat et al. 2007; Sedaghat et al. 2008; Rong and Perelson 2009a, b). In the absence of the immune response, which we include later, and for large cell populations, the model is described by the ordinary differential equations

$$\begin{aligned}
\frac{dT}{dt} &= b - d_T T - kVT \\
\frac{dI}{dt} &= (1 - p_{\text{lat}})kVT - d_I I - kVT \\
\frac{dL}{dt} &= p_{\text{lat}} kVT - d_L L - rL \\
\frac{dV}{dt} &= n d_I I - cV
\end{aligned} \tag{3.20}$$

Here uninfected target cells, T, are produced at linear rate b, die at rate d_T, and can be infected by virus particles, V, at rate k. Infected cells face two possible fates, as follows. With probability p_{lat}, they become latently infected, L, and with probability $1 - p_{\text{lat}}$, they become actively infected and virus-producing, I. Latently infected cells are either reactivated at rate r or die at rate d_L. Actively infected cells produce n virus particles each and die at rate d_I. The virus particles lose infectivity at rate c, which is

Tab. 3.2: Parameters of the basic model.

Notation	Definition	Units	Value	References
R_0^{LT}	Reproduction ratio in systemic infection	1	10 in Fig. 3.5 and 3.6 15* in Figs. 3.8 and 3.9	Nowak et al. (1997)
R_0^{muc}	Reproduction ratio in early mucosa	1	0.1 in 3.6A, B 0.25 elsewhere	Li et al. (2005) and Miller et al. (2005)

Tab. 3.2 (continued)

Notation	Definition	Units	Value	References
b_{muc}	Replenishment rate of infectable cells in mucosa	cell/day	$R_0^{muc} = b_{muc}\, nk/cd_T$	****
b_{LT}	Linear replenishment rate of infectable cells in lymphoid tissue	cell/day	$2 \cdot 10^{11}\, d_T$	**
k	Virus infectivity parameter	1/day/ RNA copy	$R_0^{LT} = \frac{b_{LT}nk}{cd_T}$	Nowak et al. (1997)**** Perelson et al. (1996) and Zhang et al. (1999)
n	Number of virions from an infected cell	RNA copy/cell		
c	Virion clearance rate	1/day		
				Nowak et al. (1997)
d_T	Death rate of target cells	1/day	0.1	Stafford et al. (2000)
d_I	Death rate of virus-producing cells	1/day	1.0	Klatt et al. (2010)
d_L	Death rate of latent cells	1/day	$10^{-3} - 10^{-4}$	Finzi et al. (1997) and Sedaghat et al. (2008)
r	Activation rate of latent cells	1/day	Mucosa: 2×10^{-3} LT: $r(0) = 10^{-3} - 10^{-4}$ $r(E)$: eq. (3.26)	Finzi et al. (1997) and Sedaghat et al. (2008)
p_{lat} or p	Probability of latency	1	Free	N/A
f_{act}	Fraction of nonlatent transfer at $p_{lat} = 0.5$	1	Free	N/A
f_{nonlat}	Fraction of nonlatent transfer at $p_{lat} = p_{opt}$	1	Free	N/A

****Specific values of parameters b_{muc}, k, c, n chosen in simulation do not matter provided the two experimental values of the R_0 are correct.

****Results for optimal p (Fig. 3.5C) depend on a single parameter, R_0^{LT}.

**b_{LT} corresponds to the total CD4 cell count of 2 10^{11} in an uninfected individual (Murphy 2011).

*In the model with the immune response, R_0^{LT} is increased by a factor of 1.5 in order to compensate for the viral eclipse phase, which slows down the predicted viral expansion (Klenerman et al. 1996; Sergeev, Batorsky, Coffin et al. 2010; Sergeev, Batorsky and Rouzine 2010). Parameter values obtained from data in human patients are the same for the lymphoid tissue and the mucosa, except the replenishment rate of target cells, b, which is assumed to be much less for the mucosa to account for the lower R_0 in the same (Zhang et al. 1999; Li et al. 2005; Miller et al. 2005; Haase 2011).

assumed to be very fast. All parameter values are given in Tab. 3.2. Parameters for the extended model including the immune response are in Tab. 3.3. Equation (3.20) is solved numerically using program code 15Ss.m in a MATLAB™ package.

3.2.2 Effect of latency depends on both mucosal and systemic infection

To determine $p_{\text{lat}}^{\text{opt}}$, the two-compartment model calculates the effect of latency in both mucosal and systemic infection. The total transmission rate between individuals is proportional to the product of two factors, the initial inoculum of infected cells in the mucosa, I_0, and the probability that an initially infected cell establishes systemic infection, p_{estab}, as given by

$$p_{\text{trans}} = p_{\text{estab}}\ I_0 \tag{3.21}$$

where p_{estab} is assumed to be small, because $<1\%$ of sexual contacts result in HIV infection (Gray et al. 2001; Wawer et al. 2005; Fraser et al. 2007). Given eq. (3.21), the overall goal is to determine the effect of latency on I_0 and p_{estab}.

3.2.2.1 Latency increases the probability of systemic infection

To shed light on the effect of latency on the establishment probability p_{estab}, consider mucosal infection. As mentioned above, the first five days are characterized by a lack of detectable virus replication (Li et al. 2005; Miller et al. 2005), proving that the reproduction number in the mucosa, R_0^{muc}, is initially far below unit (Section 3.2.9). Condition $R_0^{\text{muc}} \ll 1$ is also consistent with the low frequency of successful transmission and the week of delay before the systemic phase of infection (Miller et al. 2005).

Both deterministic (Fig. 3.5A) and stochastic version of the model (Fig. 3.6A, B), predict a fitness advantage for latency, because, in its absence, all infected cells become extinct during mucosal infection (Fig. 3.5A inset, 3.6C). Computer simulations were performed with $R_0 \leq 0.25$ assuming a few dozen of initially infected cells, as in the animal mucosal experiments (Zhang et al. 1999; Miller et al. 2005; Haase 2011). In contrast, even a low probability of latency enables viral survival (Fig. 3.5A, 3.6D). To test the robustness of these predictions across all values $R_0 < 1$ and $I_0 < 100$, these findings have been repeated analytically by a branching-process technique. Instead of averaging thousands of Monte-Carlo simulations for each R_0 and I_0, the branching-process approach (Section 3.2.7) derives the expression for the extinction probability as a function of time, in the general form. Results show that, in the absence of latency, the probability of extinction approaches 1 by day 5 of mucosal infection, unless R_0 is close to 1 (Fig. 3.6C). Animal mucosal challenge experiments suggest $R_0 < 0.25$ (Miller et al., 2005).

The average surviving number of infected cells in mucosa derived analytically for the Wright–Fisher process (Section 3.2.7) increases approximately linearly with p_{lat},

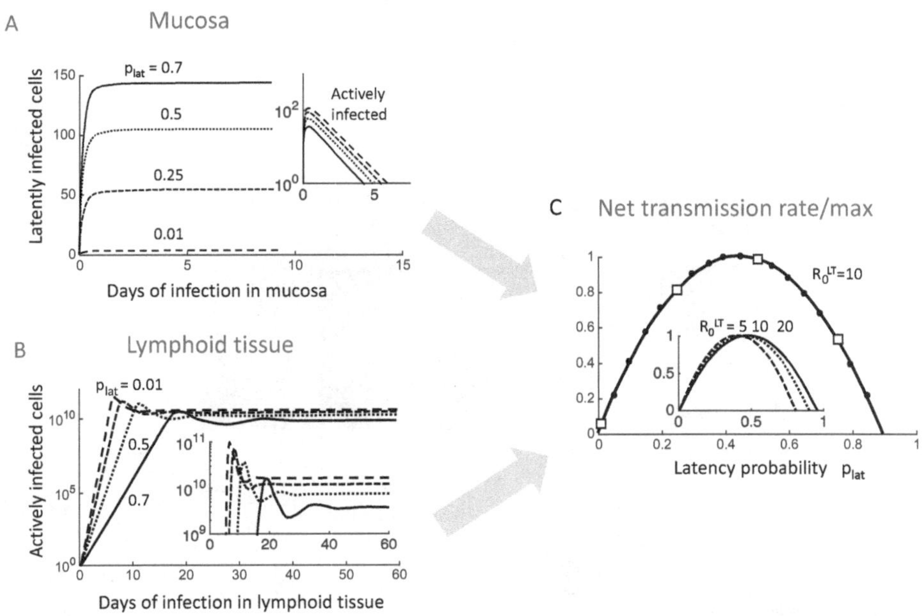

Fig. 3.5: Optimal latency probability optimizes the gain in the establishment rate of infection against the decrease in inter-host inoculation.
(A) Dynamics of latently infected cells in early mucosal infection ($R_0^{\text{muc}} = 0.25$). The number of surviving latently infected cells increases with p_{lat}. Inset: When p_{lat} increases, actively infected cells get extinct more rapidly. **(B)** In a systemic infection ($R_0^{\text{LT}} = 10$), the virus load (and the viral dose transmitted to the next host) decreases with p_{lat}. (A–B) are obtained numerically from eqs. (3.20) and parameters in Tab. 3.2 at $r = 0$. **(C)** Normalized probability of host-to-host transmission, p_{trans}, as a function of $1 - p_{\text{lat}}$, obtained analytically from eq. (3.25) (solid line) and numerically using the plateau levels of actively infected cells, I, and latently infected cells, L, from (A, B) (dots). Based on Rouzine et al. (2015).

for each value of inoculum I_0 (Fig. 3.6E). Given that latency increases initial viral survival in the mucosa, we tested whether latency increases the probability that an initially infected cell seeds a systemic infection in the lymphoid tissue, p_{estab}. To do so, the Wright–Fisher model was extended into a two-compartment mucosa-to-lymph model that directly connects the two typical stages of lentiviral infection (Section 3.2.8). Only a single parameter value is assumed to differ between the mucosal and lymphoid compartments: $R_0^{\text{muc}} < 1$, and $R_0^{LT} > 1$ (Tab. 3.2). The two-compartment model directly confirms (Fig. 3.6F) that increasing p_{lat} increases the establishment probability of systemic infection.

3.2.2.2 Latency decreases the inoculum

While increasing p_{lat} increases the probability of systemic infection given an infected cell inoculum, I_0, which was considered, so far, fixed, the probability of transmission also depends on I_0 itself, which is proportional to the viral load of the transmitting

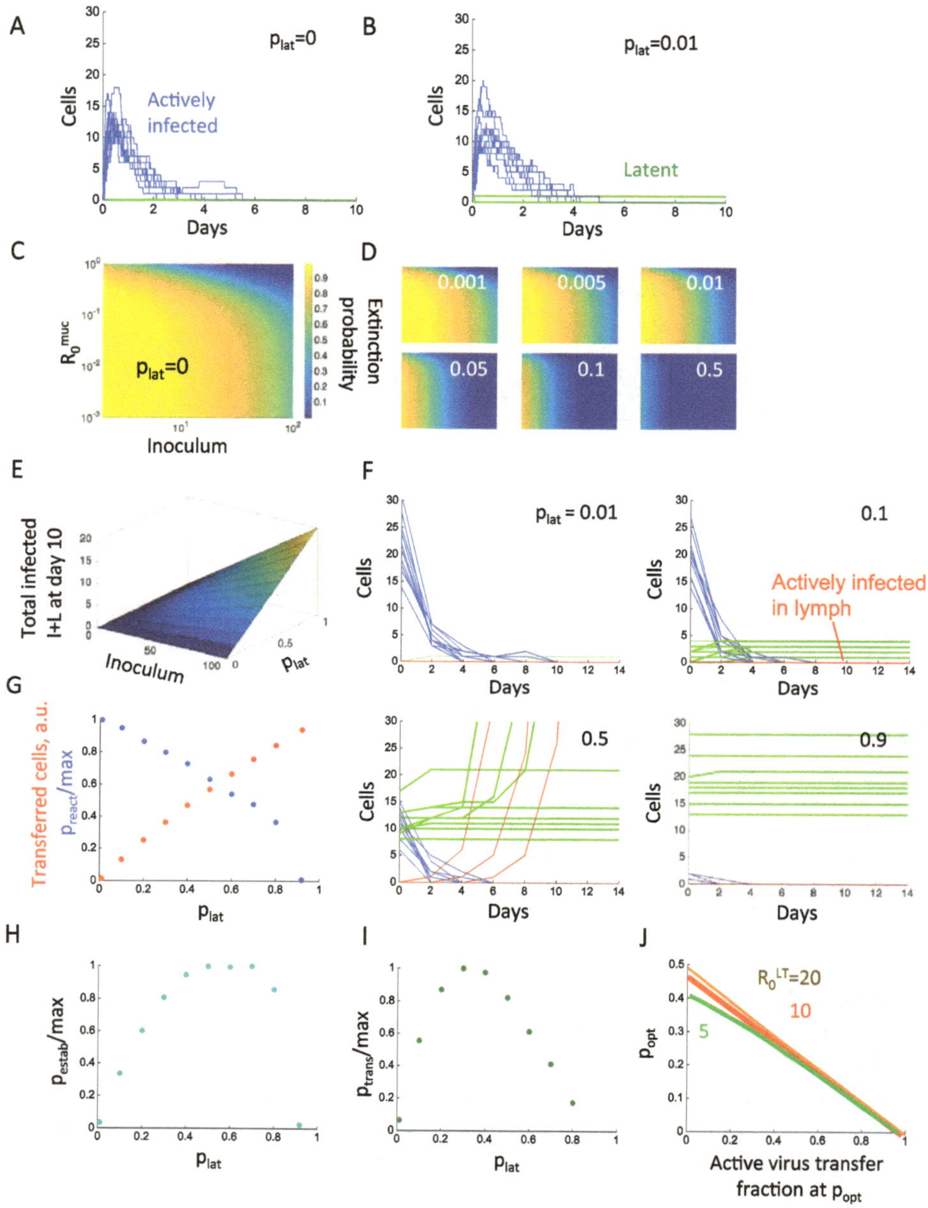

Fig. 3.6: Stochastic simulation demonstrates that HIV latency increases the chances of virus survival if $R_0 < 1$.

(**A**) 10 representative runs of Gillespie simulation of the one-compartment model [eqs. (3.28); for parameters, see Section 3.2.14]. (**B**) Same with $p_{lat} = 0.01$. The difference between the longest runs in A and B is due to random fluctuations. (**C**) Analytical extinction probability derived using branching- process technique, in absence of latency ($p_{lat} = 0$) at day 5 post infection, as a function of R_0 and the number of initially infected cells, I_0 (Section 3.2.7). (**D**) Extinction probability for $p_{lat} = 0.001,\ 0.005,\ 0.01,\ 0.05,\ 0.1, 0.5$ at day 5 post infection, as a function of R_0 and I_0. (**E**) The number of surviving cells (at day 10 and $R_0 = 0.25$), patient [eq. (3.33) in Section 3.2.8]. Thus, one has to find out how the virus load in a steady-state infection depends on latency probability.

For this end, the deterministic model in the lymphoid tissue compartment, eqs. (3.20), was solved numerically at $R_0 = 10$ and a single founder infected cell (Keele et al. 2008; Kearney et al. 2009; Maldarelli et al. 2013). The results demonstrate a decrease in the steady-state viral load with p_{lat} (Fig. 3.5B). As expected, increasing latency decreases inoculum received by the next individual in the transmission chain.

3.2.3 Probability of latency near 50% optimizes transmission

With eq. (3.21) for the transmission rate, which serves as the proxy of virus fitness, the goal to find out whether the increase of p_{estab} due to latency exceeds the decrease of I_0, and to calculate the value of p_{lat} that maximizes the product $p_{\text{estab}} I_0$ (and hence virus fitness). For this aim, p_{estab} and I_0 are derived below analytically, as functions of p_{lat}.

Let us denote the number of latently infected cells survived after mucosal infection $L_{\text{init}}^{R_0>1}$. As shown above, the majority of latently infected cells emerge in the first generation of infection, which implies $L_{\text{init}}^{R_0>1} \approx p_{\text{lat}} I_0$ (Figs. 3.5A and 3.6F). To establish systemic infection, at least, one of latent cells must be reactivated, which happens with probability p_{react}. Thus, the probability of establishment per initial infected cell is

$$p_{\text{estab}} = \frac{L_{\text{init}}^{R_0>1}}{I_0} p_{\text{react}} \approx p_{\text{lat}}\ p_{\text{react}} \tag{3.22}$$

Here all active cells are assumed to be extinct, and only latently infected cells seed systemic infection. To demonstrate robustness of results to this assumption, the possibility of a nonlatent seeding, including Langerhans dendritic cells, actively infected cells, and free virions, is considered later on (Section 3.2.10).

Fig. 3.6 (continued)
as a function of p_{lat} and I_0, from the analytic Wright–Fisher approach (Section 3.2.7). **(F)** Dynamics in two compartments, target-cell poor ($R_0^{\text{muc}} < 1$) and target-cell rich ($R_0^{\text{LT}} > 1$) at four different values of p_{lat}, for 10 random runs of stochastic Wright–Fisher simulation for actively infected cells in mucosa (blue), latent cells in lymphoid tissue (green), and actively infected cells in lymphoid tissue (red) (Section 3.2.8). **(G–I)** Coupled-compartments Wright–Fisher simulation with nonconstant p_{react} (Section 3.2.10.8). **(G)** Fraction of the initial infected cells in mucosa that become latent (and are transferred to lymphoid tissue) and the reactivation probability p_{react}, as a function of p_{lat}. **(H)** The establishment probability per initially infected cell, $p_{\text{estab}} = p_{\text{react}} \cdot L_{\text{init}}^{R_0>0}/I_0$, has maximum at $p_{\text{lat}} \approx 0.6$. **(I)** The net transmission probability including inoculum dependence $I_0(p_{\text{lat}})$ (Section 3.2.15) has a maximum at $p_{\text{lat}} = 0.35$. Each point in (G–I) is average of 10^4 runs with parameters: $R_0^{\text{muc}} = 0.25$, $R_0^{\text{LT}} = 10$, $dI = 0.5$, $r = 0.006$. **(J)** Inclusion of nonlatent transfer between mucosa and LT preserves high values of p_{opt}. The dependence of p_{opt} on the fraction of active-infection transfer f_{latent} recalculated a posteriori at $p_{\text{lat}} = p_{\text{opt}}$ is given by eqs. (3.38), (3.39). Based on Rouzine et al. (2015).

The next task is to find the average infectious dose I_0 transmitted to a new patient, as a function of p_{lat}. The average dose is directly proportional to the average viral load over the systemic infection (Nowak and May 2000). The systemic infection includes the initial viral expansion, the peak, and the steady-state attained after the peak. The result has the form (Section 3.2.8)

$$I_0 \approx \text{const}(p_{\text{lat}}) \cdot \left[(1 - p_{\text{lat}})R_0^{\text{LT}} - 1\right] \tag{3.23}$$

The first term in eq. (3.23) is constant in p_{lat} but may depend on other parameters. (Note that eq. (3.23) is obtained under the assumption that latently infected cells rarely reactivate prior to cell death, $r \ll d_L$, see Sections 3.2.3.1 and 3.2.10.)

From eqs. (3.21)–(3.23), the transmission probability, as a function of p_{lat}, is given by

$$p_{\text{trans}} \approx \text{const}(p_{\text{lat}}) \cdot p_{\text{lat}} \cdot \left[(1 - p_{\text{lat}})R_0^{\text{LT}} - 1\right] \tag{3.24}$$

This function shown in Fig. 3.5C has a maximum at $p_{\text{lat}} \equiv p_{\text{lat}}^{\text{opt}}$. Differentiating with respect to p_{lat} yields

$$p_{\text{lat}}^{\text{opt}} \approx \frac{1}{2} - \frac{1}{2R_0^{\text{LT}}} \tag{3.25}$$

Strikingly, for a typical value of $R_0^{\text{LT}} = 10$, the probability of latency that maximizes lentiviral transmission is quite large, close to a half (Fig. 3.5C). In agreement with the analytic result, the numerical averaging of inoculum over time also show an optimum at $p_{\text{lat}} \approx 0.5$ (Fig. 3.5C). This optimum at $p_{\text{lat}} \approx 0.5$ exists across the entire observed range of R_0^{LT} values (Fig. 3.5C, inset). Here the probability of reactivation, p_{react}, is assumed to not depend on p_{lat}; lifting this assumption slightly lowers $p_{\text{lat}}^{\text{opt}}$ (Section 3.2.10 and Fig. 3.6G).

3.2.3.1 Result is robust to the presence of nonlatent routes

So far, no additional channels of seeding infection exist except latent cells were allowed. To test robustness to that assumption, the optimal latency probability was rederived allowing a fraction of successful infections to arrive via nonlatent routes (Section 3.2.10). As long as the fraction stays less than 100%, the optimal latency probability remains surprisingly large (Fig. 3.6J). For example, even if 80% of lentiviral transmissions are established by nonlatent routes, such as Langerhans dendritic cells, one still has $p_{\text{lat}}^{\text{opt}} = 0.1$ (Fig. 3.6J).

Some changes in the assumptions make $p_{\text{lat}}^{\text{opt}}$ even larger than in eq. (3.25). Relaxing the previous assumption that most latently infected cells die before reactivation, $r \ll d_L$, reduces the cost of latency to inoculum and, hence, increases the optimal probability of latency. As a result, if $r > d_L$, we have $p_{\text{lat}}^{\text{opt}} = 1$ (Section 3.2.10). Next, if we lift the assumption of a linear relation between the virus load and transmission rate and allow it to saturate at high viral loads (Fraser et al. 2007), then I_0 becomes less sensitive to latency, and $p_{\text{lat}}^{\text{opt}}$, again, increases toward 1 (Section 3.2.10).

3.2.3.2 Experiments in cell culture and primates confirm predictions

The predicted latency probability of 50% (Dahabieh et al. 2013) or higher (Calvanese et al. 2013) was actually observed in cell culture. This value is also consistent with a study in *rhesus macaques*, where a large reservoir of latently infected cells was documented on day 3 of mucosal infection (Whitney et al. 2014). However, the 50% of latency differs from the low frequency of latent cells observed in chronically infected patients under retroviral treatment. Only 1 in $10^6 - 10^7$ patient $CD4^+$ T cells is latently infected (Chun, Carruth et al. 1997; Sedaghat et al. 2007), which corresponds to $p_{lat} \sim 10^{-5} - 10^{-4}$ (Sedaghat et al. 2007; Rong and Perelson 2009b). Some other studies suggest a 60-fold higher latent cell frequency (Ho et al. 2013), but this is still far below 50%. The real problem is in explaining the difference between the high frequency in early infection (or cell culture) and the low frequency in chronic infection. What could be the factor absent from the first case but present in the second case?

3.2.4 Why latency in patients is rare: including the immune response

In contrast to an early or a cell-culture infection, a chronic infection elicits a vigorous HIV-specific immune response, including T-cell and B-cell response (Turnbull et al. 2009). As evident from the experiments on CD8 T-cell depletion in monkeys (Jin et al. 1999; Schmitz et al. 1999) and the rapid genetic variation in CD8 T-cell epitopes (Goonetilleke et al. 2009; Ganusov et al. 2011; Batorsky et al. 2014), the cytotoxic T-cell response strongly limits HIV replication in a host. As a result, the immune response has to be included into the basic model of viral dynamics (De Boer and Perelson 1998; Nowak and May 2000) to explain the 100 – 1000 fold decrease in the viral load between the peak of acute infection and the chronic infection (Stafford et al. 2000). CD8 T cells kill actively infected cells, which they recognize by specific TCR to MHC-I-virus-peptide binding, and suppress viral replication by secreting cytokines. Neither mechanism, however, concerns latently infected cells that do not express virus proteins and, hence, cannot be recognized by CD8 T cells.

The critical change will be in the assumption that p_{lat} and r are constant. In reality, both parameters are sensitive to cell cycle phase and, hence, to cytokines secreted by T cells that may activate cells. The activation levels of $CD4^+$ T cells increase during chronic HIV infection, as witnessed by the expression of three activation markers (Li et al. 2005) and the increased turnover rate of $CD4^+$ T cells (Mohri et al. 1998). One mechanism of cell activation is a homeostatic response to the depletion of $CD4^+$ T cells by HIV infection. Another potential mechanism is secretion of TNF-α by killer $CD8^+$ T cells (Murphy 2011), and accumulation of transcription factors, such as NF-κB that activates the HIV LTR promoter and suppresses the provirus entry to latency. In any case, there is solid evidence that cellular activation factors can reduce p_{lat} sharply and thus activate HIV transcription (Chun et al. 1998; Siliciano and Greene 2011; Calvanese et al. 2013). Razooky et al. (2015) demonstrated that cellular activation affects the

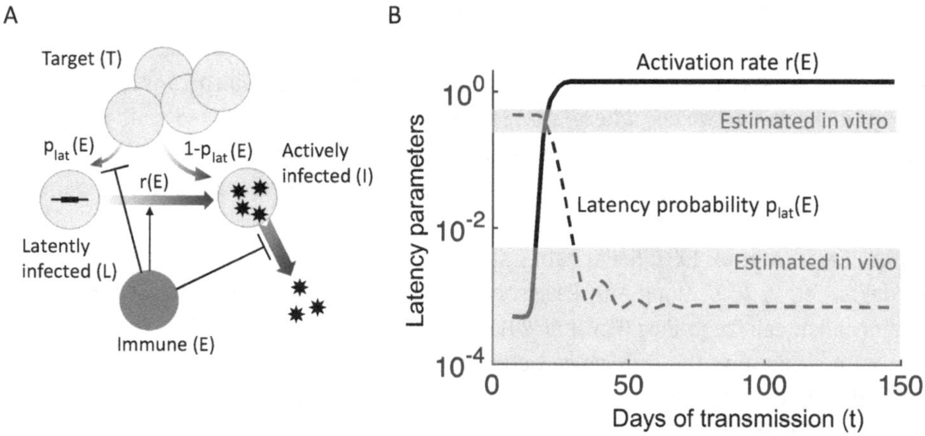

Fig. 3.7: Including the immune response into the model of systemic HIV infection explains the difference in the frequency of HIV latent cells between experimental models (cell-culture and nonhuman primates) and chronically infected humans.
(**A**) Extended model includes CD8⁺ T cells, E, which kill actively infected cells (or suppress viral replication) and activate latently infected cells, eqs. (3.26)–(3.27). (**B**) Latency probability p_{lat} and reactivation rate r change strongly after the viremia peak due to bystander cytokine activation by immune cells, eqs. (3.26). Based on Rouzine et al. (2015).

expression of HIV provirus, and hence decreases p_{lat} and increases r, by changing the relative time intervals of latent and active states.

Because the adaptive immune response secretes cellular activation factors (Chun et al. 1998; Li et al. 2005; Siliciano and Greene 2011; Calvanese et al. 2013), this effect may explain the depressed latency levels in chronically infected patients. Because cytokines secreted by $CD8^+$ T cells are short-lived and act on a short distance, one can assume that p_{lat} and r are functions of the effector $CD8^+$ T-cell concentration, $E(t)$, as follows (Fig. 3.7B)

$$p(E) = p(0)\frac{E_{0L}}{E_{0L}+E}, \quad r(E) = r(0) + d_I \frac{E}{E_{0L}+E} \tag{3.26}$$

Here E_{0L} is the characteristic number of immune cells at which a half of infected cells receives a cytokine signal above a proviral activation threshold, and the values $p(0) \approx 0.5$ and $r(0) = 10^{-4} - 10^{-3}$ (Tab. 3.2) correspond to the absence of the HIV-specific immune response.

With latency parameters defined in eq. (3.26), the model of systemic infection given by eqs. (3.20) needs to be upgraded to include the dynamics of antigen-specific CD8 T-cell response (De Boer and Perelson 1998; Davenport et al. 2004; De Boer 2007). In a systemic infection, the cell numbers in all compartments are very large, so we can safely use the deterministic approximation and ODE, as follows:

$$
\begin{aligned}
\frac{dT}{dt} &= b - d_T T - kVT \\
\frac{dI_E}{dt} &= (1 - p_{\text{lat}})kVT - d_{IE} I_E + rL \\
\frac{dI}{dt} &= d_{IE} I_E - d_I \left(1 + \frac{E}{E_0}\right) I \\
\frac{dL}{dt} &= p_{\text{lat}} kVT - d_L L - rL \\
\frac{dV}{dt} &= n d_I I - cV \\
\frac{dE}{dt} &= \frac{aI(E_N + E)}{I + I_{av}} - d_E E \\
\frac{dE_N}{dt} &= \frac{aIE_N}{I + I_{av}}
\end{aligned}
\tag{3.27}
$$

where the additional model parameters are in Tab. 3.3. As compared to the model in eqs. (3.20), which assumes that the death rate of virus-producing cells d_I is constant, in the expanded model in eqs. (3.27), cytotoxic CD8 T cells E can kill actively infected cells (term E/E_0 in line 3). Two new equations describe the activation of naive CD8 T cells E_N by virus, as well as proliferation and the decay of effector cells. As is always the case with models including cytotoxic response, there are also infected cells in the eclipse phase between infection and virus production, I_E. The eclipse phase, which does not exceed a day, is known to exist and is necessary to prevent the artifact of a sharp drop of viremia and giant oscillations (Rouzine et al. 2006; De Boer 2007; Sergeev, Batorsky, Coffin et al. 2010; Sergeev, Batorsky and Rouzine 2010). Additional parameters in eqs. (3.27) are given in Tab. 3.3.

Tab. 3.3: Additional parameters for the model including the immune response, Fig. 3.7 and eqs. (3.26) and (3.27), to explain data in Fig. 3.8.

Notation	Definition	Units	Value	References
a	Maximum proliferation rate of CTLs	1/day	$a - d_E = 0.5$	De Boer et al. (2003)
d_E	CTL death rate	1/day	0.75	Data fitting*
d_{IE}	Inverse length of eclipse phase	1/day	1.4/day	Markowitz et al. (2003) and Brandin et al. (2006)
E_0	CTLs that halve the infected cell lifetime	Cell	$2 \cdot 10^8$	Data fitting*
I_{av}	CTL avidity threshold in the number of infected cells	Cell	10^8	Data fitting*

Tab. 3.3 (continued)

Notation	Definition	Units	Value	References
$E_N(0)$	Initial naive CTL number	Cell	$2\cdot10^6$	Murphy (2011)
E_{0L}	CTL # that decreases p_{lat} by 50%	Cell	$4\cdot10^6$	Data fitting*

*Out of the seven parameters, four parameters, d_E, E_0, E_{0L}, I_{av}, are adjusted to fit the four experimental plateaus in patients (Fig. 3.8), and three are fixed and cited from the literature. Specifically, E_0, E_{0L}, I_{av} are estimated from the predicted steady-state levels of E, I, and L, compared to their experimental estimates (Fig. 3.8): $E = 10^9$ cells (Ogg et al. 1998; Turnbull et al. 2009), $I = 10^8$ cells (Haase 1999), and $L = 10^6$ cells (Chun, Carruth et al. 1997). d_E is estimated to fit L under ART: $L = 10^5$ cells (Finzi et al. 1997). The total cell counts are assumed $T(0) = b/d_T = 2\cdot10^{11}$ for both $CD8^+$ T and $CD4^+$ T cells (Murphy 2011).

Equation (3.27) is a relatively simple model of CD8 cell response. More complex models involve helper-cell dependent activation (Rouzine et al. 2006; Sergeev, Batorsky, Coffin et al. 2010; Sergeev, Batorsky and Rouzine 2010; Rouzine 2022) or multiple CTL clones recognizing different epitopes (Batorsky et al. 2014). Yet, even this simplest model works here, because the effects of CD8 response on latency are sufficiently strong. Robustness to model variation is analyzed in Section 3.2.10.

Before the adaptive-immune response and before the viral peak, the model predicts high latency probabilities of ~ 0.5 and low reactivation rates, as in the simplified models above. However, after the peak, the increase in cellular activation (Li et al. 2005) and cell death due to abortive infection (Doitsh et al. 2010) decreases $p_{lat}(E(t))$ to low levels and increases $r(E(t))$ to high levels, as shown in Fig. 3.7B and eqs. (3.26). Thus, the immune model explains the difference between latency frequencies in culture and in systemic infection.

3.2.4.1 Models incorporating the immune response fit patient data

The next task is to test the model on available patient data, as follows:

(i) viral loads before ART (Fraser et al. 2007),
(ii) effector T-cell concentrations before ART (Turnbull et al. 2009),
(iii) latently infected cells before ART (Chun, Carruth et al. 1997), latently infected cells after ART (Finzi et al. 1999).

The immune-response model, eqs. (3.27), is able to fit these 4 data plateaus (Fig. 3.8A) using well-known parameter estimates (Tab. 3.3).

The dynamics of systemic infection is calculated numerically from eqs. (3.26) and (3.27) with the use of program ODE15S.m implemented in MATLAB™ (Fig. 3.8A). Model parameters are in Tab. 3.3 (see also Section 3.2.9). Initially, we have $p_{lat}(0) \sim 0.5$ and $r(0) \ll 1$ (Fig. 3.7B), hence the high levels of latent cells similar to active cells (Fig. 3.8A). After the viremia peak, the immune response increases r and decreases p_{lat} by several orders of magnitude and thus depletes the latent reservoir. The population of latent

A

B

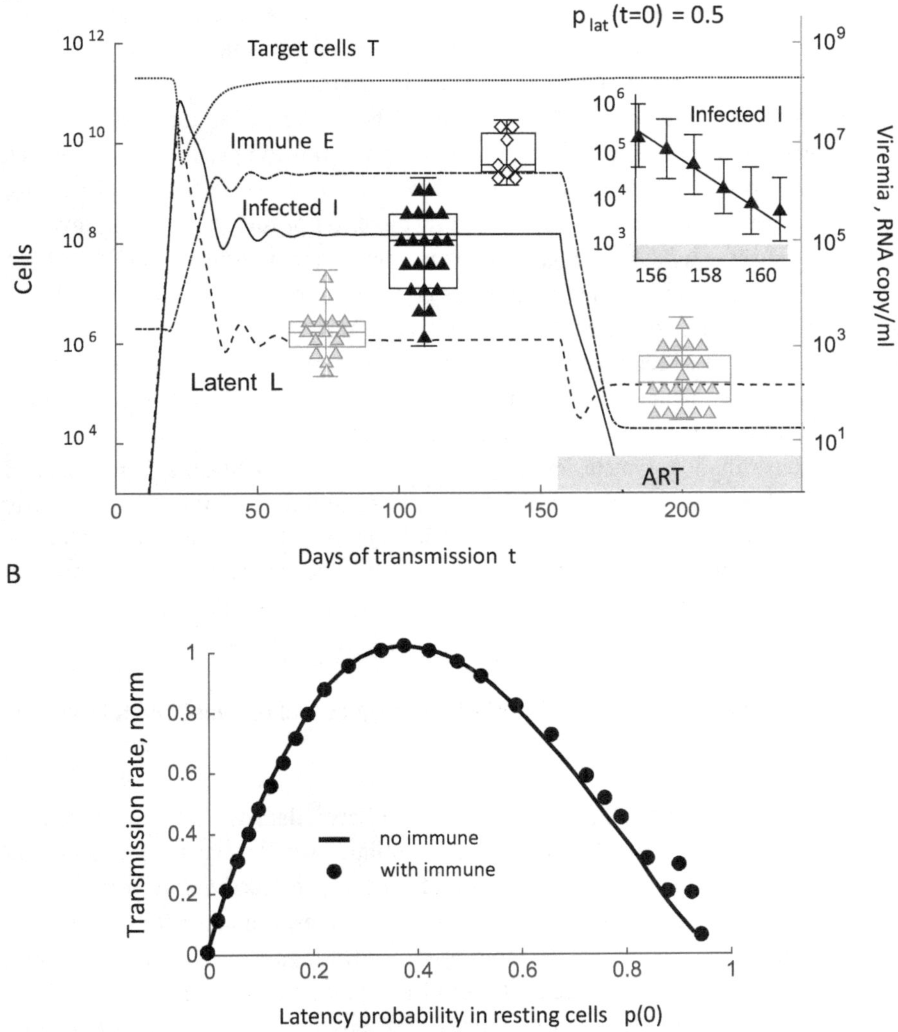

Fig. 3.8: The extended immune-response model fits the available *in vivo* data, and does not change the optimal latency probability for resting cells, $p_{lat}^{opt}(0)$.
(**A**) Dynamics of cell compartments during systemic infection calculated from eqs. (3.26) and (3.27). Antiretroviral therapy (ART) initiated during steady-state infection causes a decline of the latent reservoir, L. The saturation of the fall in the latent reservoir is due to the decline in the immune cells, E, during ART. Data points across human patients: virus load linked to infected cell number prior to ART (Fraser et al. 2007) (black triangles); latent cells prior to ART (Chun, Carruth et al. 1997) and after highly active ART (Finzi et al. 1999) (gray triangles); effector CD8 T cells (Turnbull et al. 2009) (open diamonds). Inset: Virus load after the onset of ART (Markowitz et al. 2003) (error bars show standard deviation). (**B**) Normalized transmission rate p_{trans}, as a function of $p_{lat}(0)$ calculated from the dynamics in panel (A) and eq. (3.21). Two cases are shown for comparison: with immune cells (E, filled circles) and without immune cells ($E = N = 0$, solid curve). Inclusion of immune cells into the model only weakly affects the prediction of a

cells, L, saturates at a low level, in which the arrival of new latently infected cells from target cell compartment and the activation of latent cells cancel (Fig. 3.8A). Hence, latent compartment, L, is dynamically coupled to the actively infected compartment, I.

The majority of infected individuals receive anti-retroviral therapy (ART). The onset of ART blocks new cell infections. The model predicts a decline of all infected cells, first, virus-producing cells and then, latent cells, because the two compartments are coupled (Fig. 3.8A). The predicted decay of the latent reservoir observed in patients (Fig. 3.8A) occurs as fast as the cells are activated, $r(t)$. At the onset of ART, due to high levels of CD8 T cells, the activation rate is high, and latent compartment declines rapidly. Antigen depletion causes the population of immune cells, E, to contract. As a result, reactivation rate $r(t)$ decreases and soon returns to its low background level, $r(0)$, eq. (3.26). Under long-term ART, any further decrease in the latent reservoir occurs very slowly, as it is actually observed in patients (Fig. 3.8A). The model explains the 10-fold decrease in the number of DNA positive cells observed after ART (Chun, Carruth et al. 1997).

Most interestingly, the inclusion of the immune response and time dependence of latency parameters does not change the above prediction that the optimal *initial* latency probability $p_{\text{lat}}(0)$ is large, $p_{\text{lat}}^{\text{opt}}(0) \sim 0.5$ (Fig. 3.8B). The prediction is similarly robust to epidemiological assumptions, such as the monotonic dependence of transmission on viral loads, or whether chronic or acute infection dominates transmission (Section 3.2.10).

3.2.5 Prediction: depletion of $CD8^+$ T cells in macaques will increase the latent reservoir

Testable predictions of the above model are, as follows. Because CD8 T cells reduce latency, depletion of $CD8^+$ T cells with anti-CD8 antibodies will cause the latency probability p_{lat} to return to its initially high value of 0.5, and the reactivation rate r to drop to its initially low value. In SIV-infected rhesus macaques, the artificial depletion of $CD8^+$ T cells causes a rapid (10 – 100)-fold rise in viral load (Jin et al. 1999; Schmitz et al. 1999). The present model predicts that CD8 depletion should boost the small latent reservoir even higher, up by 4 – 5 orders of magnitude (Fig. 3.9). At the same time, CD8 depletion during early prepeak infection (Matano et al. 1998), would increase the latent reservoir size 2 – 3 fold only [Fig. S4 in Rouzine et al. (2015)]. Testing these predictions could be very illuminating for the dynamic nature of the formation of latent reservoir.

Fig. 3.8 (continued)
large optimal latency probability for resting cells, $p_{\text{lat}}^{\text{opt}}(0) \sim 0.5$. Model parameters in panels A–B are in Tabs. 3.2 and 3.3, with $p_{\text{lat}}^{\text{opt}} = 15$ and $p_{lat}(0) = 0.5$ in panel A. See also Fig. S3 in Rouzine et al. (2015). Based on Rouzine et al. (2015).

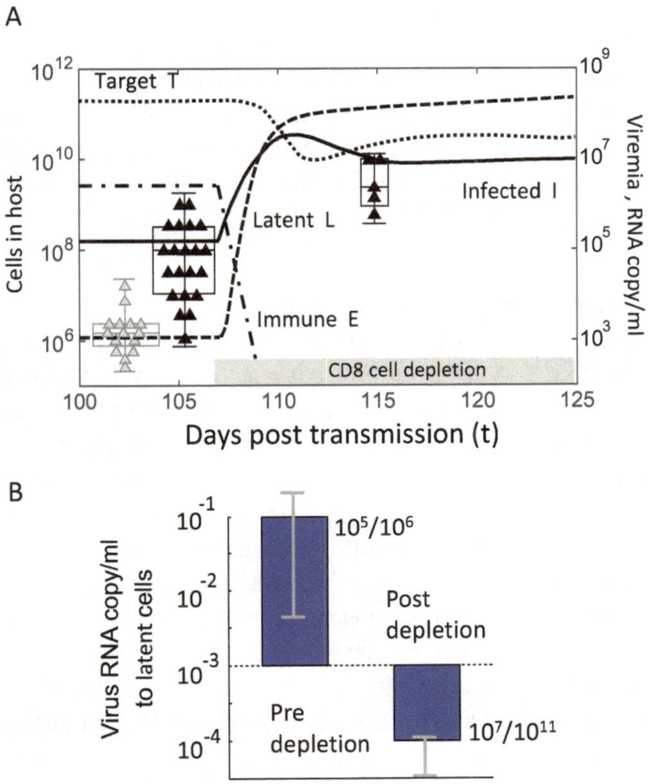

Fig. 3.9: Testable prediction: depletion of CD8$^+$ T cells in SIV-infected macaques will boost the latent reservoir much more than viremia.
(A) Dynamics in systemic infection for the model with the immune response, eqs. (3.26) ands (3.27). Data points and parameters are as in Fig. 3.8. (B) The ratio of actively to latently infected cells will be inverted by CD8$^+$ T-cell depletion ("postdepletion" is day 125 in A). The increase in viremia has been observed in Jin et al. (1999); Schmitz et al. (1999). The dotted horizontal line at 10^{-3} RNA/ml/cell corresponds to a 1:1 ratio of latently and actively infected cell counts. Blue bars correspond to the parameters and compartment sizes in (A). The maximal expected errors (vertical bars) are estimated from the whisker box borders in (A) (the two middle quartiles). See also fig. S4 in Rouzine et al. (2015). Based on Rouzine et al. (2015).

Molecular engineering of recombinant SIV strains with strongly reduced latency probabilities could offer another experimental test in animals. If these strains established systemic infections less frequently than wild-type after mucosal infection, this result would directly demonstrate that latency is of an advantage to fitness.

3.2.6 Discussion

Present modeling informed by data strongly suggests that HIV latency serves as a Trojan horse that allows the virus to get inside the body and start a systemic infection.

3.2.6.1 Latency and other mechanisms of initial viral survival

In principle, systemic infection could be seeded by a route alternative to latently infected $CD4^+$ T cells. One possibility is migration of dendritic cells (Kahn and Walker 1998; Wu and KewalRamani 2006). Langerhans dendritic cells express CD4 and can be infected by HIV. Even if in the presence of nonlatent seeding, the principal result of this study about the large latency probability remain robust provided latency is responsible for, at least, a fraction of systemic infections (Fig. 3.6J).

3.2.6.2 Potential therapy: suppressing latent reactivation in mucosa

If confirmed, the model offers a potential therapeutic strategy, as follows (Fig. 3.10). During the early infection (i.e., first 3 – 5 days), the efficient reactivation of latent cells starts systemic infection and increases the size of the latent reservoir (Fig. 3.6F, Section 3.2.8). This reactivation is the weak link, because it depends on the initial immune stimulation, for example, caused by antigen-presenting cells such as Langerhans dendritic cells (Peressin et al. 2014). If the early reactivation were decreased, for example, by suppressors of HIV transcriptional reactivation (Weinberger et al. 2008) or dendritic cell migration control, systemic infection would be made less likely (Fig. 3.10C). The caveat of this proposed approach is the need of early detection of HIV during the first week of infection. Nevertheless, early detection and early treatment of HIV is done (Haase 2011). A substantially smaller latent reservoir $\sim 10^2$ cells in an infected individual would require fewer cells to be reactivated by subsequent "shock-and-kill" strategies (Archin et al. 2012; Deeks 2012). This combined therapy could enhance the chances of HIV eradication.

3.2.6.3 Implications for alternate antiviral therapy approaches

Shock-and-kill therapies try but fail to clear latent reservoir, due to the existence of cells that cannot be activated (Siliciano and Greene 2011). Instead of reactivating all 10^5 latently infected cells, a completely opposite approach would be to *increase* the fraction of latent cells (Weinberger et al. 2008; Dar et al. 2014). A nontoxic drug that switched 90 – 95% of actively infected cells to latency would drive basic reproductive ratio R_0 below 1, making HIV infection to decay. Although this approach is purely hypothetical, the more comprehensive virus-cell models could help with the identification of such latency-stimulating drugs. If "shock-and-kill" latency eradication strategies will be eventually abandoned, alternative approaches to viral latency could be considered.

3.2.7 Derivations for stochastic dynamics in mucosa ($R_0^{\mathrm{muc}} < 1$)

Sections 3.2.7–3.2.10 serve as a mathematical appendix. Below notations p and p_{lat} are used interchangeably. Dynamics of infection in the mucosal compartment is described

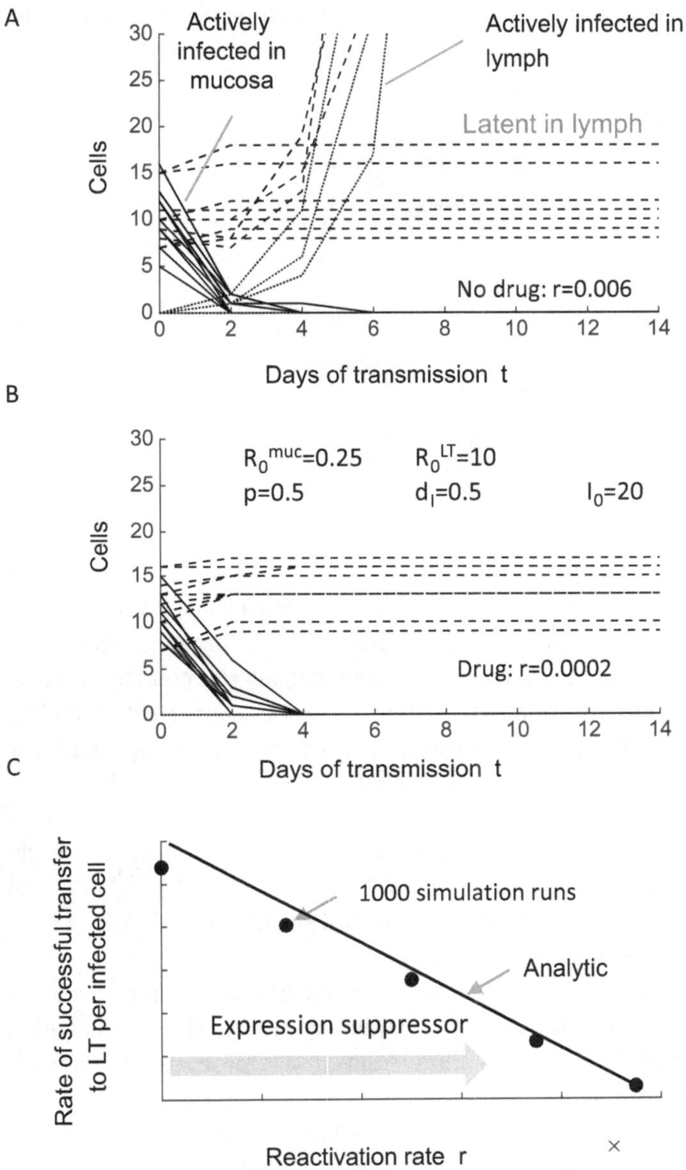

Fig. 3.10: Decreasing reactivation in the post-mucosal phase of infection will reduce latent reservoir size, enhancing the effectiveness of a subsequent 'kick-and-kill' therapy.
(A) Establishment of a systemic infection between days 4 – 6. Stochastic simulation of Wright–Fisher model (Section 3.2.17), $r(0) = 6 \times 10^{-3}$/day. **(B)** The same, when reactivation rate is reduced 30-fold.
(C) The remaining unactivated latent compartment is relatively small. Based on Rouzine et al. (2015).

by a stochastic model, which corresponds to the ODE system in eqs. (3.20) (Pearson et al. 2011).

$$\begin{aligned}
&\xrightarrow{b} T\\
&T \xrightarrow{d_T} \emptyset\\
&T + V \xrightarrow{(1-p)k} I\\
&T + V \xrightarrow{p\,k} L\\
&L \xrightarrow{r} I\\
&I \xrightarrow{nd_I} I + V\\
&I \xrightarrow{d_I} \emptyset\\
&V \xrightarrow{c} \emptyset\\
&L \xrightarrow{d_L} \emptyset
\end{aligned} \tag{3.28}$$

Here state variables denoted by capital letters and model parameters denoted by lower case letters are the same as in eqs. (3.20) and Tab. 3.2. Target cell dynamics, $T(t)$, is either considered explicitly or, when virus load is small, represented by a constant, $T = \frac{b}{d_T}$. The only difference from eqs. (3.20) is that each state variable is no longer a continuous variable, but an integer number, which can change by 1 with the probability proportional to the rate of each "chemical" process in eq. (3.28) and the small time interval chosen to make this probability small.

Gillespie simulation of eqs. (3.28) was carried out using the standard xSSA package in Mathematica™ and parameters from Tab. 3.2: $R_0^{\text{muc}} = 0.1$, and $k = \left(\frac{cd_T}{bn}\right) R_0^{\text{muc}}$. In addition, experimental values $c = 5/\text{day}$ (Perelson et al. 1996) and $n = 500$ (Zhang et al. 1999) were used. Results are shown in Fig. 3.6A, B.

To calculate the probability of extinction at fixed time as a function of R_0 and inoculum I_0, below analytic method based on a branching process (Markov chain) is used (Fig. 3.6C, D). Indeed, Gillespie algorithm is too slow to produce ~10^8 runs required for a high-quality heatmap.

The faster approach requires simplifications, as follows. Firstly, for a small inoculum, $I_0 < 100$ infected cells, reactivation $L \xrightarrow{r} I$ can be neglected ($r < 10^{-3}$, Tab. 3.2). In this case, the dynamics of mucosal active infection $I(t)$ is found first, and $L(t)$ is found after. Further, because HIV virions are cleared rapidly, $c \gg d_I$ and d_T (Perelson et al. 1996), their concentration follows in time the infected cell number. Hence, we can exclude virus load as a free variable. Lastly, the total number of target cells is assumed constant, because viral load is small, and target cells are not depleted.

After these simplifications, the probability to find I infected cells at time t denoted $P(I, t)$ is found from the Markov chain in continuous-in-t limit

$$\frac{\partial P(I,t)}{\partial t} = (I-1)(1-p)R_0^{\text{muc}}P(I-1,t) + (I+1)P(I+1,t) - I\left[1+(1-p)R_0^{\text{muc}}\right]P(I,t)$$

where $P(I,t)$ is the probability of having $I = 0, 1, 2, 3, \ldots$ actively infected cells as time t (measured in units $1/d_I = 1$ day). The equation was solved numerically in MATLAB^{TM} by iterating in discrete time with small steps and the initial Poisson distribution

$$P(I,0) = \frac{((1-p)I_0)^I}{I!} e^{-(1-p)I_0}$$

Now, to calculate the stochastic dynamics of latent cells, let us introduce the probability $G(L,t)$ of having L latent cells at time t averaged over all possible random trajectories described by distribution $P(I,t)$. We get

$$\frac{\partial G(L,t)}{\partial t} = [G(L-1,t) - G(L,t)] \sum_{I_{\text{muc}}=1}^{\infty} pIP(I,t)$$

with the initial Poisson distribution

$$G(L,0) = \frac{(p\, I_0)^L}{L!} e^{-p\, I_0}$$

The probability of zero latent cells at time t, $G(0, t)$, is calculated numerically from

$$G(0,t) = \exp\left[-p\, I_0 - p\int_0^t dt' \sum_{I_{\text{muc}}=1}^{\infty} I_{\text{muc}}\, P(I_{\text{muc}}, t)\right]$$

Similarly, the probability of active-cell extinction at time t [condition $I(t) = 0$] is given by $P(0,t)$ calculated numerically, as explained above. The probability of complete virus extinction at time t, including actively producing and latent cells, $I(t) = 0$ and $L(t) = 0$, is given by the product of the two probabilities, $P(0,t)G(0,t)$, plotted in Fig. 3.6C, D.

It is also instructive to estimate the average number of latent cells surviving mucosal infection (Fig. 3.6E). This can be done analytically in a closed form, if we use a discrete-time Markov process for the average numbers of cells. As before, we neglect the reactivation of latent cells in mucosa. Time $t = 1, 2, 3, \ldots$ is discrete and expressed in units of the infected-cell lifetime, $1/d_I = 1$ day (Tab. 3.2), that is, generations of actively infected cells are nonoverlapping. This represents a generalized version of the Wright–Fisher process. Actively infected and latently infected cells change between generations, as given by

$$\bar{I}(0) = (1-p)I_0$$
$$\bar{I}(t) = R_0^{\text{muc}}(1-p)\bar{I}(t-1), t = 1, 2, 3 \ldots$$
$$\overline{L'}(t) = R_0^{\text{muc}} p\bar{I}(t-1)$$
$$\bar{L}(t) = pI_0 + \sum_{i=1}^{t} \overline{L'}(i)$$

where the ensemble averaging is denoted by overline, and $\overline{L'}(t)$ is the fraction of latent cells born in generation t. The initial latent cells, $\bar{L}(0) = pI_0$, come directly from the infecting virus. Iterating these relations, one gets

$$\begin{aligned}\bar{I}(t) &= I_0(1-p)a^t \\ \bar{L}(t) &= I_0\, p\left[1 + a\frac{1-a^{t-1}}{1-a}\right] \\ a &\equiv R_0^{\text{muc}}(1-p)\end{aligned} \tag{3.29}$$

which is smaller than unit, because $R_0^{\text{muc}} < 1$. From these expressions, we can obtain the number of surviving infected cells $\bar{I}(t) + \bar{L}(t)$ (Fig. 3.6E). At infinite time, the average number of surviving infected cells is given by

$$L_{\text{init}}^{R_0>1} = I_0 \frac{p}{1-a} \tag{3.30}$$

These cells initiate systemic infection.

In addition, the extinction time of actively infected cells $t_{\text{act\_extinct}}$ is estimated from the condition $\bar{I}(t_{\text{act\_extinct}}) = I_{\text{th}}$. To match Gillespie simulation results (Fig. 3.6A, B), the extinction threshold I_{th} is chosen to be 0.3 cell. As a result

$$t_{\text{act extinct}} = \frac{\log\left(\frac{I_0}{I_{\text{th}}}\right)}{\left|\log\left[(1-p)R_0^{\text{muc}}\right]\right|} \tag{3.31}$$

If latent probability p is very small, all cells can get extinct with an appreciable probability. The time when it happens is found from the condition $\bar{I}(t) + \bar{L}(t) = I_{\text{th}}$.

$$t_{\text{all extinct}} = \frac{\log\left(\dfrac{\frac{I_{\text{th}}}{I_0} - \frac{p}{1-a}}{1 - \frac{p(2-a)}{1-a}}\right)}{|\log a|}$$

where a is defined in eq. (3.29) and is much smaller than unit. When p approaches the threshold value equal to $I_{\text{th}}/I_0 \ll 1$, this time diverges to infinity, because the latent cells number surviving active infection exceeds the extinction threshold I_{th}.

3.2.8 Two-compartment model ($R_0^{\text{muc}} < 1$ and $R_0^{\text{LT}} > 1$)

Next, to understand the time delay between transmission to the mucosa and the systemic infection in lymphoid tissue (LT), a two-compartment model is considered (Fig. 3.6F–I). As compared to the simulation in Fig. 3.6A, B, the transfer between compartments and reactivation of latent cells, which occurs before the full extinction of actively infected cells in mucosa, are now included explicitly.

For simplicity, time scale separation is used. In the initial phase of the systemic infection in lymphoid tissue (LT), the virus load is still very low, and target cells $T(t)$ are at their uninfected level. As above in Section 3.2.7, short-lived virions are not considered explicitly. The model, whether deterministic, eqs. (3.20), or stochastic, eqs. (3.28), can be simplified to exclude variables $T(t)$ and $V(t)$ and keep p, r, R_0^{muc}, and R_0^{LT} (Tab. 3.2) as the only input parameters. The resulting Markov process with discrete generations has a form

$$\begin{aligned} I_{\text{muc}}(t+1) &= \text{Poisson}\left[(1-p)R_0^{\text{muc}} I_{\text{muc}}(t)\right] \\ L_{\text{LT}}(t+1) &= L_{\text{LT}}(t) + \text{Poisson}\left[pR_0^{\text{muc}} I_{\text{muc}}(t) + pR_0^{\text{LT}} I_{\text{LT}}(t)\right] - L_{\text{react}} \\ I_{\text{LT}}(t+1) &= \text{Poisson}\left[(1-p)R_0^{\text{LT}} I_{\text{LT}}(t)\right] + L_{\text{react}} \end{aligned} \tag{3.32}$$

Here I_{muc} and I_{LT} are numbers of virus-producing cells in the two compartments and L_{LT} is the latently infected cell number in the lymphoid tissue. Poisson[X] is a random integer chosen from a Poisson distribution with average X.

In eqs. (3.32), actively infected cells are assumed to remain in their respective compartments. In contrast, latently infected cells (resting memory CD4 T cells) can circulate freely between local mucosa and LT (Murphy 2011). The number of latent cells reactivating in LT is given by

$$L_{\text{react}} = \text{Poisson}[r(t)L_{\text{LT}}(t)]$$

Here $r(t)$ is the reactivation rate, which, as discussed above, can depend on time. For example, the reactivation of CD4 T cells can occur via T-cell receptor when exposed to macrophages and dendritic cells, which express HIV peptides in MHC-II context and migrate from the mucosal infection site (Murphy 2011).

The form of $r(t)$ is chosen to be step-like to generate a 5–7 delay observed in an animal host (Haase 2011). Specifically, rare activation of latent cells is assumed to occurs at maximal rate $r(t) = r$ during mucosal infection (Tab. 3.2), and then $r(t) = 0$ for the period of several days, until the adaptive immune response kicks in, as described by eqs. (3.26). In real life, $r(t)$ has a finite value $r(t) < 10^{-3}$ which causes eventual cell reactivation even after years of therapy (Tab. 3.2). However, on the time scale of weeks considered here, this fact is negligible. The transfer of virions or actively infected cells is considered in Section 3.2.10.

The Wright–Fisher model for two compartments was simulated in MATLAB™ using the "broken-stick" method (Macarthur 1957) to generate Poisson-distributed random numbers around the respective average values (Fig. 3.6F–I). Parameter values are given in Tab. 3.2. This simulation explains why the peak of viremia, which occurs at 10 – 12 days post-transmission, has a rather well-defined timing, despite very stochastic nature of mucosal infection. According to simulation (Fig. 3.6F), although latent cells can be reactivated anytime with the same probability during the initial period, the cells that are activated the latest have the highest chance to survive until reaching LT rich in target cells, $R_0^{\mathrm{LT}} > 1$. Therefore, the virus expansion after transfer to LT (red lines in Fig. 3.6F) follows the extinction of mucosal infection tightly.

Let us now return back to deterministic equations, eq. (3.20), and calculate the average inoculum per unprotected sexual encounter I_0 proportional to the average virus load

$$I_0 = \mathrm{const}(p) \frac{1}{t_{\mathrm{inf}}} \int_0^{t_{\mathrm{inf}}} dt\ V(t) \tag{3.33}$$

In eq. (3.33), const(p) does not depend on p but may depend on other model parameters, and $V(t)$ is calculated numerically from eqs. (3.20), with some initial condition $V(0)$, which only shifts the process in time. In this basic model without the immune response, the peak and the steady-state values of $V(t)$ are comparable, but the average steady state is much longer, ~ 10 years, and hence dominates the integral in eq. (3.33).

Because the net transmission probability p_{trans} is very small (Fraser et al. 2007), its value can be approximated by

$$p_{\mathrm{trans}} \approx p_{\mathrm{estab}} I_0 \tag{3.34}$$

Here $p_{\mathrm{estab}} = (L_{\mathrm{init}}^{R0>1}/I_0)\ p_{\mathrm{react}}$ is the probability of infection transfer from mucosa to LT, which is the product of the latent cell fraction in mucosa $L_{\mathrm{init}}^{R0>1}/I_0$ and the probability of cell reactivation, p_{react}. Assuming that reactivation probability does not depend on p (we relax this assumption for the coupled-compartment model considered below), and $R_0^{\mathrm{muc}} \ll 1$, p_{estab} is proportional to the number of latent cells formed in mucosa. Calculating $V(t)$ numerically from eqs (3.20), we obtain normalized $p_{\mathrm{trans}}(p)$ from eqs. (3.33) and (3.34) (Fig. 3.5C).

We can also obtain $p_{\mathrm{trans}}(p)$ analytically. From eqs. (3.20), assuming $r \ll d_L$ (Section 3.2.10), steady-state viremia is given by

$$V = \frac{d_T}{k}\left[(1-p)R_0^{LT} - 1\right] \tag{3.35}$$

From here and eq. (3.33), we arrive at eq. (3.23) for I_0. Assuming $R_0^{\mathrm{muc}} \ll 1$, the final level of latent cells formed in mucosa is approximately linearly proportional to p, eq. (3.30). (Finite value R_0^{muc} creates a correction in p_{estab}; for example, $R_0^{\mathrm{muc}} = 0.25$ increases it by 14%.) Hence, from eq. (3.34), $p_{\mathrm{trans}}(p)$ is the quadratic dependence on p, eq. (3.24), which is almost identical to the numeric dependence (Fig. 3.5C). The latency probability is maximal at the value given by eq. (3.25).

As we show in Section 3.2.10, a similar result for $p_{\text{trans}}(p)$ is obtained when the model includes the immune response against HIV transmission, steady-state viremia is low, and transmission often occurs during the acute peak of infection. In general, the form of $p_{\text{trans}}(p)$ is quite robust to variations of the model (Section 3.2.10).

3.2.9 Parameters sensitivity analysis

The next task is to examine robustness to parameter values.

HIV demographics in early mucosa ($R_0^{\text{muc}} < 1$). The central assumption $R_0^{\text{muc}} < 1$ during the first 5 – 6 days of infection is derived from the small probability of HIV infection and experimental data in Miller et al. (2005); Haase (2011). Miller et al. (2005) inoculated female macaques with a 1mL viral dose with the high concentration of 10^9 RNA copy/mL (during a typical transmission, it is ~10^5 RNA copy/mL sperm). The authors failed to detect any consistent evidence for active infection in the interval 0 – 5 days postinfection (see their fig. 1A). Following day 6 and on, a local RNA expansion in mucosa is observed, concurrent with the RNA expansion in lymph nodes, implying $R_0^{\text{LT}} > 1$ starting from that time (Miller et al. 2005, fig. 1A–C). Therefore, our compartment termed "mucosa" means early mucosa, while our compartment termed "LT" (Fig. 3.3) includes late mucosa, lymph nodes, spleen, and gastrointestinal LT.

Large early latent reservoir, and the estimate of inoculum I_0 from the count of HIV DNA$^+$ cells in early mucosa. Experiments demonstrate the existence of a large latent compartment in mucosa early after infection. The observed frequency of SIV DNA$^+$ cells inside a mesenteric lymph node of a monkey on day 3 postinfection is 200 per 10^6 CD4$^+$ cells (Whitney et al. 2014, fig S5). Estimating CD4 T-cell density at 300 CD4 cell/mkg by extrapolating from mice to monkeys (Zhang et al. 1998) for a 2 g lymph node in a monkey which weighs 10 mg in a mouse (Kim et al. 2008), latent reservoir in monkeys is $L_{\text{init}}^{R_0>1} = 10^5$ SIV DNA$^+$ cells. This results corresponds to a high dose of virus, 1ml supernatant with the concentration 10^9 SIV RNA/ml. A typical transmission has a 10^4-fold less concentrated virus, hence, the result rescales down to $L_{\text{init}}^{R_0>1} \sim 10$ cells, which corresponds to our choice of inoculum $I_0 = 20 - 30$ cells in our simulations (Figs. 3.5 and 3.6).

Parameter choice and parameter sensitivity in LT. For the basic model (Figs. 3.3 and 3.5), typical parameters from the literature are given in Tab. 3.2. The prediction for optimal latency, eq. (3.25), is affected by only one composite parameter, R_0^{LT}. The term in eq. (3.25) depending on R_0^{LT} represents a small correction (Fig. 3.5C), with the measured range $R_0^{\text{LT}} = 5$ to 20 (Nowak et al. 1997). Other parameters, such as death rates d_T, d_I (Tab. 3.2) affect the shape of the acute viremia peak, but not the value of p_{opt}. For the extended model with the immune response, eqs. (2.27), additional parameters are defined in Tab. 3.3. Out of the seven parameters, four parameters (d_E, E_0, E_{0L}, I_{av}) are adjusted to fit the four observed plateaus (Fig. 3.8A), while the other three are fixed and taken from the literature. The predicted steady-state levels of E, I, and L are compared to their experimental estimates (Fig. 3.8A): $E = 10^9$ cells (Ogg et al. 1998; Turnbull et al. 2009), $I = 10^8$ (Haase 1999), and $L = 10^6$ (Chun, Carruth et al. 1997). This procedure allows to estimate fitting parameters E_0, E_{0L}, I_{av}. The remaining fitting parameter d_E is estimated from the value of L under ART, $L = 10^5$ cells (Finzi et al. 1997). For both CD8⁺ T and CD4⁺ T cells in uninfected individuals, the total cell count is $T(0) = b/dT = 2 \times 10^{11}$ (Murphy 2011).

3.2.10 Robustness to model variations

To test the robustness to model changes, the model architecture was modified in four ways: (i) extended the model to include the immune response, as discussed in detail, (ii) took into account the contribution of acute phase to transmission, (iii) allowed saturation of the transmission rate in the doze, and (iv) included actively producing cells into the mucosa-lymph transfer. As shown below, although dynamics of infected cells is generally sensitive, the form of the dependence of the transmission probability on p either changes little or is skewed toward even larger p, provided model parameters stay within the order of magnitude shown in Tab. 3.2. Model variation affects only the constant prefactor in eq. (3.24).

High latent cells and sensitivity to r/d_L. For the sake of simplicity, inequality $r \ll d_L$ was assumed above. In real life, the two parameters are within the same range, and both $r < d_L$ and $r > d_L$ are possible (Tab. 3.2). In the more general case, as can be shown from eqs. (3.20), the steady-state virus load is given by

$$V = \frac{d_T}{k}\left[R_0^{LT}\frac{1-p+\dfrac{r}{d_L}}{1+\dfrac{r}{d_L}} - 1\right]$$

Thus, the virus load in steady state V can be sensitive to r/d_L even though both r and d_L are small. As a result, in the opposite extreme scenario, $r \gg d_L$, eq. (3.23) for $I_0(p)$ loses factor $1-p$ before R_0^{LT}, so that $I_0(p)$ calculated in the basic model becomes independent on p. As a result of the change, the optimum value for the initial p_{opt} becomes 1 instead of 0.5. Thus, p_{opt} is bound the interval $[0.5,\ 1]$ depending on the ratio $r/d_{L.}$

However, this sensitivity of V to r/d_L is an artifact of the basic model, eqs. (3.20). As it is well-known, a model without the immune response cannot explain a high, variable peak-to-steady-state ratio for $V(t)$. Another artifact is a very high number predicted for latent cells, L. The steady-state values for T and L are

$$T = \frac{b_{LT}}{R_0^{LT} d_T}\frac{1+\dfrac{r}{d_L}}{1-p+\dfrac{r}{d_L}}$$

$$L = \frac{b_{LT}p}{(r+d_L)R_0^{LT}}p\left[R_0^{LT} - \frac{1+\frac{r}{d_L}}{1-p+\frac{r}{d_L}}\right]$$

Therefore, at $p \sim 0.5$, the ratio $L/T \sim R_0^{\mathrm{LT}} d_{\mathrm{T}} / (d_{\mathrm{L}} + r) \sim 10^3 - 10^4$ (Tab. 3.2). The unrealistically high L (and V) predicted by the basic model, eqs. (3.20), indicate that it is not able to explain the low levels of latency in chronic infection (Fig. 3.8A), as we already discussed above. Hence, it must be upgraded to a model including immune response, eqs. (3.26) and (3.27), as it was done.

Factors affecting the transmitted dose. The doze was assumed to be proportional to the average virus load in steady state. The real infectious doze depends on various epidemiological factors, such as an infection risk group and the variation of host's infectivity depending on infection stage (Baggaley et al. 2006). Here only high-risk groups of hosts that dominate propagation of HIV are taken into account. Next, a course of an HIV infection comprises a highly infectious acute phase with a tall viremia peak (1–2 months) and a much longer but much less infectious chronic stage (~100 months). In high-risk groups, the acute phase is responsible for a half of transmissions (Wawer et al. 2005; Fraser et al. 2007). A similar contribution from acute phase is expected in a natural host population.

Basic model and acute-stage transmission. Consider now the contribution of the acute phase to the transmission [see the time integral in eq. (3.31)], as a function of p. The initial expansion slope $d\log I(t)/dt$ is equal to the initial reproduction number $(1-p)R_0^{\mathrm{LT}}$, which is smaller than the raw value in the absence of latency, R_0^{LT}, due to the diversion of infected cells from virus production. As the population of infected cells expands, uninfected target cells T are depleted proportionally to the viral load (Rouzine et al. 2015, fig S3A). The expansion is stopped, and the viral load reaches its maximum, when the reproduction ratio is decreased from the initial value $(1-p)R_0^{LT}$ to 1 due to this depletion. To stop virus expansion, the target cells must be depleted by a factor of $(1-p)R_0^{LT}$. Since the degree of depletion is proportional to the virus load, the virus peak height must be proportional to $1-p$. Using this proportionality in eq. (3.33), where the time integral is contributed mostly from the peak region, one again arrives at eq. (3.23) for I_0 and at the same result for p_{opt} as for the chronic-phase transmission, eq. (3.25).

Immune response and acute-stage transmission. In the model including the immune response, eqs. (3.26) and (3.27), CD8 T cells becomes numerous after the viremia peak and lower the steady-state virus load by orders of magnitude below the value predicted by eqs. (3.20) of basic model (Rouzine et al. 2015, fig. S3A). However, the shape of the summit of viremia peak, which determines I_0 in the time integral in eq. (3.33), is barely changed by the immune response. This is because the height of the peak $I(t)$ is limited by the depletion of target cells, which occurs right before the immune cells rise to prominence. The shape of the summit is also robust, because the decay $I(t)$ after the summit is determined by the lifetime of the eclipse phase cells I_E, eqs. (3.27) (Klenerman et al. 1996;

Rouzine et al. 2006). Hence, assuming the main role of the acute phase in transmission, the result for the transmission rate as a function of p is almost the same as for the basic model, eqs. (3.20). Numeric simulation confirms this analytic estimate. Equations (3.26) and (3.27) are computed, and the time integral in eq. (3.33) is evaluated over the time interval of 20 days centered at the viremia peak. The result for $p_{\text{trans}}(p)$ (Fig. 3.8B) is similar to that with no immune response (Fig. 3.5C).

Immune response and mixed acute-chronic-stage transmission. Suppose, a half of transmission comes from acutely infected individuals and another half from the chronic phase (Fraser et al. 2007). Assume also that the preimmune-response value $p(0)$ is at the evolved optimum, p_{opt}. The immune response is present, as in eqs. (3.26) and (3.27). As we just discussed, the acute infection part of the integral for the average dose in eq. (3.33) is proportional to $1 - p$, but the steady-state virus load is pinned near the CTL avidity threshold, $I \sim I_0$, and does not depend on p. Hence, the bracketed expression in eq. (3.23) for the dose I_0 is replaced with

$$\frac{R_0^{\text{LT}}(1-p)-1}{2}+\frac{R_0^{\text{LT}}(1-p_{\text{opt}})-1}{2} \tag{3.36}$$

The second term is due to steady state, which is constant in p and calculated from the condition that both terms are equal at $p = p_{\text{opt}}$. The best net transmission rate $p\ I_0(p)$ will be attained at $p = p_{\text{opt}} = (2/3)(1 - 1/R_0^{\text{LT}})$, which is larger than our base result, eq. (3.25). Thus, the inclusion of the two infection stages into transmission in the presence of the immune response will only increase the optimum latency rate.

Nonlinear dependence of the transmission rate on the viremia. The probability of transmission was assumed to be linear in infected cell counts, eq. (3.33). In real life, it may saturate at high viremia levels, as it does in HIV-status discordant couples (Fraser et al. 2007) and the other high-risk groups of individuals, where transmission rates are much higher (May 2004). Again, this factor may only increase p_{opt}. A slower dependence of transmission rate on viremia would translate to a slower dependence of transmission rate on p, so that p_{opt} would only increase. Thus, as long as we assume that only latently infected cells seed HIV infection in lymph from mucosa, the prediction of large optimal p remains robust to the other assumptions. What happens if this central assumption is relaxed?

Transmission in the presence of non-latent virus transfer (Fig. 3.6J). Suppose, actively infecting cells (including CD4 T cells and dendritic cells) can also seed infection in lymph. Regardless of a model variant considered above, the expression for the effective transmission rate, eq. (3.21), takes a more general form

$$p_{\text{trans}} = p_{\text{estab}}(p)\ I_0(p) \tag{3.21}$$

$$p_{\text{estab}}(p) = \text{const}(p)\ [(1-p\ )f_{\text{act}} + (1-f_{\text{act}})p] \tag{3.37}$$

where p is the value before the immune response, $p \equiv p(0)$, and $I_0(p)$ is given by eq. (3.35). The first term in eq. (3.37) is any nonlatent virus transfer proportional to the probability of nonlatency, $1-p$, and a new constant, f_{act}. By definition, f_{act} is the relative fraction of nonlatent transfers at $p=0.5$. At fixed f_{act}, we calculate the new optimum in p

$$p_{opt}=\frac{1}{2}\left(1-\frac{1}{R_0^{LT}}-\frac{f_{act}}{1-2\,f_{act}}\right) \tag{3.38}$$

As we see from eq. (3.38), the non-latent transfer decreases p_{opt} compared to the case $f_{act}=0$. The optimal value p_{opt} given by eq. (3.38) is positive at

$$f_{act}<\left(1-1/R_0^{LT}\right)/\left[1+2\left(1-1/R_0^{LT}\right)\right]\approx 1/3$$

However, f_{act} is not directly observable. The observable parameter is the fraction of non-latent transfer rate f_{nonlat} obtained as the relative weight of the first term in eq. (3.37) at $p=p_{opt}$

$$f_{nonlat}=\frac{\left(1-p_{opt}\right)f_{act}}{\left(1-p_{opt}\right)f_{act}+p_{opt}\left(1-f_{act}\right)} \tag{3.39}$$

As one check, $f_{nonlat}>f_{act}$. As f_{act} increases to about $1/3$, p_{opt} vanishes, and the optimal transfer switches completely to the active-cell component, $f_{nonlat}=1$. Until then, the optimal latency probability is not zero (Fig. 3.6J). For example, for $f_{nonlat}=0.9$, which is 90% of active transfer and 10% of latent transfer, eqs. (3.38) and (3.39) predict $p_{opt}\sim 0.05$. Thus, to render latency useless to virus, active virus must completely dominate the transfer between mucosa and LT.

Dependence of establishment probability p_{estab} and reactivation probability p_{react} on p in decoupled and coupled models (Fig. 3.6G–I). In the main derivation, the probability of reactivation of a latent cell in LT, p_{react}, was assumed fixed and did not depend on p. This is because the basic ODE model, eqs. (3.20), considers two compartments separately (Fig. 3.5). Hence, it does not describe the dynamics of reactivation of latent cells explicitly, assuming a single reactivated cell in LT.

To test the effect of dynamic coupling of two compartments more explicitly, a Wright–Fisher simulation of the stochastic coupled model (Section 3.2.8) was carried out. The result is that the value of p_{react} decreases with p by 60% compared to its maximum at $p=0$ (Fig. 3.6G). Indeed, the reactivation of latent cells occurs all the time during mucosal infection. As latency parameter p increases, the active infection burns out more rapidly, so that the time of activation decreases. This effect causes a modest decrease in p_{opt} from 0.45 (Fig. 3.5C) to 0.35 (Fig. 3.6I). Again, the prediction of large p_{opt} is robust.

The peak of latent cells is sensitive to the details of latency control by the immune response. Unlike a large value of p_{opt}, the other predictions are sensitive to model variations. For example, the latent cell peak in acute infection in LT (Fig. 3.8A) is sensitive to the parameters of the latency control by CD8 T cells, eqs. (2.26) (data not shown). The important latency parameters include: (i) minimal possible value of $p(E)$ (currently set at zero); (ii) maximum of $r(E)$ (set at d_I = 1/day), (iii) the characteristic level of CD8 T cells E_{0L}, at which CD8 T cells reactivate latent cells significantly (assumed $E_{0L} = 4 \times 10^6$ cells per human), and (iv) inhomogeneity of latent cell population due to methylation and the variation in the HIV gene integration site (Dar et al. 2012). Especially important are latent cells resisting activation (Eisele and Siliciano 2012; Ho et al. 2013). Therefore, it is difficult to make a positive prediction regarding the size (or even the existence) of the peak of latent cells during acute infection.

Simplified immune models fail to predict the realistic viral dynamics. To test whether the extended immune model in eqs. (2.26), (2.27) is the simplest useful model, two even simpler versions were investigated. In the first version, cells in the eclipse phase, I_E, were eliminated by replacing the first three equations in eqs (3.27) with only two

$$\frac{dT}{dt} = b - d_T T - kVT$$

$$\frac{dI}{dt} = kVT - d_I I\left(1 + \frac{E}{E_0}\right)$$

The removal of eclipse phase leads to a few unobserved artifacts, including the sharp decrease in infected cells at the onset of ART within less than a day, by more than 10-fold (Rouzine et al. 2015, fig. S3B, C). Nothing of a sort is observed in real patients, where viremia decays smoothly at the rate of ~ 1/day (Fig. 3.8A inset).

In another simple version of the model, the first equation in eqs. (3.27) was eliminated, and T was set at a fixed value for uninfected patients, $T = b/d_T$. In acute infection, this version predicts an extremely high peak of infected cells overshooting the normal level of target cells, followed by strong oscillations of virus load (Rouzine et al. 2015, fig. S3D). This feature is not observed.

3.3 Recombination and the optimal mutation rate of polio virus

As discussed in Chapter 2, genetic evolution occurs due to the combined action of several factors. New genetic variants are created by random mutation and recombination of different genomes. Their subsequent dynamics is decided by the factors of natural selection including epistasis (Chapter 2), genetic drift, and linkage effects (clonal interference, genetic background, and enhanced accumulation of deleterious alleles). Random mutation carries costs and confers benefits to an organism. While a higher mutation probability (mutation rate) allows faster adaptation, it also produces larger numbers of deleterious mutations. In keeping with this fact, modeling studies in multi-locus evolution have predicted the existence of an optimal mutation rate (Rouzine et al. 2003; Gerrish et al. 2013). Theoretically, the role of recombination is thought to decrease the negative effects of linkage, accelerate adaptation, and eliminate deleterious mutations (Muller 1932; Fisher 1958; Maynard Smith 1971; Felsenstein 1974; Otto and Barton 1997; Rouzine and Coffin 2005).

RNA viruses adapt rapidly to the environment, which fact is linked to many diseases and their resistance to antiviral drugs and the immune response (Holland et al. 1982; Steinhauer and Holland 1987; Elena and Sanjuan 2005b). Therefore, understanding the acting factors of virus adaptation are critical to the development of antivirals and vaccines, as well as to understanding the mechanism of virulence. The adaption of a population occurs by random mutation and subsequent selection of new beneficial alleles (Domingo and Holland 1997; Elena and Sanjuan 2005b; Hartfield et al. 2010). High mutation rates of RNA viruses compared to DNA organisms are one of the reasons for their evolutionary plasticity (Domingo and Holland 1997; Elena and Sanjuan 2005a). At the same time, most mutations in a well-evolved virus are deleterious (Domingo et al. 1996; Holmes 2009; Barlukova and Rouzine 2021). Theoretical and experimental studies demonstrate that, in populations lacking recombination, beneficial alleles compete with each other creating the effect of clonal interference (Fisher 1930; Muller 1932; Fisher 1958; Gerrish and Lenski 1998; Worobey and Holmes 1999; Cooper 2007; Hartfield et al. 2010). At the same time, accumulation of deleterious mutations is enhanced by linkage, the process known as "Muller's ratchet" (Felsenstein 1974; Chao 1990; Duarte et al. 1992). Interference of deleterious alleles with the amplification of linked beneficial alleles is another effect of linkage (Roze and Barton 2006; Hartfield et al. 2010). Experimental and theoretical studies demonstrate that recombination partly alleviates these negative effects of linkage, by combining beneficial mutations within the same genome and filtering out deleterious mutations (Zeyl and Bell 1997; Worobey and Holmes 1999; Burch et al. 2003; Bonhoeffer et al. 2004; Rouzine and Coffin 2005; Roze and Barton 2006; Cooper 2007; Neher et al. 2013).

Poliovirus (PV) is a positive-sense (similar to mRNA in a cell), single-stranded RNA virus that causes poliomyelitis (Racaniello and Baltimore 1981; Agol 2006). As it is known from tissue culture experiments, poliovirus genome undergoes frequent recombination mediated by template switching during replication (Kirkegaard and Baltimore 1986; Runckel et al. 2013; Lowry et al. 2014). The steps of poliovirus recombination include the premature termination of RNA synthesis, disassociation of the polymerase-nascent strand complex from the template, its re-association to another genome, and then completion of replication, resulting in a chimeric genome. In agreement with this scheme, the nucleotide homology between parental sequences is important for the recombination frequency and location (Kirkegaard and Baltimore 1986; Sztuba-Solinska et al. 2011).

The effects of mutation and recombination combined with natural selection and random drift have been studied in multi-locus models (Maynard Smith 1971; Gerrish and Lenski 1998; Rouzine et al. 2003; Rouzine and Coffin 2005; Brunet et al. 2007; Gheorghiu-Svirschevski et al. 2007; Rouzine and Coffin 2007; Brunet et al. 2008; Rouzine et al. 2008; Rouzine and Coffin 2010; Hallatschek 2011; Good et al. 2012; Walczak et al. 2012; Neher and Hallatschek 2013), and so was the synergy of mutation and recombination (Neher et al. 2010; Neher et al. 2013; Weissman and Hallatschek 2014). However, the interaction between these factors is insufficiently studied experimentally, in particularly, during an acute viral infection.

This section is based on a study, which conducted experiments on virus variants with different mutation and recombination rates in both cell culture and animals (mice) and then used mathematical modeling to interpret them (Xiao et al. 2016). Because the results of experiments are transparent for a nonvirologist, their results will be described below in full. The details of the protocols can be found in the original work (Xiao et al. 2016).

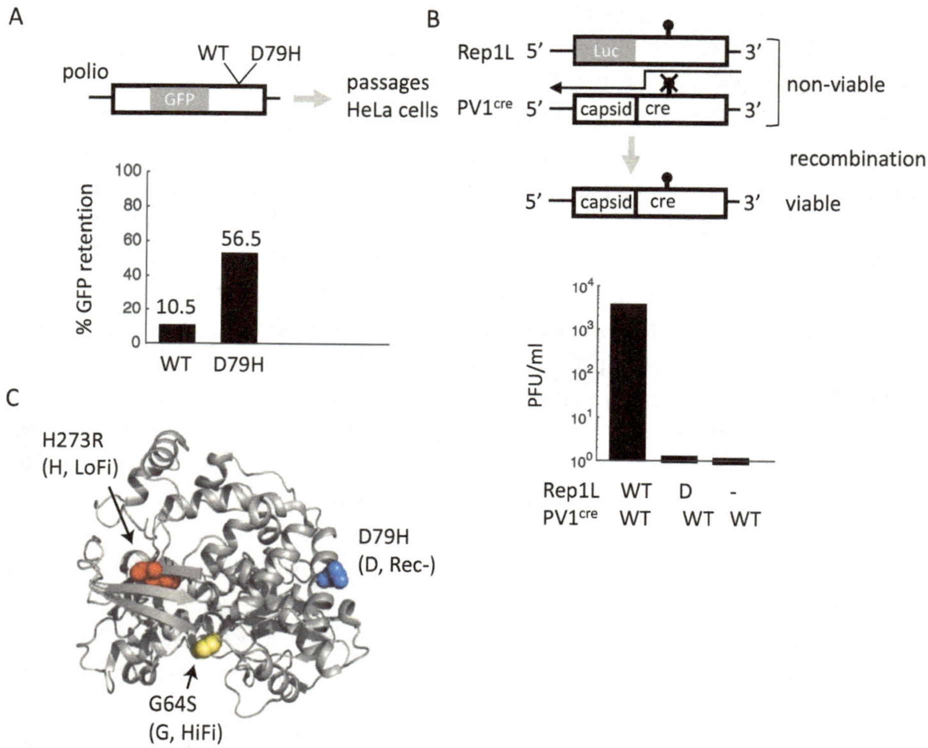

Fig. 3.11: Isolation of a recombination-deficient poliovirus variant.
(A) GFP was cloned into poliovirus genome, the resulting recombinant virus was used to infect HeLa cells and GFP retention was measured between passages by limited dilution. A mutation, D79H within $3D^{pol}$ confers an increase in GFP retention. **(B)** CRE-REP recombination assay (Lowry et al. 2014) confirmed that D79H mutation reduces the recombination rate. *In vitro* transcript (IVT) RNAs, $PV1^{cre}$, Rep1L, were cotransfected into HeLa cells. $PV1^{cre}$ contains a mutant *cis*-acting replication element (CRE), which prevents positive strand viral RNA synthesis. Sub-genomic replicon (Rep1L) does not encode structural protein, therefore, neither of them produce viable progeny. Following co-transfection viable progeny is produced only if recombination of the two defective IVT RNAs takes place at any site between the structural proteins (capsid) and CRE. Introduction of the D79H (denoted also D) mutation into Rep1L dramatically reduces the

The team produced virus variants with a high and low mutation rate, and a high and low recombination rate, in any combination. The accumulation of beneficial and deleterious mutations in culture was measured by a precise deep-sequencing approach. This information was compared with the viral adaptation in an animal host (mice), as measured by tissue tropism and animal mortality. The experimental results on virus adaptation were then interpreted with a computational model of stochastic multi-locus evolution (Batorsky et al. 2011). Using that model, the previous analytic predictions for the long-term asexual evolution (Rouzine et al. 2003; Gerrish et al. 2013) were generalized for the short-term viral evolution during an acute infection.

3.3.1 Experiments on strains with altered recombination and mutation rates

3.3.1.1 Mutation D79H decreases recombination rate 10-fold

The mutation rate of poliovirus depends on its polymerase protein (Steinhauer et al. 1992; Vignuzzi et al. 2006). A recombination-deficient mutant of polymerase that replicates well in a cell culture was identified. To its genome, by means of recombination, the green fluorescent protein (eGFP) was added to visualize the cells infected with this variant. The eGFP sequence was inserted between the structural and nonstructural coding regions separated by artificial 2A protease cleavage sites (Fig. 3.11A). This construct was used to infect a cell culture (HeLa cell line), virus was passed several times between plates, and the viral clones that did not delete the eGFP coding sequences were selected. Then, the process was repeated until a virus variant that retained eGFP for a much higher number of passages than the wild type virus was obtained [for details see Xiao et al. (2016)]. The genome of that variant had a single amino-acid substitution within the viral polymerase protein (RdRp), specifically, an aspartate to histidine mutation at the aminoacid position in that protein with number 79, denoted D79H. The probability that virus deletes GFP from its genome after 10 passages was quantified. The variant carrying the D79H allele retained the GFP code in 56% of the virus clones compared to only 10% retained by the initial strain, wild type (Fig. 3.11A).

To confirm by another method that D79H decreases poliovirus recombination, the process proposed by Lowry et al. (2014) was used as well. In this assay, two defective viral RNA sequences, denoted $PV1^{cre}$ and Rep1L, can be used to estimate recombination rate, as follows. $PV1^{cre}$ contains a mutant domain (Goodfellow et al. 2003), which prevents positive-strand RNA synthesis. Another RNA defective viral sequence

Fig. 3.11 (continued)
number of recombinant viable progeny. The titer of viable progeny (TCID50/ml) was measured by the standard TCID50 assay (*Y*-axis). **(C)** The structure of RNA-dependent RNA polymerase of the type 1 poliovirus Mahoney strain. Colors show locations of mutations affecting fidelity and recombination. Red: low-fidelity mutant, H273R (denoted also LoFi), yellow: high-fidelity mutant, G64S (HiFi); blue: recombination-deficiency mutant, D79H (Rec-). Based on Xiao et al. (2016).

(Rep1L) does not encode any structural proteins. Therefore, neither construct alone can leave viable progeny. Only together, they can make viable virus. After co-transfection (a method of laboratory infection bypassing the use of cell-surface receptor normally necessary for virus entry) of both constructs into permissive cells, viable progeny is produced only by recombination of the two defective RNAs (Fig. 3.11B). Introduction of the D79H mutation into Rep1L dramatically reduces the viable progeny number (Fig. 3.11B bottom). Thus, these results by two independent methods demonstrate that D79H mutation significantly decreases the rate of recombination.

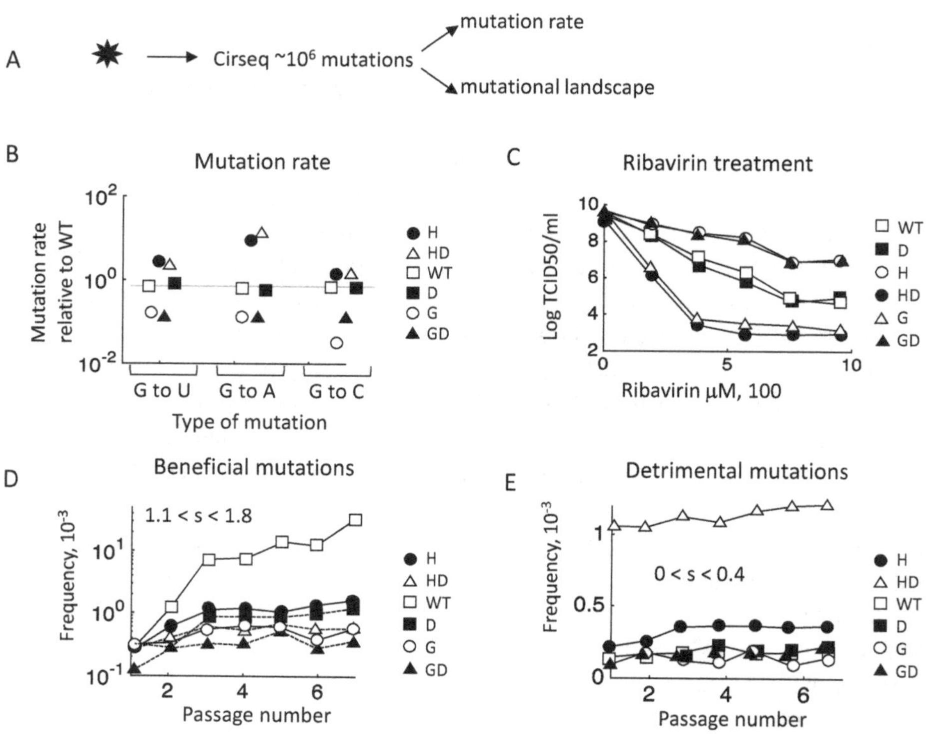

Fig. 3.12: Recombination-deficient mutation does not modify mutation rate but significantly affects the accumulation rate of beneficial and deleterious mutations.
(A) Series passage of viral strains in HeLa cells. For each passage, 10^7 monolayer HelaS3 cells in were infected by viral strains at MOI = 0.1, with population size 10^6 PFU. The next-generation CirSeq libraries were made followed the protocol, mutation rate and mutation landscape of viral strains were calculated (Acevedo and Andino 2014; Acevedo et al. 2014). Wild-type (WT), recombination-deficiency mutant D79H (denoted D), high-fidelity mutant G64S (denoted G), low-fidelity mutant H273R (H), and double mutants G64S/D79H (GD) and H273R/D79H (HD). **(B)** The frequency of mutations that result in ssense codons or catalytic site substitutions was used to measure mutation rates for each mutation type (Acevedo and Andino 2014; Acevedo et al. 2014). The frequency of deleterious mutations at mutation-selection balance is the mutation rate, μ, over the deleterious selection coefficient, s, defined here as the virus fitness (exponential growth rate) relative to WT minus 1. For lethal mutations, $s = -1$, thus, their frequencies equal the mutation rate. **(C)** RNA virus mutagen ribavirin sensitivity assay of viral strains. HelaS3 cells

The discovery of mutation D79H that prevents recombination offered a chance to study the simultaneous variation of recombination and mutation rate. For this purpose, D79H was introduced into a previously found high-fidelity mutant virus with mutation G64S and low-fidelity mutant H273R (Pfeiffer and Kirkegaard 2003; Vignuzzi et al. 2006; Korboukh et al. 2014). Mutation G64S increases polymerase fidelity, that is, decreases mutation rate relative to the wild-type poliovirus (WT). In contrast, H273R confers to virus more frequent mutations. The three mutations are located at different sites of polymerase: G64S is situated at the "finger" of polymerase, H273R at its "palm," and D79H on its surface (Fig. 3.11C). In order to connect the adaptation rate to recombination and mutation rates, the team compared double mutants G64S/D79H (denoted for brevity GD) and H273R/D79H (HD) with single mutants G64S (G, HiFi), D79H (D, Rec-), and H273R (H, LoFi). Together with wild type, these five mutants form a set of six strains that correspond to any combination of the normal or low recombination rate with the normal, high, or low mutation rate.

3.3.1.2 Mutation D79H does not alter mutation rate

The first task was to test whether mutation D79H turning off recombination has the secondary effect of affecting mutation rate. Direct measurements of mutation rate are challenging, because the accuracy relies on observing rare mutation events (Sanjuan et al. 2010). The team used an indirect method (Acevedo et al. 2014) based on a high-fidelity sequencing tool, CirSeq (Fig. 3.12A). The idea is to observe the frequency of very deleterious mutations, which is well known to be on the order of the mutation rate. Each of the five mutants was passed seven times between plates with cells, keeping the multiplicity of infection (frequency of infected cells) low, MOI = 0.1. Populations resulting from each passage were analyzed using CirSeq (Acevedo et al. 2014). A large dataset covering more than 95% of possible mutations in the poliovirus genome was obtained. Confirming previous findings, polymerase fidelity mutations G64S and H273R displayed an increase (H273R) or decrease (G64S) in mutation rate for nucleotide replacements G to U, G to A and C to U (Fig. 3.12B) (Acevedo et al. 2014). In contrast, D79H did not change the mutation rate (Fig. 3.12B).

To confirm this result by another method, the sensitivity of the maximal virus amount in cell culture to single and double mutants was checked in the presence of different concentrations of ribavirin, an RNA virus mutagen (Crotty et al. 2001). The

Fig. 3.12 (continued)
were pretreated with the indicated concentrations of ribavirin (shown in *X*-axis) for 4 h. Cells were infected by viruses at MOI = 0.1 for 40 min and were covered with fresh ribavirin media for additional 24 hours. Viral production was measured by standard TCID50. **(D)** The accumulation of beneficial mutations across passages for viral strains. The beneficial mutations are in interval $1.1 < 1 + s < 2.3$. **(E)** The accumulation of deleterious mutations across passages for viral strains. Deleterious mutations in the range of fitness 0.6 to 1 of WT fitness are selected (average $s = -0.2$, $0 < |s| < 0.4$). In simulation below, we define s through relative fitness $\exp(s) - 1$, which is $\approx s$ for small s. Based on Xiao et al. (2016).

wild type is suppressed by high drug concentrations by 4–5 orders of magnitude, as opposed to the low-mutation mutant G64S that decreases by only two orders. This mutation not only decreases mutation rate, but also confers resistance to the mutagen (Pfeiffer and Kirkegaard 2003; Vignuzzi et al. 2006). As expected, high-mutation H273R variants were hypersensitive (Fig. 3.12C). What is important is that adding recombination-deficient mutation D79H to any of these strains did not have any effect on the virus sensitivity to mutagen. Thus, it was confirmed that mutation D79H decreases recombination rate, but does not change mutation rate.

3.3.1.3 Mutation D79H slows down viral adaptation in culture

The next task was to use CirSeq to measure the substitution rate of beneficial mutations caused by virus adaptation to a tissue culture. During serial passages, a steady increase in the frequency of beneficial mutations was observed only in the wild-type virus (Fig. 3.12D). The frequency of best-fit mutations (in the top 5% of fitness gain) increased, over 7 passages, by a factor of 100. No such increase was observed in the double mutants with a high or low mutation rate, HD or GD respectively, and only modest increase occurred for single mutants D79H (low recombination) and H273R (high mutation rate) (Fig. 3.12D). This striking result indicates that the ability of selection to adapt has optimally evolved in WT and can only be reduced either by alteration of polymerase fidelity in any direction or by the decrease in recombination. The combination of the two is especially detrimental for adaptation (Fig. 3.12D).

In contrast, accumulation of deleterious mutations is enhanced in high-mutation mutants (H and HD), especially, in the recombination-deficient variant with a high mutation rate (HD) (Fig. 3.12E). Such an enhanced accumulation of deleterious mutations at low recombination reveals a linkage effect called "Muller's ratchet" well studied before experimentally and theoretically (Felsenstein 1974; Chao 1990; Duarte et al. 1992; Rouzine et al. 2003; Rouzine et al. 2008).

3.3.1.4 Mutation D79H does not affect fitness in culture

Next, the effect of the single and double mutants on viral fitness was tested in cell culture. Cells (HeLaS3 line) were infected at a high multiplicity, MOI = 10, and virus was allowed to grow for one replication cycle, 8 h (Fig. 3.13A). No significant change in the replication speed from wild type was observed for any of the five mutant viruses, neither in virus particle production (Fig. 3.13B) not in viral RNA synthesis (Fig. 3.13C).

To make the test more sensitive, the relative fitness of pairs of mutants was measured using a growth competition assay (Clarke et al. 1994). Mutation D79H alone did not reduce replication fitness, regardless of the other mutations (Fig. 3.13D and 3.13E). However, the presence of either a high-fidelity G64S or low-fidelity H273R mutation was accompanied by a fitness reduction compared to WT regardless of presence of D79H (Lauring et al. 2012; Korboukh et al. 2014; Xiao et al. 2016, fig. S1).

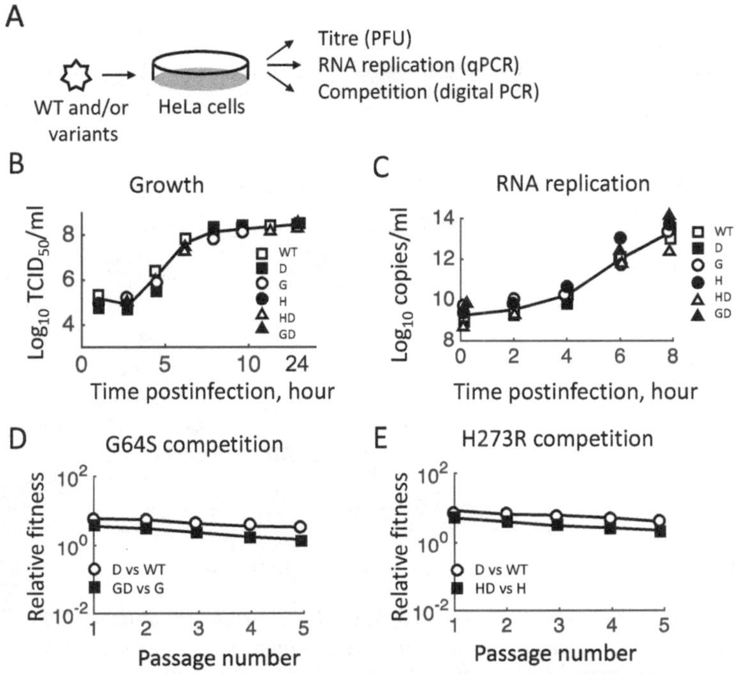

Fig. 3.13: Recombination-deficient mutation does not affect viral fitness in cell culture.
(A) Single-step growth curve in HelaS3 cells for viral strains: 5×10^5 monolayer HelaS3 cells were infected by WT or mutant strains at multiplicity of infection, MOI ~ 10. **(B)** Viral titer was measured by the standard TCID50 assay for single-step growth curve. Viral titers were shown TCID50/ml. **(C)** Copy numbers of the positive strand of viral RNA genome measured by qRT-PCR. **(D, E)** Growth competition assay of viral strains. Competition assay were performed on HelaS3 cells with each pair of viruses at total MOI = 0.01 as described in Xiao et al. (2016) (Experimental Procedures). RNA genome copies were detected with a pair of Taqman primers by digital droplet PCR (table S1 in Xiao et al. (2016). Relative fitness values of pairs of viral strains were measured as a function of passage numbers. D79H mutation does not affect fitness (fig. S1 in Xiao et al. (2016). Based on Xiao et al. (2016).

One can conclude that combining a recombination-deficient mutation with the mutations altering mutation rate does not observably decrease the virus fitness in culture. The availability of such mutations that reduce recombination rate and increase or decrease mutation rate but do not affect viral fitness directly, provided an opportunity to investigate experimentally how mutation and recombination interact driving adaptation in animals.

3.3.1.5 Double mutants impair viral adaptation in mice

When infecting an animal, virus is exposed to selection pressures caused by the host immune response (Section 4.2) and the need to adapt to various tissues and cell types (Whitton et al. 2005). Thus, an infection in vivo provides an ultimate test for the roles

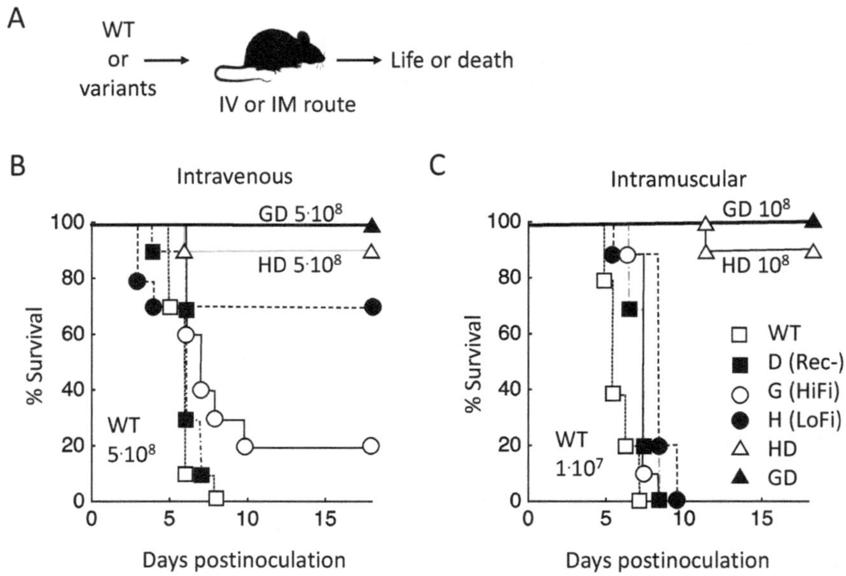

Fig. 3.14: Defects in both recombination and mutation rate significantly attenuates poliovirus virulence. **(A)** Survival of mice infected with WT or mutants by two routes was measured. **(B)** Percentage survival of susceptible mice infected through the tail vein (IV) with $5 \cdot 10^8$ PFU with wild-type virus (WT), double mutant viruses, including high fidelity/recombination-deficient (G64S/D79H, denoted GD) and low fidelity/ recombination-deficient variant (H273R/D79H, denoted HD), and single mutant viruses, including D79H (denoted D or Rec-), G64S (G or HiFi), and H273R (H, LoFi). **(C)** Percentage survival of susceptible mice infected by intramuscular route (IM) with 10^7 PFU per mouse for WT, D79H, G64S). Susceptible mice were infected with 10^8 PFU per mouse for H273R/D79H (HD) and G64S/D79H (GD). Based on Xiao et al. (2016).

of mutation and recombination in viral adaptation. The 5 mutants and wild type were used to infect mice through two routes: a systemic infection by intravenous injection or by a more local intramuscular route that leads to the central nervous system (CNS) (Fig. 3.14A).

Tab. 3.4: 50% lethal doses for intramuscular inoculation.

Mutant	Fidelity	Recombination	LD_{50}
WT	Wild type	Wild type	1.0×10^6
D	Wild type	Reduced	2.3×10^6
G	High	Wild type	2.3×10^6
H	Low	Wild type	1.9×10^6
GD	High	Reduced	$>1.0 \times 10^8$
HD	Low	Reduced	$>1.0 \times 10^8$

After intravenous infection, wild-type virus rapidly propagated through an animal, invaded CNS, and killed all mice within 6 days (Fig. 3.14). Recombination-defective strain D79H was just as lethal, and the low-mutable strain G64S also killed 80–100% of mice (Fig. 3.14B and 3.14C). The highly mutable strain H273R was less pathogenic when inoculated intravenously, but still lethal in the intramuscular inoculation route leading to CNS (Fig. 3.14B and 3.14C). Importantly, combining the two mutations affecting mutation rate with the recombination-reducing mutation decreased pathogenicity dramatically. Regardless of the route, 90–100% of mice inoculated with double mutants GD or HD lived. Measuring the 50% lethal dose (LD_{50}) of each virus variant for the intramuscular route demonstrated that the pathogenicity of double mutants was diminished by the factor of 100 in terms of LD_{50} relative to the wild-type (Tab. 3.4). To exclude a fitness effect (tested in a cell culture but not in mouse neurons), intracranial inoculation of double GD variant into the brain was used. High amounts of virus attested to the replication ability of GD in the brain tissue.

One must conclude that the decrease in recombination rate combined with a strong change in the mutation rate sharply reduced the ability of the virus to adapt and establish infection in the central nervous system, while allowing normal (wild type) levels of viral replication in spleen and muscle (Fig. 3.15).

3.3.2 Mathematical modeling and fitting data

3.3.2.1 Adaptation rate is maximal near wild-type mutation rate

As predicted by analytic and computational models, the adaptation rate of an asexual population has a maximum in mutation rate due to the balance between accumulation of beneficial and deleterious mutations. Adding the process of recombination accelerates adaptation. Depending on the time scale of evolution and the initial population diversity, recombination and mutation are predicted to affect adaptation either independently or in synergy. On the short time scale and in the presence of standing variation, recombination and natural selection act independently of mutation de novo (Rouzine and Coffin 2005; Gheorghiu-Svirschevski et al. 2007; Rouzine and Coffin 2007; Dutta et al. 2008; Rouzine and Coffin 2010). In the long-term evolution, or in the absence of standing variation, the two processes act in synergy, and the genome evolves as the system of partly overlapping asexual blocks (Neher et al. 2010; Neher et al. 2013; Good et al. 2014; Weissman and Hallatschek 2014).

Hence, the next task was to test, based on the experiments discussed above, whether adaptation has indeed an optimum in mutation rate, and whether the optimal mutation rate depends on recombination. Two alternative hypotheses were considered in this regard: either recombination accelerates adaptation without affecting the optimal mutation rate, or the interplay between recombination and mutation shifts the optimal mutation rate (Fig. 3.16). To find out which hypothesis fits the experimental system, the team used a general computational model (Batorsky et al. 2011)

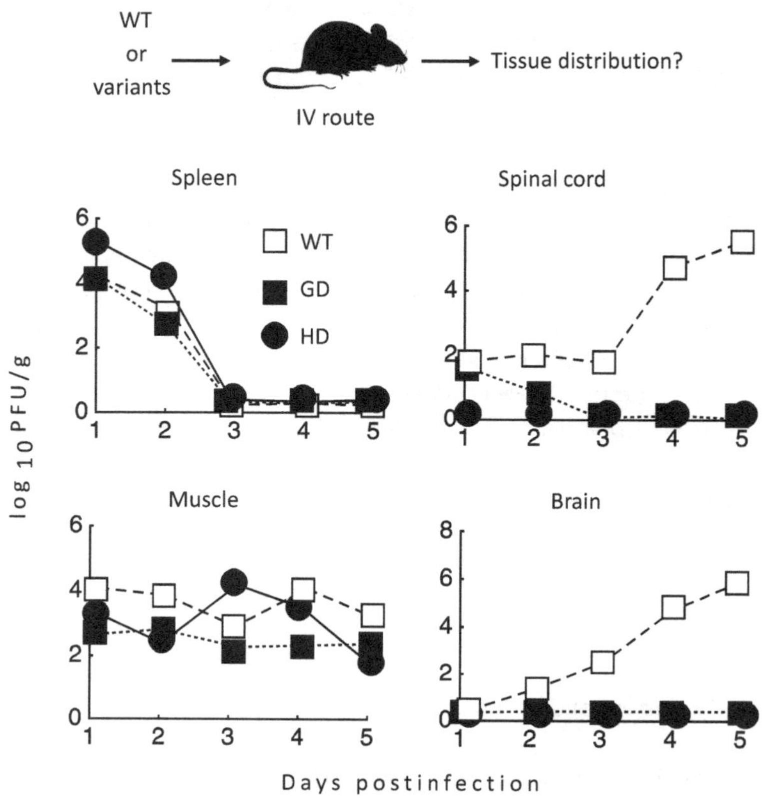

Fig. 3.15: The interplay between recombination and mutation rate determines viral tissue tropism. Six-week-old mice were injected with $2.5 \cdot 10^7$ PFU intravenously. Virus titers of different organs from susceptible mice were measured by plaque assay. Animals were inoculated with wild type virus, high fidelity/recombination-deficient GD mutant-, or low fidelity/recombination-deficient HD. The number of mice is 5 animals per time point, which was 1 day, per virus type. Based on Xiao et al. (2016).

(Section 1.3), which can reproduce either behavior depending on the initial state of a population, model parameters, and the time scale.

The starting point was an analytic model without recombination but including beneficial and deleterious mutation of a fixed fitness effect, random genetic drift, and all linkage effects (Rouzine et al. 2003). The model tracks the evolution of N genomes comprised of L sites (nucleotide positions), where both $L$and N are assumed to be much larger than unit. The evolutionary factors in the model include natural selection, random genetic drift due to random sampling of progeny, and forward and back mutations with identical probability μ per site per generation. The effect of mutation on log fitness is either s (beneficial) or $-s$ (deleterious), depending on a site. The fraction of sites that can have beneficial mutation is denoted α. Thus, beneficial and deleterious mutation rates per genome are given by $\alpha\mu L$ and $(1-\alpha)\mu L$. The target quantity is the average

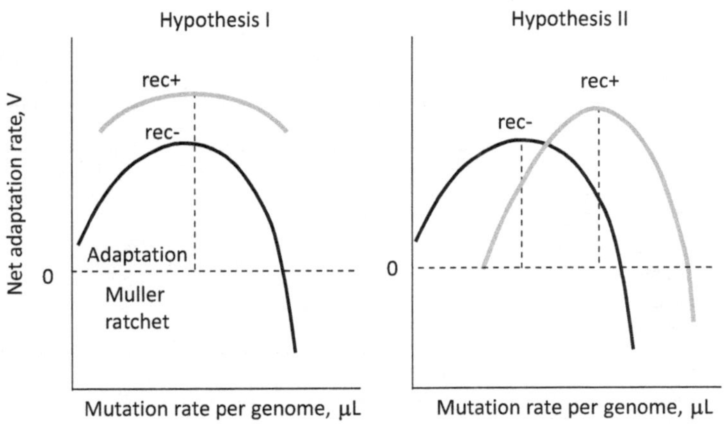

Fig. 3.16: Two hypotheses about the dependence on viral adaptation rate on mutation and recombination rate. Hypothesis 1: Adaptation rate has a maximum in the mutation rate which is elevated by recombination independently on mutation rate. Black and gray lines: schematic adaptation rate in the absence and in the presence of recombination, respectively. The optimum represents a trade-off between the selection of beneficial mutations and accumulation of deleterious mutations due to Muller's ratchet. Hypothesis 2: Recombination and mutation act synergistically, so that recombination shifts the maximum in mutation rate. Based on Xiao et al. (2016).

substitution rate $V = V_{\text{ben}} - V_{\text{del}}$, where V_{ben} and V_{del} denote the average substitution rates of beneficial and deleterious mutations per generation. Model parameters are given in Tab. 3.5.

The method of "traveling wave" was previously developed to derive the speed of the asexual evolution with multiple linked sites (see Rouzine 2020b, chapter II). A model including both deleterious and beneficial mutations and covering a broad range of parameters was studied in Rouzine et al. (2003) and (2008). Assuming small selection coefficient $s \ll \mu L$ and large population size, $Ns \gg 1$, the substitution rate V was calculated analytically [see Rouzine et al. (2003), Appendix, eqs. (15), (16), (19)–(21), Fig. 2A–B]. The normalized substitution rate $V/\mu L$ is a function of the composite parameter $(s/\mu L)\log(N/N^*)$ and the frequency of less-fit sites, α, where $N^* \sim 1/[\alpha\sqrt{s\mu L}]$. Using parameters in Tab. 3.5, the values of $s\log(N/N^*)$ and α can be calculated. Plotting the substitution rate V as a function of the genomic mutation rate μL for parameters from Tab. 3.5 demonstrates the existence of a maximum at $\mu L \approx 1.5$ (Fig. 3.17A inset). Thus, the model of asexual population (Rouzine et al. 2003) predicts the existence of a mutation rate that optimizes the rate of adaptation (Gerrish et al. 2013).

The cited derivation describes the time range corresponding to dozens or hundreds of generations, when a traveling wave in fitness is well-established and moves in a stationary fashion (Rouzine et al. 2003). This is not the case in an acute infection in the experimental mouse model, where the number of generations does not exceed $t \approx 20$, which corresponds to 7 days and 3 cycles per day (Fig. 3.13). A realistic model of viral adaptation

Tab. 3.5: Input and output model parameters.

Notation	Definition	Range	Source
L	Number of nonconserved nt positions	3,000	a b
α	Fraction of sites with less-fit alleles	0.08	a b d
μ	Average mutation rate per nt for wildtype: low-fidelity strain H273R (Fig. 3.11C: high-fidelity strain G64S (Fig. 3.11C):	10^{-4} 3×10^{-4} 0.15×10^{-4}	a b
s	Fitness gain/loss	0.2	b
N	Population size (number of infected cells)	$10^3 - 10^4$	a
r	Recombination rate per genome	$0 - 1$	a b
M	Number of crossovers per genome	1	a b
$2d$	Preexisting genetic distance	3	d
T	Total time (number of generations)	20	a b c
V_{ben}, V_{del}	Beneficial (deleterious) substitution rate per genome per generation	Output	
$V = V_{ben} - V_{del}$	Net substitution rate	Output	

[a]Present experiments
[b]Acevedo et al. (2014) and Stern et al. (2014)
[c]The total of 7 passages, with ~5 replication cycles per 8 h passage, was used. The last 2 cycles in each passage are excluded due to high MOI, because natural selection is suppressed by protein complementation between virus variants in a cell.
[d]Parameters d and α are adjusted to fit data in Fig. 3.14. Other parameters are estimates from the present or published data.

in acute infections must take into account the short period of the process. On such short time scales, it is much easier to use direct Monte-Carlo simulation.

We used a Monte-Carlo simulation algorithm implemented in MATLAB™ (Batorsky et al. 2011) to simulate the stochastic evolution of N individual genomes (Section 1.3). The algorithm simulates a Wright–Fisher process with nonoverlapping generations described in Chapter 2 (Nielsen and Slatkin 2013). The model parameters are listed in Tab. 3.5. Each genome is represented by a binary string $t_i = 0$ or 1, $i = 1, \ldots, L$. Genome fitness is determined by the sum $\exp(\sum_i s_i t_i)$, where the selection coefficient s_i at site i may be either positive or negative. The values s_i are chosen randomly from a distribution $g(s)$. The simplest case is a fixed absolute value of s with a variable sign, as given by

$$g(s) = \alpha\delta(s - s_0) + (1 - \alpha)\delta(s + s_0) \tag{3.40}$$

where $\delta(x)$ is a Dirac delta-function. At each generation step, the "broken-stick" method (Macarthur 1957) is used to generate the progeny number for each genome, with the average progeny number given by genome fitness, and total progeny equal to N. A fraction μ of all sites randomly mutates, $t_i \to 1 - t_i$. The substitution rate V is given by the rate of increase of allelic averaged over the second half of the simulated period.

Simulation confirms the existence of an optimal adaptation rate at genomic mutation rate $\mu L \approx 0.3$ (Fig. 3.17A), which is 5-fold smaller than for the long-term evolution (Fig. 3.17A inset). Interestingly, this optimal value is similar to the mutation rate of wild-type polio determined experimentally (Fig. 3.12B). Strikingly, the mutation rate of wild-type poliovirus has evolved to optimize viral adaptation *within host* during the course of infection in the host.

3.3.2.2 Recombination and mutation affect adaptation independently

Then, the effect of recombination was added to the model. Recombination was predicted to dramatically alter the viral adaptation of a multi-locus system (Rouzine and Coffin 2005; Gheorghiu-Svirschevski et al. 2007; Rouzine and Coffin 2007; Neher and Leitner 2010; Neher et al. 2010; Neher et al. 2013). At each generation step, a fraction of genomes r is randomly chosen to undergo M recombination crossovers with another genome, also chosen randomly. One of two possible recombinants replaces one parental genome. From the data in the present work, on the average, one crossover per genome was assumed, $M = 1$. The case $r = 1$ is the limiting case of strong recombination characteristic for sexually reproducing populations, and $r = 0$ represents the asexual case above (Model 1). RNA viruses that have recombination, including poliovirus, are somewhere in between. The value of r depends on the frequency of co-infected cells and, hence, on the multiplicity of infection (Rouzine and Coffin 2005).

The results demonstrate that the optimal mutation rate μL does not change due to recombination, but all adaptation rates increase uniformly for all mutation rates (Fig. 3.17A). Thus, either the effects of mutation and recombination are independent, as expected in short-term adaptation with standing variation, or, at least, their synergy does not affect the optimal balance between adaptation and Muller's ratchet (Fig. 3.16, Hypothesis I). The results also confirm that, for the short-term evolution, the initial diversity is critical for the beneficial effect of recombination. [In viruses that do not have recombination, the initial diversity is also critically important in a short-term evolution, more important than mutation de novo (Dutta et al. 2008).] The initial distribution of the allele number among genomes was taken to be Poisson distribution. The pairwise genetic distance of $2d = 3$ substitutions was adjusted to fit the virulence data in Fig. 3.14. This value falls within the range observed for poliovirus in culture (Fig. 3.12D) and not far from the range observed for vesicular stomatitis virus in culture, where $2d = 0.8 - 1.4$ was predicted (Dutta et al. 2008).

If initial diversity is absent, $d = 0$, the effect of recombination at small mutation rates disappears (Fig. 3.18), because it has no alleles to combine. Another model parameter tuned to fit the data is $\alpha = 0.08$, the less-fit site frequency. Only these sites can

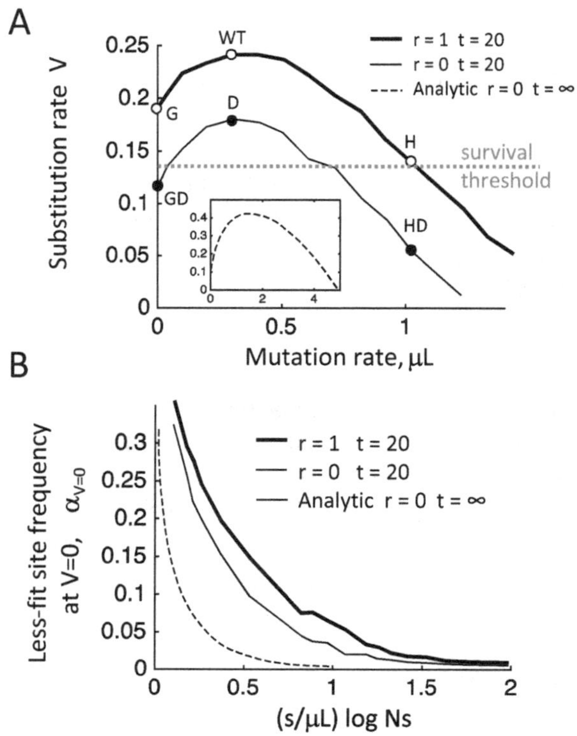

Fig. 3.17: Evolution rate predicted by the model has a maximum in optimum mutation rate, which explains the variation of mice mortality between viral strains.
(A) Curves: Net adaptation rate V defined as the number of beneficial minus detrimental mutations per genome per replication cycle is shown as a function of genomic mutation rate μL, for sexual ($r=1$, thick curve) and asexual ($r=0$, thin curve) evolution in $t=20$ generations. Monte-Carlo simulation of evolution is performed on a symmetric double-peak ($+s$, $-s$) fitness landscape using a modified code from (Batorsky et al. 2011). Circles: experimental virus strains that are able (to invade the CNS (above dotted line), unable to invade (below dotted line), or are at the "invasion threshold" (dotted line) chosen to explain the observation that the low-fidelity variant's (H) behavior depends on an inoculation route (see data in Fig. 3.14). Filled and open circles correspond to recombination-competent and recombination-deficient strains, respectively. **Inset:** Adaptation rate in asexual populations after a very large time (Rouzine et al. 2003, appendix, eq. 20). **(B)** Frequency of less-fit sites at which adaptation rate V becomes zero, $\alpha_{V=0}$, calculated at different mutation rates and plotted as a function of composite parameter $(s/\mu L)\log(Ns)$ for sexual (thick solid curve) and asexual (thin solid curve) case. Dashed curve: asexual equilibrium at infinite time according to Rouzine et al. (2003, appendix, eq. 31 therein) and Goyal et al. (2012, appendix, eq. 13 therein). Fixed parameters: (A) $\alpha=0.008$; (A, B) $N=1000$, $s=0.2$, $d=1.5$, $M=1$, $T=20$. Based on Xiao et al. (2016).

have beneficial mutation. Two model parameters, d, α were found from fitting. The other parameters, $N=10^3-10^4$, $T=20$, $s=0.2$ (Tab. 3.5), were determined from experimental data.

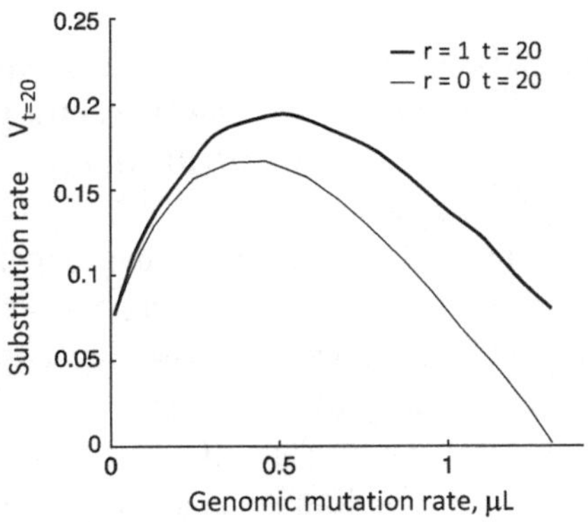

Fig. 3.18: The predicted adaptation rate of a virus that is initially genetically uniform is not affected by recombination at small mutation rates (compare to Fig. 3.17A).
Y-axis: adaptation rate *V* at $t = 20$. *X*-axis: genomic mutation rate μL. Red and blue curves correspond to the presence and absence of recombination, respectively. Fixed model parameters: $N = 1000$, $\alpha = 0.05$, $s = 0.2$, $M = 1$. Based on Xiao et al. (2016).

3.3.2.3 Mathematical modeling fits the mice survival data

To determine the roles for mutation and recombination in experimental pathogenesis, one can assume that the virus invades the brain only when it adapts to the tissue fast enough, before it is curbed by the immune system, that is, when the adaptation rate *V* exceeds a threshold (Fig. 3.17A, dashed horizontal line). Under this assumption, the model predicts the correct order of the pathogenesis for all strains. In particular, the two recombination-capable strains (WT and G) have *V* far above the threshold and hence they kill all mice, as observed experimentally (Fig. 3.17A). The only recombination-deficient strain that is above the threshold in adaptation rate and can invade the brain is the one with the with the wild type (optimal) mutation rate, variant D. The two recombination-deficient double mutants, with either high or low fidelity (variants GD and HD), adapt too slowly to invade the brain (Fig. 3.17A) and are benign, as observed experimentally. The highly mutable recombination-competent mutant that does intermediate damage (H), is chosen to be near the threshold in *V* for pathogenesis, and this is how the threshold for *V* is estimated (Fig. 3.17A). Thus, the model explains the pathogenicity variation across six strains with different rates of recombination and mutation from a single predicted quantity, their adaptability. Also, the model confirms the existence of an optimal mutation rate, explains why pathogenic effects for viruses with nonoptimal mutation rates are weaker and why the presence of recombination partly compensates for a nonoptimal mutation rate.

3.3.2.4 Sensitivity to parameters

The sensitivity of the predicted adaptation rate to variation in parameters *s*, *d*, and *α* was tested (Fig. 3.19), of which *d* and *α* were the fitting parameters. Unlike the best-fit values which match data (Fig. 3.17A), a two-fold change in any of these parameters results in an incorrect prediction for, at least, one strain. Either it contradicts to mice survival data, or predicts that WT and variant G evolved at a similar rate, in contrast to the data in culture (see Fig. 3.19A–F and the caption). Thus, the specific choice of model parameters d and α (Fig. 3.17A) is necessary to fit data.

Further, two limiting cases for recombination (outcrossing) rate r were considered, assuming either 100% or no recombination ($r = 1$ or $r = 0$). Although this assumption appears to be supported by GFP-retention data (Fig. 3.11), the outcrossing rate depends also on the cell coinfection rate and hence MOI (Rouzine and Coffin 2005). To investigate the cause of intermediate r, simulation was repeated for several values of r (Fig. 3.19G). As expected, the adaptation rates predicted at intermediate r values are bound between the predictions made for the two limiting cases ($r = 0$ and 1), and nothing new is learned in the process. As long as r is not much smaller than 1, predictions match the observed pathogenesis of 6 strains as well as at $r = 1$.

3.3.3 Optimal mutation rate replaces the concept of "error catastrophe"

In summary, combining experimental data on mice survival with mathematical modeling explains the narrow range of genomic mutation rates, 0.1 – 1, observed for many RNA viral species. The model does not support the old paradigm of "error catastrophe" predicting the "meltdown" of genome above a threshold for mutation rate (Steinhauer and Holland 1987; Domingo and Holland 1997) based on some deterministic models of evolution making unrealistic assumptions about fitness landscape (Eigen 2002; Biebricher and Eigen 2005). Instead, more realistic models of stochastic evolution offer an alternative explanation of the narrow range of genomic mutation rates, specifically, that a virus has an optimal adaptation rate tailored to the competing evolutionary factors acting during adaptation to a host (Rouzine et al. 2003; Gerrish et al. 2013).

This result is quite nontrivial due to a potential evolutionary conflict between selection forces acting on the host scale and population scale (Section 4.4). Indeed, the evolutionary optimization of virus occurs in the natural host population, which raises interesting questions about the link between pathogenic effects and reproduction number in a population. Addressing this issue in a host population requires a very different study. In Section 4.4, such a multi-scale model is studied for another system.

The combined effect of mutation and recombination is what allows viruses to exploit the few beneficial mutations without experiencing overload by many deleterious mutations, Muller's ratchet (Felsenstein 1974; Chao 1990; Duarte et al. 1992). The presented experimental and theoretical work assesses whether the experimentally observed optimal mutation rate depends on the presence of recombination. Interestingly, recombination augments viral adaptation and pathogenesis without shifting the optimal mutation rate (Fig. 3.17A). In fact, simulation demonstrates that recombination accelerates adaptation uniformly, but in a manner that depends on mutation rate. At low

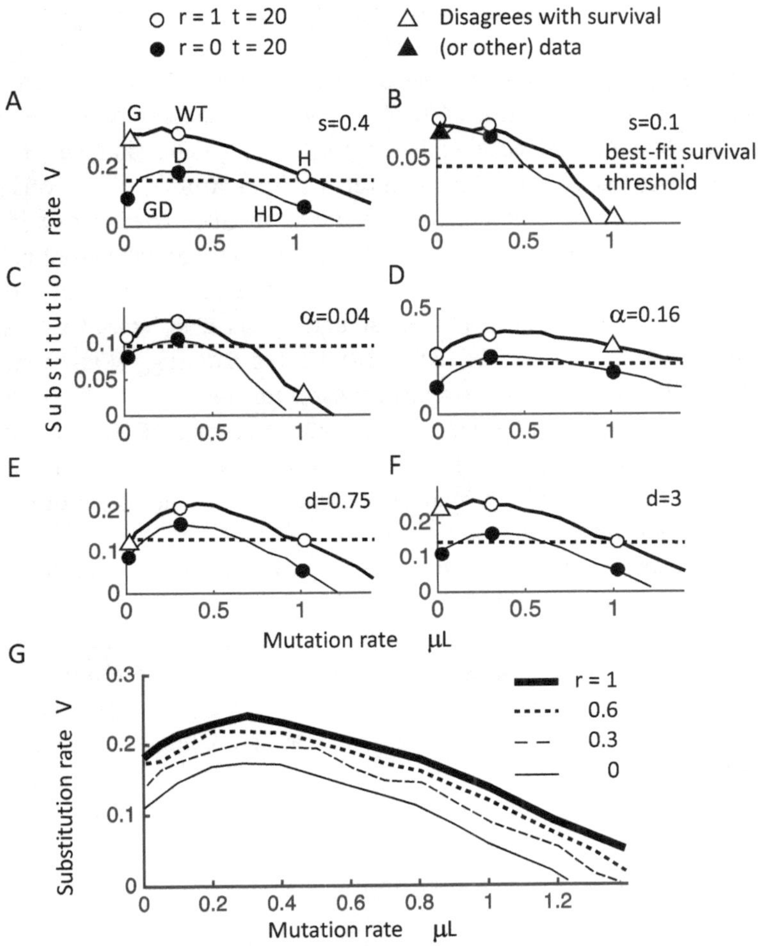

Fig. 3.19: Robustness test.
(A–F) The dependence on the net adaptation rate on the genomic mutation rate as in Fig. 3.17, varying one fixed model parameter at a time (shown). Open and closed circles show correct predictions for asexual and sexual strains, respectively. Open and closed circles show predictions that disagree with data in Fig. 3.14 (B–D) or Fig. 3.12D (A, F). Other parameters and notation are as in Fig. 3.17A and Tab. 3.5. Thin curves show simulation results: Blue: average over random runs, red: the average minus SD, green: the average plus SD. **(G)**. Dependence on the net adaptation rate on the genomic mutation rate (as in Fig. 3.17A) for different recombination rates per genome, *r* (shown). Other parameters and notation are as in Fig. 3.17A and Tab. 3.5. Based on Xiao et al. (2016).

mutation rates, its main role is to combine beneficial mutations together and, at high mutation rates, its main role is to purge deleterious mutations. Simulation results also stress the importance of small preexisting genetic variation. In the absence of initial diversity and in the limit of rare mutation, the effect of recombination is zero (Fig. 3.18).

Indeed, recombination can only bring together the already-existing alleles, it cannot create diversity from none.

These results can help to explain previous experimental observations connecting standing variation with pathogenicity (Vignuzzi et al. 2006) and adaptability (Lauring et al. 2012; Stern et al. 2014). An RNA virus evolves at the level of allelic combinations rather than single-site mutants, so that adaptation rate depends strongly on the various types of interaction between mutations in the same genome. Experiment and models demonstrate that various types of linkage (Fisher-Muller-Hill-Robertson) effects impair adaptive selection in a multi-locus system (Fisher 1930; Muller 1932; Felsenstein 1974; Chao 1990; Duarte et al. 1992; Gerrish and Lenski 1998; Worobey and Holmes 1999; Roze and Barton 2006; Cooper 2007). Models and data suggest also that even very rare recombination is able to partly compensate for these effects (Rouzine and Coffin 2005; Neher and Leitner 2010; Neher et al. 2010; Batorsky et al. 2011; Neher et al. 2013). The presented combination of animal experiments and mathematical modeling confirms these earlier findings by directly demonstrating the effect of recombination on adaptation and links it to virulence.

In conclusion, this experimental and mathematical study (Xiao et al. 2016) discovered how optimal mutation rate, recombination, and preexisting genetic variation affect the short-term adaptation of virus in acute infection. As follows from this study, viruses do not exist on the brink of "catastrophe" by evolving the highest mutation rate they can take, as previously proposed (Eigen 2002; Biebricher and Eigen 2005), but that their mutation rate is optimized to balance the accumulation of beneficial against detrimental mutations.

3.3.4 Steady-state derivation

The remaining two subsections serve as a mathematical appendix. In addition to the optimal mutation rate that maximizes adaptation rate, there is another important value of mutation rate worthy of study. Sooner or later, an evolving system arrives at an equilibrium state where natural election compensates for the combined effect of mutation and linkage, and population fitness no longer changes in time (Fig. 3.17B). If a population has a higher or lower fitness than its equilibrium value, it will accumulate deleterious or beneficial alleles, respectively, until it eventually reaches the equilibrium where the adaptation rate is zero, $V = 0$ (Rouzine et al. 2003; Goyal et al. 2012). At equilibrium, the fraction of deleterious alleles α stabilizes at a certain value, denoted as $\alpha_{V=0}$, which depends on a single composite parameter, $(s/\mu L)\log(Ns)$, which represents natural selection, mutation, and random drift (Fig. 3.17B) (Rouzine et al. 2003; Goyal et al. 2012). Accumulation of more frequent deleterious mutations is arrested by natural selection of a few beneficial mutations. The equilibrium condition has the form

$$(s/\mu L)\log(N/N^*) = 1 - 2a - \sqrt{a(1-a)}\log\frac{1-a}{a} \tag{3.41}$$

where the characteristic population size N^* is estimated by different methods to be either $N^* = a\sqrt{s\mu L}$ (Rouzine et al. 2003, appendix, eq. (31) and fig. 2B) or $N^* = s$ (Goyal et al. 2012, appendix, eq. (13) and figs. 4B and S1). Because N^* is in the argument of a large logarithm, the difference between the two estimates is minor in the relevant parameter range, and one can use, for example, the simpler estimate $N^* = s$. The validity of eq. (3.41) has been confirmed by Monte-Carlo simulation (Goyal et al. 2012). Thus, neither adaptation nor Muller's ratchet continue indefinitely but end in a stable state where the population no longer gains or loses fitness.

3.3.5 The point of no adaption in a short-term evolution

These modeling predictions can be generalized to the case of short-term evolution relevant for these experiments ($t = 20$ gen), when equilibrium is not reached yet (Fig. 3.17B). The adaptation rate at final point $t = 20$ is equal to zero, at some value of $a = a_{V=0}$, which depends on time t. Using Monte-Carlo simulation, $a_{V=0}$ was calculated as a function of $x \equiv (s/\mu L)\log(Ns)$, both in the presence and in the absence of recombination (Fig. 3.17B). For which value of x, the frequency of less-fit sites $a_{V=0}$ is larger than in the long-term equilibrium. Indeed, in the short-term evolution, natural selection has less time to act, so that the virus population tolerates more deleterious alleles.

So far, the absolute value of selection coefficient s was assumed the same for all sites. In real life, it varies among sites (Section 2.4), which makes multi-locus theory more complex (Good et al. 2012). The simplification, in particular, neglects genetic hitchhiking, that is, the amplification of a deleterious allele linked to a beneficial allele with a larger $|s|$, which also affects the fixation probability of the beneficial mutation (Johnson and Barton 2002). To test robustness of this simplification, the model was generalized to include the distribution of fitness effects measured experimentally for WT poliovirus (Stern et al. 2014) (Fig. 3.20A). The results shown in Fig. 3.20B are very similar to the fixed-$|s|$ model (Fig. 3.17A).

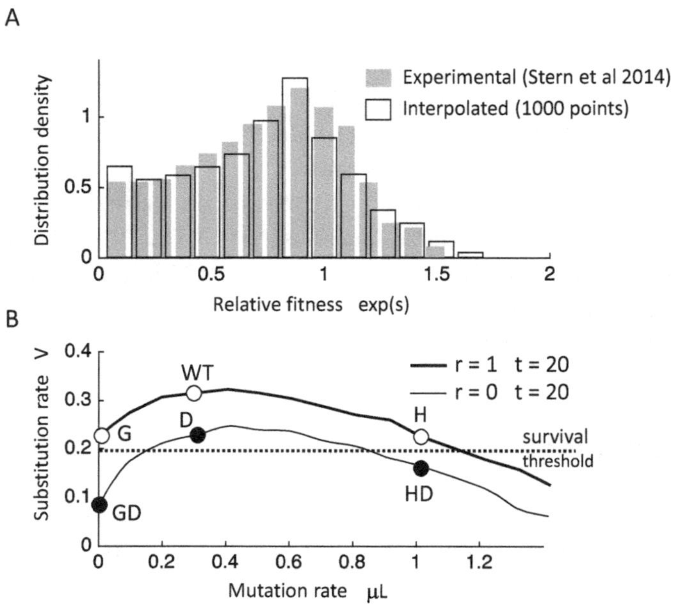

Fig. 3.20: A generalized model with variable fitness effect of mutation among genomic sites predicts substitution rates similar to the +s, −s model (results in Fig. 3.17A).
(A) Grey histogram: Experimental distribution of mutation fitness effect exp(s) for polio (Stern et al 2014). Clear histogram: Distribution of 1000 points simulated according to an interpolation formula representing a sum of a uniform and lognormal distributions. (B) Predicted substitution rate V as a function of mutation rate μL using the distribution in (A). Parameters a and s are no longer used. Other parameters and notation are as in Fig. 3.17A and Tab. 3.5. Thus, results of simulation are robust with respect to the change in the distribution of s. Based on Xiao et al. (2016).

Chapter 4
Evolutionary escape from an opposing species (Red Queen effect)

In ecological systems, a species has to adapt to the environment containing a hostile (and often evolving) species. This situation occurs in a broad variety of ecological systems, organisms, and viruses (Clarke et al. 1994). Because wolves and lynxes hunt hares, the hares have to adapt to run faster, use evasive maneuvers, and confuse their tracks. Such an adaptation of an organism facing an opposing species and continuously evolving to survive is termed "Red Queen effect" (Fig. 4.1). This chapter contains detailed analysis of such an evolutionary competition for a pathogen escaping either from an adapting immune system or its secondary parasite. Several biological scales, from a cell to an individual host to a host population, are considered.

Fig. 4.1: The Red Queen effect.
The effect bears the name of the Red Queen's race from the novel *Through the Looking-Glass* by Lewis Carroll. As the Red Queen told Alice: "Now, here, you see, it takes all the running you can do, to keep in the same place." Similarly, the evolution of influenza virus, at the population level, is driven by the immune response in individuals recovered from natural infection or vaccinated, whose number gradually accumulates in the population. In order to avoid extinction, the virus has to mutate perpetually to distance itself genetically from this immune response accumulating in the population. Based on an illustration by Sir John Tenniel from Lewis Carroll's *Through the Looking-Glass*, 1871.

4.1 Evolution of antibody epitopes of influenza virus

Propagation of many RNA viruses depends on the outcome of the race between the immune response and viral evolution. To escape immune recognition in hosts previously exposed to infection, virus accumulates mutations in immunologically important genomic

https://doi.org/10.1515/9783110697384-004

regions. The immune system follows the path of virus evolution by producing lymphocytes that recognize these new antigenic variants. The development of antiviral treatment and effective preventive measures depends on evaluating the details of viral evolution on the scale of a population. To evade immune recognition by hosts previously exposed to infection, in a never-ending chase, viruses accumulate mutations in immunologically relevant regions of the genome (Smith et al. 2004).

For example, influenza virus infects 5 – 15% of the world population annually. The global persistence of the virus due to the reinfection of previously infected individuals is caused by the rapid evolution of antibody-binding regions in the hemagglutinin protein serving as the viral receptor (Smith et al. 2004). The information available on influenza virus is quite large and includes its worldwide circulation (Rambaut et al. 2008; Russell et al. 2008; Bedford et al. 2015), genetic mapping of antibodies and virus variants, molecular structure and virology cycle, and fitness effect measured for some specific mutations (Smith et al. 2004; Koel et al. 2013; Fonville et al. 2014; Neher et al. 2016).

Computer simulation combined with data analysis sheds light on mechanisms of virus evolution and allows to predict short-term evolution (Lin et al. 2003; Bedford et al. 2012; Strelkowa and Lassig 2012; Bedford et al. 2014; Luksza and Lassig 2014; Bedford et al. 2015). Until Rouzine and Rozhnova (2018) published a work reviewed in this section, the general connection between the population-scale parameters, immunology, and evolutionary behavior of the virus was unclear. On their own, genomic and epidemiological data do not tell how the immune response, molecular, and evolutionary factors interact to produce the observed viral evolution and infection incidence. Later, the same question was addressed by Yan et al. (2019); Marchi et al. (2021), whose work reproduced the main results of Rouzine and Rozhnova (2018), although differed in some important details (Section 4.1.6).

The general analytic approach described in this section combines a susceptible-infected-recovered (SIR) framework (Gog et al. 2003; Lin et al. 2003) with standard immunobiology (Murphy 2011) and the stochastic theory of asexual evolution (Rouzine et al. 2003; Desai and Fisher 2007; Brunet et al. 2008; Rouzine et al. 2008; Hallatschek 2011; Good et al. 2012; Desai et al. 2013; Neher and Hallatschek 2013). Analysis reveals that the continuous escape of the virus occurs in the form of a traveling wave in a fitness landscape created by the immune memory in a population and moving behind the wave. The fitness landscape can be expressed in terms of the population-level virus reproduction number and the cross-immunity distance defined as the number of mutations required to change the transmission rate by the factor of 2. The problem is then reduced to the standard asexual theory, which allows to express observable parameters (viral evolution rate, its genetic diversity, infection incidence, and the average time to the most recent common ancestor) in terms of the reproduction number, population size, and the cross-immunity distance. These predictions are fit to the available data on influenza A virus to estimate two unknown parameters of the model and compare them with observations in animals and other modeling studies. The model and the results of this work might be relevant for the evolution of the immunologically-important genomic regions of SARS-CoV-2 (Rouzine and Rozhnova 2023).

4.1.1 Model of influenza transmission in a population

4.1.1.1 Strain-structured epidemiological model

The goal is to cross the models of evolution that focus on the dynamics of genetic variants with the epidemiological models that study pathogen transmission in a population. Both types of dynamics are coupled for the viruses that change genetically to evade immune memory of previously infected individuals (Grenfell et al. 2004). Let us assume that all individuals are either infected now or were infected previously. The infected individuals are classified according to the RNA sequence of the antibody-binding region of the virus located in the hemagglutinin gene for influenza virus. The infecting strain is labeled by antigenic coordinate x defined as the genetic distance in terms of amino acid changes from the original strain of 1918. After infection is cleared, the individual retains the immunological memory that provides full protection against the same virus variant and partial probabilistic protection against infection by genetically close variants. For the most of this section, one-dimensional space x that represents the trunk of the phylogenetic tree is considered. For each recovered individual, only the memory of the most recent infection is tracked (Lin et al. 2003; Bedford et al. 2012). The last-memory approximation is justified in Section 4.1.9. Multidimensional versions of this strain-structured epidemiological model are considered in Section 4.1.8. Because the focus is on long-term evolution, the factors causing seasonal oscillations are neglected. The additional assumption is that the immunologically-relevant regions of virus genome are located in the same spot for individual hosts, and that the cost of mutation there is negligible. This assumption based on the structure of hemagglutinin is valid for influenza and some other respiratory viruses controlled by neutralizing antibodies (Rouzine and Rozhnova 2023).

Let $i(x,t)dx$ be the fraction of population taken by the individuals currently infected with variants in the interval $[x, x+dx]$, and $r(x,t)dx$ is the fraction of the individuals previously infected with variants $[x, x+dx]$ and then recovered. The dynamics of the distributions $i(x,t)$ and $r(x,t)$ is described by PDEs of a form

$$\frac{dr(x,t)}{dt} = -r(x)R_0 \int_x^{\infty} dy\, K(x-y)i(y,t) + i(x,t)$$

$$\frac{di(x,t)}{dt} = i(x,t)\left[R_0 \int_{-\infty}^{x} dy\, K(y-x)r(y,t) - 1\right] + (\text{mutation term}) \tag{4.1}$$

Each individual is either infected or recovered, as given by

$$\int_{-\infty}^{\infty} dx\, [r(x,t) + i(x,t)] = 1 \tag{4.2}$$

The treatment of mutation in eq. (4.1) is described below. The time in eqs. (4.1) is measured in the units of the average time between consecutive transmissions, which is similar to the average recovery time, t_{rec} (Tab. 4.1).

Epidemiological processes are included, as follows. Firstly, the individuals recovered from strain x can be infected with strain y, with the probability proportional to the cross-immunity function $K(x-y)$, which depends on the genetic distance $u = x - y$ and y, such that

$$K(u) > 0, u < 0; \; K(u) \equiv 0, u > 0; \; K(-\infty) = 1$$

Here individuals recovered from variant x are assumed to be infectable only by variants ahead of x, as given by $y > x$. This approximation has a minor effect on the results (Section 4.1.9). Secondly, infected individuals recover with rate 1. Thirdly, the individuals infected with variant x may produce a mutant, x'. The probability of this even is assumed small. The maximal transmission rate equals the basic reproduction number in a naive population R_0.

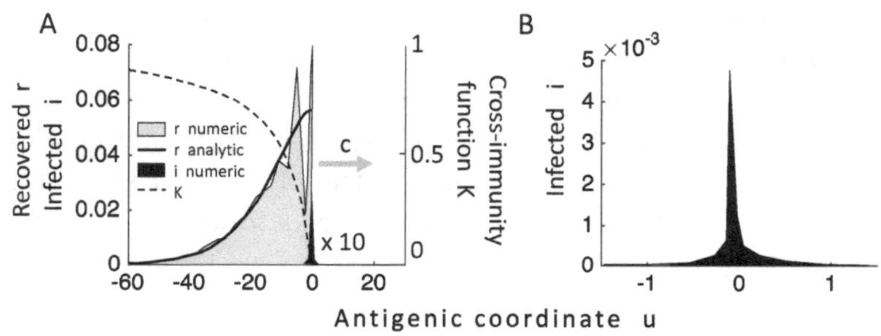

Fig. 4.2: One-dimensional epidemiological model predicts a steady traveling wave along fitness axis. (A) Frequencies of recovered individuals and the infected individuals in population in the reference frame moving with the wave. Here *X*-axis is the antigenic coordinate in that reference frame, $u = x - ct$. Solid line shows analytic prediction for $r(u)$, eq. (4.5). Small black peak is the distribution of the infected individuals $i(u)$. Gray area shows the result of a full stochastic simulation of the epidemiological model, eqs. (4.1) and (4.2). Dashed line is the cross-immunity function $K(u)$ (Tab. 4.1). (B) Infected individual density $i(u)$. Parameters: $R_0 = 2, a = 9, U_b = 5.8 \times 10^{-6}$ per epitope per transmission, $N = 10^8$. Based on Rouzine and Rozhnova (2018).

4.1.1.2 Including mutation and random genetic drift

So far, dynamics of already existing variants was considered. In fact, the antigenic evolution is driven by the emergence of new viral variants. Variant x occasionally undergoes a random mutation event forward, $x \to x + \Delta x$, which helps it to decrease antibody binding energy and, hence, recognition. The new influenza strain with a new antigenic coordinate, $x + \Delta x$, can be transmitted to another person, with some probability.

The relevant model parameters are the average mutation rate U_b per genome per infectious period (Tab. 4.1) and the distribution of the mutation effect, Δx, among sites. The model assumes probability density of the form

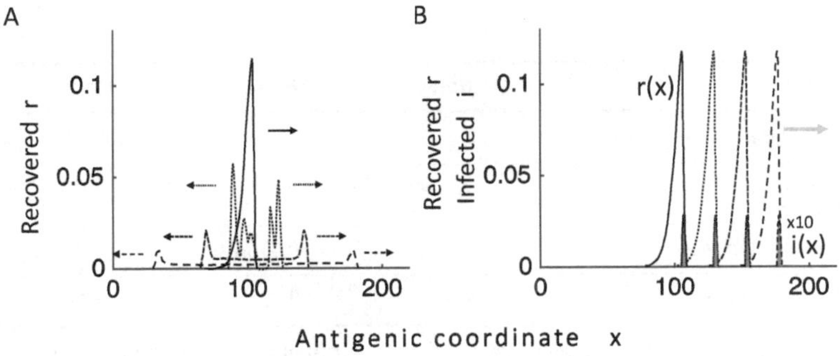

Fig. 4.3: Finite population size N eliminates the artifact of "mirror wave".
(A) The recovered individual density at times equal to 1, 300, 800, 2,000 transmission intervals. Deterministic limit $N=\infty$. (B) Both the infected (small gray peaks) and the recovered (solid, dotted, and dashed curves) are shown at times 1, 1,000, 2,000, 3,000. Population size $N=10^8$. Parameters in (A, B): symmetric immunity function $K(u)=|u|/(|u|+a)$, $R_0=2.6$, $a=7$, $U_b=5.8\times10^{-5}$. Predicted quantities: speed $c=0.023$ substitutions per transmission interval t_{rep}, average selection coefficient $\sigma=0.155$, infection incidence 0.285. Thus, finite population size (just as the asymmetry condition $K(u)=0, u<0$, Fig. 4.2) eliminates the artifact of mirror wave, a consequence of the single-memory approximation. Based on Rouzine and Rozhnova (2018).

$$\rho(\Delta x)=\frac{e^{-(\Delta x)^\beta}}{\Gamma\left(1+\frac{1}{\beta}\right)},\qquad \Delta x>0 \tag{4.3}$$

where β is a fixed parameter. Below cases $\beta=1$ and $\beta=2$ are considered. A detailed discussion of fitness landscape is given in Section 2.4. Model parameters are listed in Tab. 4.1.

Tab. 4.1: Model parameters: input (seven upper rows) and output (four lower rows).

Notation	Name	Unit	H3N2	H1N1
R_0	Basic reproduction number	1^b	1.8^a	1.46^a
t_{rec}	Recovery time	Day	5^a	5^a
U_b	Mutation rate per genome	$1/t_{rec}\|1/\text{year}$	$510^{-4}\|0.036^c$	$810^{-4}\|0.058^c$
$a=1/\|K'(0)\|$	Cross-immunity distance	AA	15^c	14^c
$K(u)$	Cross-immunity function	1	$\frac{\|u\|}{a+\|u\|}, u<0$	$\frac{\|u\|}{a+\|u\|}, u<0$
N	Population size	Human	10^8	10^8
β	Mutation distribution parameter	1	2	2

Tab. 4.1 (continued)

Notation	Name	Unit	H3N2	H1N1
σ	Average selection coefficient	1	0.048[d]	0.028[d]
$\frac{365N_i}{t_{rec}N}$	Annual incidence	1/ year	0.07[d]	0.04[d]
c	Substitution rate	$1/t_{rec}\|yr$	0.036|2.6[a]	0.031|2.26[a]
T_{MRCA2}	Pairwise coalescent time	Year	3.03[a]	4.59[a]

[a]Known from published data for influenza A strains H3N2 and H1N1 (Carrat et al. 2008; Strelkowa and Lassig 2012; Biggerstaff et al. 2014; Bedford et al. 2015).
[b]Unit 1 stands for "dimensionless."
[c]Input parameter of the model which was adjusted to fit published data.
[d]Value predicted for the best-fit parameter set.

Epidemiological and evolutionary dynamics in this model is analyzed in two steps, as follows:

(i) An ansatz is made that, in the parameter range $a \gg 1$, the infected individual density, $i(x,t)$, represents a solitary peak narrow in x compared to recovered individual density, $r(x,t)$. This ansatz is used to derive the general form of $r(x,t)$.
(ii) Standard theory of asexual evolution is applied to obtain the distribution of infected individuals, $i(x,t)$.

Detailed derivations are given in Section 4.1.7; intermediate results are outlined below.

4.1.2 Two-component traveling wave

First, one can neglect the mutation term in eq. (4.1) (it will be reintroduced later) and seek a solution in the traveling wave form

$$r(x,t) = r(x-ct)$$

$$i(x,t) = i(x-ct)$$

where $x - ct \equiv u$ is the antigenic coordinate in the moving reference frame, and c is the average substitution rate. Without the loss of generality, the peak of the wave $i(u)$ is set at $u = 0$, as given by $[di/du]_{u=0} = 0$.

4.1.2.1 Density of recovered individuals

Substituting the above *ansatz* into eqs. (4.1) and solving the resulting ODE, as explained in Section 4.1.7, for the infected and recovered density one obtains

$$i(u) = Acf(u)$$

$$r(u) \approx \begin{cases} A\exp(-AR_0 \int_u^0 dv\, K(v) & u<0 \\ 0 & u>0 \end{cases} \tag{4.4}$$

where $A = \text{const}$ is obtained from eq. (4.2), and $f(u)$ is assumed to be a narrow peak with unit area, a width much less than that of $r(u)$, and of unknown form. The speed of the wave, c, and the form of $i(u)$ will be considered in Sections 4.1.3 and 4.1.7.

At large values of R_0, $K(v)$ in the integrand of eq. (4.4) can be approximated with its linear expansion at $v = 0$. Upon integration, the recovered individual density becomes a half-Gaussian

$$r(u) \approx \begin{cases} \frac{2R_0}{\pi a} e^{-\left[\frac{R_0 u}{a\sqrt{\pi}}\right]^2} & u<0 \\ 0 & u>0 \end{cases} \tag{4.5}$$

and $A = 2R_0/(\pi a)$. Hence, the total frequency of infected individuals is given by

$$\frac{N_{\text{inf}}}{N} \equiv \int_{-\infty}^{\infty} du\, i(u) = Ac = \frac{2R_0 c}{\pi a} \ll 1 \tag{4.6}$$

and the annual incidence of infection is

$$\text{Annual incidence} = \frac{2R_0 c}{\pi a} \cdot \frac{365}{t_{rec}} \tag{4.7}$$

Equation (4.7) is directly testable in observational data.

The analytic solution, eqs. (4.4) and (4.5), was obtained from the assumption that the infected peak $i(u)$ is much narrower than $r(u)$. To test whether this is actually the case, the recovered density, eq. (4.5), is compared with Monte-Carlo simulation based on eqs. (4.1) (Section 4.1.3). The simulation confirms these analytic findings. Firstly, it shows a traveling wave with two linked components (Fig. 4.2) and, secondly, that the infected component $i(u)$ is a narrow peak compared to $r(u)$. Thirdly, the recovered curve averaged over time, $r(u)$, agrees with the analytic result, eq. (4.5). It has a sharp step where $i(u)$ is localized and a long tail at $u < 0$. The sharp step is caused by the recovery of infected individuals. The long tail is caused by reinfection of recovered individuals genetically remote from the infected peak. Interestingly, assuming that K(u) is symmetric with respect to u=0 and infinite population size creates an artifact of a mirror wave moving in the opposite direction (Fig. 4.3, Section 4.1.9)

4.1.2.2 Moving fitness landscape

The distribution of infected individuals, $i(u)$, as well as the speed of evolution are found from an independent argument, by connecting the SIR model to the standard

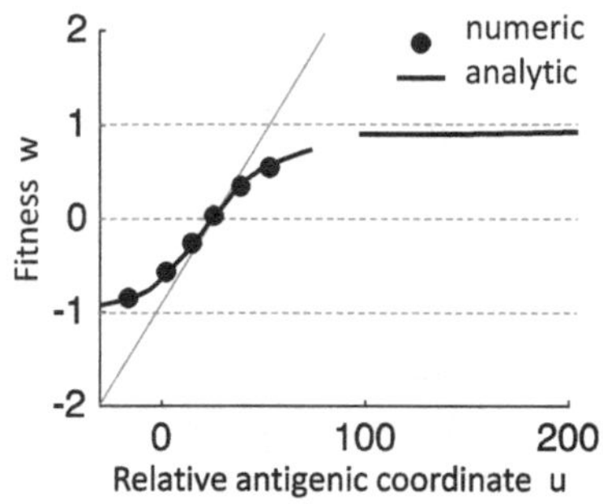

Fig. 4.4: Traveling fitness landscape and its linear approximation near the infected peak.
Solid curve: analytic result from eq. (4.8). Black circles: Monte-Carlo simulation based on eqs. (4.1).
Thin line: linear approximation with the average selection coefficient $\sigma = 0.066$ from eq. (4.10). Parameters are as in Fig. 4.2: $R_0 = 2, a = 9, U_b = 5.8 \cdot 10^{-6}, N = 10^8$. Based on Rouzine and Rozhnova (2018).

traveling wave theory (Rouzine et al. 2008; Good et al. 2012). The difference from the standard theory is that the fitness landscape is not fixed but travels with the wave. Another caveat is that fitness landscape has to be derived from the epi demiological model, eqs. (4.1). Intuitively, the distribution of memory cells in genetic space creates the fitness landscape, which pushes the entire wave forward.

Maltusian fitness of a virus in a host population (reproduction number) is defined as the average number of secondary infections per infected individual (Rice 2004; Nowak 2006; Astier 2007; Poulin 2007). The reproduction number is smaller than its value in a naive population, R_0, because it is decreased by the immune memory in a population. Virus propagates if the reproduction number is larger than unit. Equivalently, fitness $w(x,t)$ can be defined as the net exponential expansion rate of the density of infected individuals $i(x,t)$ measured per infectious period

$$w(x,t) = \frac{\partial \log i(x,t)}{\partial t} = R_0 \int_{-\infty}^{x} dy\ K(y-x) r(y,t) - 1$$

By the definition, fitness is positive (virus strains are selected for) in front of the peak of the infected density and negative (strains are selected against) behind the peak of the wave. Obviously, the fitness landscape travels in time together with the recovered density: because $r(y,t)$ represent a traveling wave, and K depends only on the difference between $y - x$, function $w(x,t)$ must be a traveling wave with the same speed. In the moving reference frame tied to the wave, for the traveling fitness landscape, one gets

$$w(u) = R_0 \int_{-\infty}^{u} dv\ K(v-u) r(v) - 1 \tag{4.8}$$

The form of $w(u)$ that can be obtained from eqs. (4.8) and (4.4) is shown in Fig. 4.4. The asymptotic cases are

$$w(u) \approx \begin{cases} R_0 - 1 & u \gg a \\ \sigma & |u| \ll a \\ -1 & u < 0,\ |u| \gg a \end{cases} \tag{4.9}$$

where the new notation

$$\sigma \equiv -R_0 \int_{-\infty}^{0} du\, r(u) \frac{dK(u)}{du} \tag{4.10}$$

represents the slope of fitness landscape. Equality $w(0) = 0$ is equivalent to the fact that the growth rate is zero at the peak of the wave, as it should; $w(u)$ has the same sign as u. For large absolute values of u, such that $|u| \gg a$, fitness landscape $w(u)$ saturates at a plateau, eq. (4.10). In the region $|u| \ll a$, where the peak of the infected is located, fitness landscape can be approximated with its linear expansion, with a positive slope σ.

If an average mutation changes the antigenic coordinate by unit, then σ is the average fitness change due to mutation event, that is, the average selection coefficient of mutation. For sufficiently large R_0, from eqs. (4.5) and (4.10), σ can be approximated by an expansion series in $1/R_0$

$$\sigma(a, R_0) = \frac{1}{a}\left[R_0 - \xi_1 + \frac{\xi_2}{R_0} + O\left(\frac{1}{R_0^2}\right)\right] \tag{4.11}$$

Here, by the definition, $a \equiv 1/|K'(0)|$, and the second and third terms are supposed to be small corrections to the first term if R_0 is large. Thus, the average selection pressure is inversely proportional to the cross-immunity distance a and increases with the basic reproduction ratio R_0. Its value is assumed to be small, $\sigma \sim 1/a \ll 1$.

The numerical coefficients in the two correction terms in eq. (4.11), ξ_1 and ξ_2, depend on the form of cross-immunity function $K(u)$ (Tab. 4.1). For the slowly decaying cross-immunity function in Tab. 4.1, they are given by $\xi_1 = 2$ and $\xi_2 = 3\pi/\sqrt{2}$. For an exponentially decaying cross-immunity function, $K(u) = 1 - \exp(-u/a)$, they are $\xi_1 = 1$ and $\xi_2 = \pi/2\sqrt{2}$. Fitness landscape $w(u)$ calculated from the simulated recovered individual density agrees with the analytic result (Fig. 4.4).

4.1.3 Connecting to the evolution theory

4.1.3.1 Antigenic diversity and the speed of evolution

So far, the speed of viral evolution c was left undetermined. The next task is to derive it simultaneously with the density of infected individuals $i(u)$. After the average selection coefficient σ is obtained, eq. (4.11), the epidemiological problem is reduced to the evolution of an asexual population with many evolving loci. For such systems, the evolution

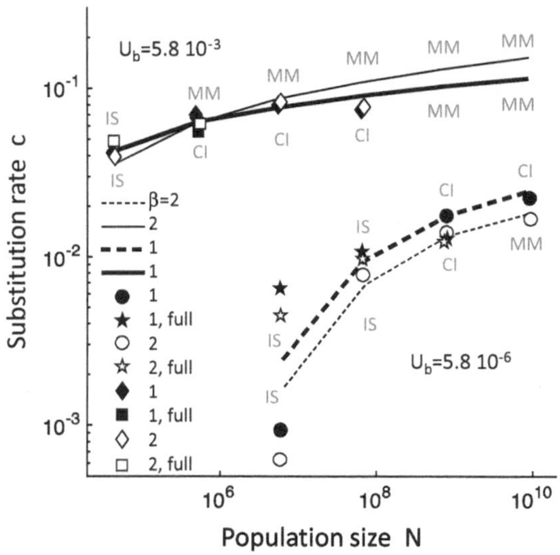

Fig. 4.5: Stochastic simulation confirms analytic results for the evolution speed.
Four broken lines are analytic results for the wave speed, *c*, from eqs. (4.12) to (4.15), at two values of mutation rate U_b, which define the broadest range of interest for RNA viruses, and two values of parameter β to test the sensitivity to the selection coefficient distribution. Symbols show results obtained by two methods of stochastic simulation shown in the legend: full stochastic simulation of the SIR model, eqs. (4.1), and reduced Moran simulation with a fixed population size N_{inf} and selection coefficient $\sigma = 0.066$. Gray letters mark the regime of evolution: isolated selection sweeps (IS), multiple mutations (MM), pairwise clonal interference (CI). Fixed parameters: $R_0 = 2, a = 9$; U_b and β are shown. Based on Rouzine and Rozhnova (2018).

rate can be expressed in terms of the population size, selection coefficient and mutation rate in the general form (Rouzine et al. 2003; Desai and Fisher 2007; Brunet et al. 2008; Rouzine et al. 2008; Hallatschek 2011; Good et al. 2012). In the general case, selection coefficient $s = \sigma\,\Delta x$ is distributed randomly across loci. Let us assume that mutational distance Δx is sampled from the distribution with a large parameter β, eq. (4.3).

The variance of antigenic coordinate $\mathrm{Var}[x] = \langle(\Delta x)^2\rangle$ and the adaptation rate v depend on cross-immunity distance a and the other parameters are given by Good et al. (2012):

$$\mathrm{Var}[x] = \frac{2\log(N_{\text{inf}}\sigma)}{\log\left(\dfrac{\beta\sigma}{U_b}\right)} \tag{4.12}$$

$$v = \sigma^2\,\mathrm{Var}[x] \tag{4.13}$$

In addition to the adaptation rate defined as the average change of fitness per unit time, another measure of the evolution rate is the average substitution rate c

$$c = \frac{v}{s*} = \frac{\sigma^2}{s*}\mathrm{Var}[x] \tag{4.14}$$

$$s* \equiv \sigma\left(\frac{\sqrt{2}}{\beta}\log\frac{\sigma}{U_b}\right)^{\frac{1}{\beta-1}} \tag{4.15}$$

where $s* \gg \sigma$ is the most probable selection coefficient of a fixed mutation (Good et al. 2012). The expressions for $\mathrm{Var}[x]$ and $s*$ are approximate, because logarithmic terms inside the large logarithms are neglected. For more accurate expressions, see Section 4.1.7.

To connect these findings to the problem at hand, the average selection coefficient σ is substituted from eq. (4.10), and the infected population size N_{inf} is substituted from eq. (4.6). As a result, the two measures of evolution speed c, v are expressed in terms of the cross-immunity distance a and the epidemiological parameters (Tab. 4.1). In the limit of very large β, eqs. (4.12)–(4.15) crossover to the results of a model with a constant selection coefficient σ (Rouzine et al. 2008).

The analytic result for the wave speed c, eq. (4.14), was tested by Monte-Carlo simulation in a wide range of N and U_b (Fig. 4.5). Two methods were used: (i) full simulation of the initial model defined by eqs. (4.1) including random mutation but not including random genetic drift, and (ii) simulation based on a Moran algorithm (Nielsen and Slatkin 2013) with a fixed population size, a linearized fitness landscape, and including random mutation and random drift (symbols in Fig. 4.5). In the second method, the selection coefficient was defined by eq. (4.10), and mutation never happened at the same site twice. In either method, the fitness effect of randomly occurring mutation was drawn from a random distribution, eq. (4.3).

Both simulation methods predicted similar time-dependences for the average infected-individual density $i(x, t)$ (Fig. 4.2). Results of simulation agreed well with analytic results and reproduced a slow increase c with N and U_b, except at the smallest U_b and N. Abbreviations IS, CI, MM in Fig. 4.5 indicate the evolution regime with respect to the number of simultaneously evolving loci. In small populations, allelic fixation occurs one at a time, as isolated sweeps (IS), or with pairwise clonal interference (CI) (Gerrish and Lenski; Schiffels et al. 2011; Good et al. 2012). In large populations, a large number of loci are evolving at the same time (Multiple Mutation, MM) (Rouzine et al. 2003; Desai and Fisher 2007; Brunet et al. 2008; Rouzine et al. 2008; Good et al. 2012). The results in eqs. (4.12)–(4.15) were derived for MM regime, which explains the discrepancy observed in simulation at the smallest U_b and N. Interestingly, the steepness β of the distribution of fitness effect of mutation weakly affects the evolution rate.

Above analysis and simulation demonstrate that the average selection coefficient, σ, is inversely proportional to a, eq. (4.11), that the rate of antigenic escape c, eq. (4.14), is also inversely proportional to a and increases very slowly (logarithmically) with population size N and mutation rate U_b, and that the adaptation rate v, eq. (4.13), is inversely proportional to a^2. The annual incidence of infection, eq. (4.7), scales as $1/a^2$ as well.

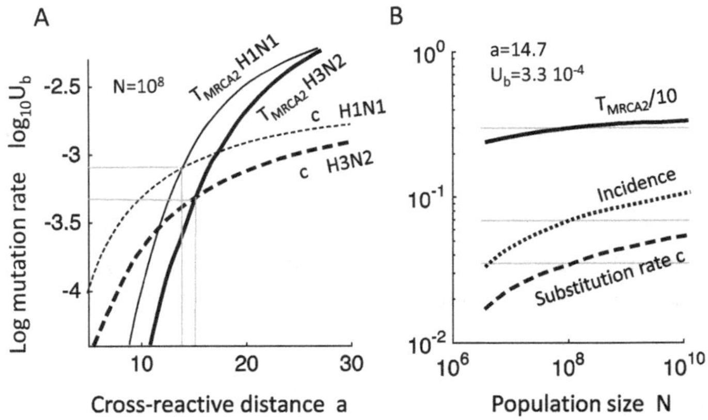

Fig. 4.6: Fitting mutation rate and cross-immunity distance to evolutionary data on influenza A. Fitting is carried out for influenza A strains H3N2 (thick curves) and H1N1 (thin curves). (A) *X*- and *Y*-axes are the cross-immunity scale, a, and the mutation rate per genome per transmission event, U_b, respectively. Analytic predictions for the evolution speed c [dashed curves, eq. (4.14)] and T_{MRCA2} [solid curves, eq. (4.16) with $z = 3$] are shown by the contours of constant heights corresponding to the observed values [(Bedford et al. 2015), extended data tab. 4.1 and the refs therein]. Population size is estimated as $N \sim 10^8$ (Biggerstaff et al. 2014). Vertical and horizontal thin gray lines show the intersection points where both parameters fit experimental values. (B) The same three quantities for H3N2 as a function of population number N, at the best-fit values of a and U_b from (A). Horizontal gray lines correspond to $N = 10^8$. (A and B) Observed values (Biggerstaff et al. 2014; Bedford et al. 2015): $R_0 = 1.8$, $c = 2.6$ AA/year, $T_{MRCA2} = 3.0$ years for H3N2 and $R_0 = 1.46$, $c = 2.3$ AA/year, $T_{MRCA2} = 4.6$ years for H1N1. Transmission period is approximated by the recovery time $t_{rec} = 5$ days. Predicted annual incidence of infection of $(4-7)$% and the cross-immunity scale $a = (14-15)$ AA are in very good agreement with independent data from horses (Park et al. 2009). Based on Rouzine and Rozhnova (2018).

4.1.3.2 Time to the most recent common ancestor

Another important observable quantity is the time to the most recent common ancestor of two coexisting viruses This parameter has been derived analytically for various asexual models (Brunet et al. 2007; Walczak et al. 2012; Desai et al. 2013; Neher and Hallatschek 2013). The results can be summarized as (Section 4.1.7)

$$T_{\text{MRCA2}} = z\sqrt{\frac{2\log(N\sigma)}{v}} \tag{4.16}$$

where numeric factor $z \sim 1$ depends on the model details, such as the form of the distribution of selection coefficient. In the case when it fixed, $z = 1.5$ is predicted. For the Gaussian distribution, eq. (4.3) with $\beta = 2$, $z = 3$. Because the Gaussian case is more realistic of the two, and because the expression for T_{MRCA2} for other forms of s-distribution is not available, $\beta = 2$ was chosen to fit data (next subsection).

4.1.4 Comparison with data on influenza A

The next task is to compare the theoretical results with available data on influenza A H3N2 and H1N1. Input model parameters and the predicted quantities are listed in Tab. 4.1. The population size N, the reproduction ratio in naive population R_0, as well as recovery time t_{rec} are found from observational data (Carrat et al. 2008; Strelkowa and Lassig 2012; Biggerstaff et al. 2014; Bedford et al. 2015). However, parameters a and U_b are difficult to measure, because they involve biological interactions at multiple biological scales: a cell, a host, and a host population. On the other hand, two quantities predicted by the model, T_{MRCA2} and substitution rate c, are known. Therefore, it makes sense to adjust the unknown model parameters a and U_b to match the known values of the two predicted quantities (Fig. 4.6A).

The annual incidence of influenza A (H3N2) varies in the interval 1 – 15% of population. The average evolution rate and T_{MRCA2} are $c = 0.036$ amino acid substitutions/genome/transmission, with a transmission period of 5 days, and $T_{\mathrm{MRCA2}} = 3.0$ years, respectively (Biggerstaff et al. 2014; Bedford et al. 2015). The relevant population size is on the order of $N = 10^8 - 10^9$ individuals, which corresponds to a large country. Naive-population reproduction ratio $R_0 = 1.8$ can be approximated by its value determined from the most rapid pandemics, such as those that occurred in 1918 and 1968. Such a major pandemic is normally caused by the reassortment of viral chromosomes or other forms of antigenic shift causing poor immune recognition, which approximates a naive population.

The total number of mutating amino acids can be estimated as $L = 120$ of nonsilent nucleotides, which are the first and second nucleotides of the 60 amino acids of the five antibody-binding regions of hemagglutinin protein (Strelkowa and Lassig 2012; Luksza and Lassig 2014). The mutation rate could be estimated from the synonymous substitution rate, which is measured to be 5.8×10^{-5} per nucleotide per transmission (Strelkowa and Lassig 2012). However, not all amino acids are equally important for antibody binding. Therefore, the effective epitope length is expected to be below 60. Plus, not all mutations are transmitted. Thus, it is difficult to determine U_b from an observation and is usually guesstimated roughly (Bedford et al. 2012; Bedford et al. 2015). Hence, parameters a and U_b are tuned to fit the predictions for c and T_{MRCA2} (Fig. 4.6A).

Influenza strain H2N3 evolves faster and has a shorter time T_{MRCA2} than strain H1N1, because of a larger value reproduction number R_0 resulting in a larger selection pressure σ. Indeed, the best-fit values of U_b and a are similar between the two strains (Fig. 4.6A).

The inferred cross-immunity distance, $a = 14 - 15$, is confirmed by the independent data on equine influenza (Park et al. 2009). The predicted annual incidence $(4 - 7)\%$ falls within the experimentally observed range 1 – 15% and previous modeling estimates (Bedford et al. 2012; Strelkowa and Lassig 2012; Luksza and Lassig 2014). The best-fit estimate of U_b at 3.3×10^{-4} is 3.3-fold larger than that in the simulation (Bedford et al. 2012), which corresponds to the effective epitope length, at the population level, of $L \sim 7$ variable amino acids.

The above results explain the inverse correlation observed between T_{MRCA2} and the substitution rate c across strains H2N3, H1N1 and influenza B (Bedford et al. 2015). The cause of the correlation is that the predicted substitution rate c is linearly proportional to the selection coefficient σ, while T_{MRCA2} is inversely proportional to σ. Broad variation of σ is caused primarily by variation in R_0. The dependence of c and T_{MRCA2} on the other parameters, U_b and N, is logarithmically slow, and a does not change much (Fig. 4.6A).

Robustness of the predicted values of c, T_{MRCA2}, and the annual incidence to changes in population size N was calculated (Fig. 4.6B) showing the usual slow logarithmic dependence predicted by the standard traveling wave theory. The sensitivity to parameters a, U_b and R_0 was studied as well (Rouzine and Rozhnova 2018, figs. S3, S4).

4.1.5 Robustness to additional dimensions and old memory

The above analysis is based on the approximations, as follows:
(i) Sigmoidal fitness landscape is approximated with its linear expansion
(ii) No infection forward in the genetic coordinate is allowed
(iii) Immune memory of only the last infecting strain is taken into account (Lin et al. 2003; Bedford et al. 2012)
(iv) One-dimensional genetic space

As demonstrated in Section 4.1.9, approximations (i)–(iii), in the relevant parameter range, have a small effect on these predictions. Antigenic space of antibody-binding region has many dimensions. For example, for $L = 7$ amino acids, with 10 chemically distinct variants per each, the total number of adjacent variants is 70 and not 2, as it were in 1D case. Strikingly, the existence of so many additional dimensions does not change the results much, because the random trajectory is quasi-one-dimensional and does not have loops. To demonstrate these facts, the model was simulated numerically on a tree of epitope variants (Rouzine and Rozhnova 2018, fig. S6) (Section 4.1.8). Real-time phylogenetic tree reveals a quasi-1D path comprising a long trunk of permanently fixed mutations and short branches representing transient virus variants, which resembles the phylogeny of influenza A (Smith et al. 2004; Bedford et al. 2012; Strelkowa and Lassig 2012; Luksza and Lassig 2014). An automatic formation of a traveling wave in the form of a snake was also observed in simulation on two-dimensional genetic space with one and two antigenic coordinates (Rouzine and Rozhnova 2018, fig. S5) (Section 4.1.8).

The actual reason why antigenic space can be reduced to a tree is the well-known fact that the return of a random walk to the origin, in a m-dimensional space, has a small probability, on the order of $1/m \ll 1$. For the same reason, the 1D topology of the virus path, the old memory cells are negligible (Section 4.1.9).

4.1.6 Discussion

The topic of this section is the evolutionary dynamics of a virus driven by the need to escape recognition by the memory B cells left in previously infected individuals. The problem is solved by expressing the fitness landscape expressed in terms of the cross-immunity function $K(x-y)$ (Fig. 4.4) and then inserting the fitness landscape into an evolutionary model. The result is a traveling wave with two components structured in the antigenic variant space: the recovered individuals and the infected individuals. The two population components have a different size and genetic diversity (Fig. 4.2). The recovered subset is genetically diverse and occupies almost the entire population. The infected subset is relatively small in size and less diverse genetically.

The speed of viral evolution, the annual incidence of infection, and the average time to the most recent ancestor are expressed in terms of model parameters $N, U_b, R_0, K(x-y)$ (Tab. 4.1). Both the selection coefficient and the substitution rate are proportional to $f(R_0)/a$, where $f(R_0)$ is a monotonously increasing function with $f(0)=0$. All parameters depend weakly on population size N and mutation rate U_b which is a universal feature of multi-locus models. Two types of Monte-Carlo simulation confirm these predictions. When compared to the epidemiological and genomic data on influenza A, these theoretical findings provide accurate estimates of four important population-scale parameters: a, L, U_b.

A much-debated aspect of influenza virus evolution is its punctuated nature (Smith et al. 2004). While most mutations have a small antigenic effect, some of them represent large jumps in antibody recognition (Bedford et al. 2014). The proposed theory naturally interprets these leaps as a consequence of the stochasticity of the traveling wave, as follows. In the traveling wave theory, the extension of the best-fit edge of a wave is made possible by adding new escape alleles to the rare best-fit sequences subject to strong stochastic fluctuations (Rouzine et al. 2003). Fitness effect of a mutation varies depending on a mutated locus, and Good et al. (2012) demonstrated that most fixed mutations have an above-average fitness effect fluctuating around the most probable selection coefficient s^* depending on model parameters σ, N, U_b. Depending on a parameter region, the results map either onto the multiple-mutation (MM) model with fixed s and a relatively smooth wave (Rouzine et al. 2008) or the two-site clonal interference (CI) model (Gerrish and Lenski 1998; Schiffels et al. 2011), where the entire "wave" is represented by one or two strongly fluctuating peaks. Next, the present work established that influenza virus typically evolves within MM regime but near the border with CI regime (Fig. 4.5). In this regime, the selection coefficients of fixed alleles are predicted to fluctuate strongly, which might explain the punctuated effect.

The analytic results agree with simulation based on a similar model (Bedford et al. 2012), which predicted the quasi-one-dimensional trajectory and the same incidence range for influenza A. (Bedford et al. 2012) assumed mutation rate $U_b \sim 10^{-4}$ and the cross-immunity distance $a = 1/0.07$ based on data in horses (Park et al. 2009). In the reviewed work, the two parameters are determined from fitting human H3N2 and

H1N1 data on c and T_{MRCA2} taken from (Bedford et al. 2015). Then, the model is tested by the comparison of the inferred a with the measured value (Park et al. 2009). Despite a different order of the two procedures, the results are similar, apart from the difference in mutation rate by the factor of 3.3.

The pioneering work by Lin et al. (2003) proposed a similar model with immune memory and 1D antigenic space, eqs. (4.1) and (4.2). Their analysis differed from the present analysis in two aspects. Firstly, the viral evolution was assumed to be completely deterministic ($N = \infty$). Therefore, the mutation term in eqs. (4.1) had a form proportional to $\frac{\partial^2 i(x,t)}{\partial x^2}$. That approximation could be correct if the front edge of the wave were a sufficiently slow function of the antigenic coordinate. Neither approximation is valid at relevant mutation rates (Section 4.1.7). As a result, Lin et al. (2003) predicted evolution speeds far below simulation results. The traveling wave theory naturally includes both the stochasticity and the sharpness of the leading edge. Future development of this model requires inclusion of finite mutation cost in the style of Batorsky et al. (2011) (Section 4.2).

Basic analytic results of the original work (Rouzine and Rozhnova 2018), including the 1D channeling and the general structure of the expression for the selection coefficient, were reproduced later by Yan et al. (2019) and Marchi et al. (2021). Both groups used a mean-field model assuming, interestingly, that all individuals in a population share their immune memory, as if they were bacteria sharing CRISPR segments (Gog et al. 2003). This mean-field model is very popular among epidemiologists due to its mathematical simplicity. Despite the bizarre description of the immune memory, the cited models produce reasonable expressions for the effective selection coefficient σ similar to eq. (4.11) but with a different function of R_0.

The artifact of the "bacterial" approximation to human immunity is the wrong predictions about the role of memory cells from the infections that occurred before the last infection. Because all individuals are assumed to share all memory cells, every individual in that models "remembers" any old infection of any other individual, greatly exaggerating the role of old memory cells (Yan et al. 2019).

To summarize, combining the standard SIR approach with the standard traveling wave approach creates a general method linking the epidemiological and immunological parameters to the observed parameters of influenza evolution. The distribution of recovered individuals in multidimensional antigenic space is shown to create a fitness landscape for the distribution of the infected individuals, the two distributions moving together along a quasi-one-dimensional. These predictions are fit to data on influenza A H3N2 producing the estimates of model parameters that agree with independent estimates. The relevance of this model to the evolution of the immunologically-important parts of SARS-CoV-2 remains to be investigated (Rouzine and Rozhnova 2023).

4.1.7 Analytic derivation for the 1D model

The remaining subsections of Section 4.1 serve as a mathematical appendix.

4.1.7.1 Traveling wave solution

One seeks a traveling wave solution of eqs. (4.1)

$$r(x,t) = r[x - x_{\max}(t)], \; i(x,t) = i[x - x_{\max}(t)], \; x_{\max}(t) = ct \qquad (4.17)$$

Here $x_{\max}(t)$ is the maximum of the infected peak. Substituting eqs. (4.17) into eqs. (4.1), one gets ODEs

$$-c\frac{di(u)}{du} = i(u)\left[R_0 \int_{-\infty}^{u} dv\; K(v-u)r(v) - 1\right] + (\text{mutation term}) \qquad (4.18)$$

$$-c\frac{dr(u)}{du} = -r(u)R_0 \int_{u}^{\infty} dv\; K(u-v)i(v) + i(u) \qquad (4.19)$$

The mutation term in eq. (4.18) denotes a random nonnegative integer divided by N, as given by $0, 1/N, 2/N, \ldots$, with the average

$$\langle \text{mutation term} \rangle = U_b[i(x+1,t) + i(x-1,t) - 2i(x,t)]$$

Because $U_b \ll 1$, this term is small and can be neglected for most values of u. Mutation becomes important at the best-fit edge of the traveling wave, $u \gg 1$, where the best-fit virus variants are produced by mutation (Section 4.1.7.3). Until then, it can be dropped from eq. (4.18).

By the definition of u, the peak of $i(u)$ is at $u = 0$, as given by $[di/du]_{u=0} = 0$. Setting $u = 0$ in eq. (4.18) and neglecting the small mutation term yields

$$R_0 \int_{-\infty}^{0} du\; K(u)r(u) = 1 \qquad (4.20)$$

meaning that the reproduction number at the peak of the infected is equal 1, as it should.

To derive $r(u)$ from eq. (4.19), let us assume that the width of $i(u)$ is much less than the width of $r(u)$. The validity of this assumption will be tested below. In this approximation, $i(u)$ in eq. (4.19) can be approximated with a delta-function

$$i(u) = Ac\delta(u) \qquad (4.21)$$

where A is a constant. Product $cA = \int_{-\infty}^{\infty} dx\; i(x,t) \ll 1$ has the meaning of the virus prevalence in a population (Tab. 4.1). Substituting eq. (4.21) into eq. (4.18), one gets

$$\frac{dr}{du} = AR_0 r(u)K(u) - A\delta(u+0), \quad u<0 \tag{4.22}$$

Integrating with the boundary condition $r(-\infty) = 0$, and using eq. (4.20), one arrives at eq. (4.4). Constant A is determined from the normalization condition, eq. (4.2)

$$\int_{-\infty}^{\infty} dx\, r(x,t) = 1 - Ac \approx 1 \tag{4.23}$$

At $R_0 \sim 1$, from eq. (4.4), $A \sim 1/a$, which suggest that the characteristic width of $r(u)$ is on the order of a, which justifies our assumption above. At $a \gg R_0 \gg 1$, one can simply eq. (4.4) to eq. (4.5) by using the Taylor expansion of $K(u)$ near $u = 0$, and the width of $r(u)$ is $\sim a/R_0$. Thus, the narrow peak of the infected at $u = 0$ is followed by a gradually decaying tail of the recovered individuals (Fig. 4.2A).

4.1.7.2 Effective selection coefficient

We use an equivalent definition of fitness, $w(u)$, as the exponential expansion rate, with the time unit being the period between transmissions. From the second of eqs. (4.1), we arrive at eq. (4.8) for $w(u)$, where $r(u)$ is substituted from eq. (4.5). Fitness landscape $w(u)$ and its linear expansion are shown in Fig. 4.4. Simplification of eq. (4.8) in three asymptotic limits is straightforward. Approximating $K(v-u)$ in the integrand of (4.8) with the asymptotic plateau, 1, one obtains the first limit in eq. (4.9), $u \gg a$. To obtain the second case in eq. (4.9), $|u| \ll a$, the RHS of eq. (4.8) is approximated with its linear expansion in u. In the third case, $u < 0$, $|u| \gg a$, the integral in eq. (4.8) is neglected altogether, because the overlap between $K(v-u)$ and $r(v)$ is exponentially small.

The value of σ in eq. (4.10) can be estimated from the scaling of $K(u)$ and $r(u)$ that follows from eqs. (4.4) and (4.23), which yields $\sigma \sim 1/a \ll 1$. Replacing fitness landscape with its linear expansion leads to accurate results when the traveling wave, including its leading edge, stays within the linear region. The corrections are calculated in Section 4.1.9.

4.1.7.3 Wave speed

The next task is to obtain the substitution rate c and the width and shape of $i(u)$ using the standard traveling wave theory (Rouzine et al. 2008; Good et al. 2012). As we mentioned above, the width of the distribution of the infected individuals, $i(u)$, is narrow compared to the width of the distribution of the recovered, $|u| \sim a$. According to these models, the shape of $i(u)$ is approximately Gaussian in the center, but ends at the leading high-fitness edge $u = u_0 \gg 1$. The speed of the wave is limited by rare stochastic events of the establishment of new mutations occurring at the edge. The exact expression for c depends on the form of distribution of selection coefficient $s = \sigma\Delta x$ among sites. A few simplest cases are considered below.

Fixed selection coefficient: In the simplest case $s = \sigma$, the wave speed c can be found from the transcendental equation (Rouzine et al. 2008)

$$\log N_{\text{inf}} = \frac{c}{2\sigma}\left(\log^2\frac{c}{eU_b} + 1\right) - \frac{1}{2}\log\frac{\sigma^2 U_b}{c\log\left(\frac{c}{U_b}\right)} \tag{4.24}$$

$$N_{\text{inf}} = cNA \tag{4.25}$$

where N_{inf} is the infected individual number, N is the total individual number, and A is found from eqs. (4.23) and (4.4). The edge location with respect to the wave peak, u_0, is given by Rouzine et al. (2008)

$$u_0 = \frac{c}{\sigma}\log\frac{c}{eU_b} \tag{4.26}$$

Exponential distribution of s: In real life, s varies randomly among sites (Section 2.4). In this case, it is better to obtain the rate of fitness increase, the adaptation rate V, and then from it calculate the substitution rate c. When s is distributed exponentially [eq. (4.3) with $\beta = 1$] with an average σ, the adaptation rate V is found from the coupled equations for V and w_0 (Good et al. 2012, eqs. S14–S15)

$$2 = \frac{U_b}{\sigma}\sqrt{\frac{2\pi\sigma^2}{V}}\left[1 + \frac{V}{w_0} + \frac{V}{\sigma w_0 - V}e^{\frac{\left(w_0 - \frac{V}{\sigma}\right)^2}{2V}}\right] \tag{4.27}$$

$$1 = N_{\text{inf}} U_b\left[-1 + \frac{(w_0)^2}{V} + \frac{2\,(w_0 + \sigma\,)}{V}\right]e^{\frac{2\sigma w_0 - V}{2\sigma^2}} \tag{4.28}$$

where w_0 is "fitness lead," that is, the fitness difference between the wave edge and the wave peak. The substitution rate c is found as

$$c = \frac{v}{s*}, \quad s* \equiv w_0 - \frac{v}{\sigma} \tag{4.29}$$

The meaning of $s*$ is the most probable selection coefficient of the new alleles that are eventually fixed in a population. The system of eqs. (4.27) and (4.28) is solved numerically for c and w_0. The resulting values of c are compared with simulation (Fig. 4.5).

Arbitrary distribution of selection coefficient: Good et al. (2012) considered also the more general case, in which the distribution density of s has an arbitrary form $\rho(s)$, $s > 0$. The values of V and w_0 are calculated numerically from two coupled equations analogous to (4.28) (Good et al. 2012, eqs. (4.17)–(4.19)). For the case $\rho(s) \propto \exp\left(-(s/\sigma)^2\right), s > 0$, which corresponds to $\beta = 2$ in eq. (4.3), the value of c is plotted in Fig. 4.5. In the case $\beta \gg 1$ in eq. (4.3), the two coupled equations simplify to eqs. (4.12)–(4.15).

4.1.7.4 Time to the most recent ancestor

An analytic expression for T_{MRCA2} for an arbitrary distribution of s is not available. However, the analytic expressions for a fixed s and a Gaussian distribution of s symmetric around 0 have been obtained and tested in simulation (Walczak et al. 2012; Neher and Hallatschek 2013). Their combined results can be written, as follows:

$$T_{MRCA2} = \frac{zw_0}{V} = z\sqrt{\frac{2\log(N_{inf}\ \sigma)}{V}} \tag{4.30}$$

which is equal to the time in which the wave moves by its lead w_0 multiplied by a numeric coefficient $z \sim 1$, and the infected population size N_{inf} is given by eq. (4.25). The value of z depends on the selection coefficient distribution $\rho(s)$, as follows

$$z = \begin{cases} 1 & s = \text{const} \quad \text{analytic} \\ 1.5 & s = \text{const} \quad \text{simulation} \\ 3 & \beta = 2 \qquad\;\; \text{simulation} \end{cases} \tag{4.31}$$

Simulation demonstrates a larger value of z than the analytic derivation.

In Sec. 4.1.4, the predicted value of T_{MRCA2} is compared with the data for influenza A H3N2 (Fig. 4.6A). Because the exact form of the distribution of s is unknown, the case $\beta = 2$ and $z = 3$ is used. Indeed, this assumption is more realistic than the case of fixed s. The reason for this choice is that we there is a published result for z at $\beta = 2$, eq. (4.31), but not, for example at $\beta = 1$ (Sec. 2.4). The evolution of HIV and influenza involves mutations with both positive and negative sign of s (Batorsky et al. 2014; Luksza and Lassig 2014). Of course, the distribution of s is not symmetric with respect to the sign (Section 2.4), but it is the positive-s part that matters for the wave progress if the wave is far from equilibrium (Rouzine et al. 2008). Hence, the result in the last line of eq. (4.31) is used for fitting data. To increase the accuracy of eqs. (4.30), $\log(N_{inf}\sigma)$ is replaced with

$$\log\left[N_{inf}\sqrt{\frac{\sigma^2 U_b}{c\log\left(\frac{c}{U_b}\right)}}\right]$$

which uses prefactor at N_{inf} from eq. (4.24) obtained for the case of constant s (Rouzine et al. 2008).

4.1.7.5 Comparison to a previous 1D model of influenza evolution

(Lin et al. 2003) proposed a similar 1D model, eqs. (4.1), predicted a traveling wave solution and calculated its speed. Their method differs from the present in two crucial aspects:

(i) The mutation term has the diffusion form proportional to the second derivative in x.
(ii) Evolution is deterministic in infinite population.

Neither approximation applies in the relevant parameter range (Tab. 4.1). Indeed, the difference between x and $x+1$ in

$$\langle \text{mutation term} \rangle = U_b[i(x+1,t) + i(x-1,t) - 2i(x,t)]$$

can be approximated with the second derivative, if and only if $|d \log i(x,t)/dx| \ll 1$, which implies $2U_b \gg R_0 - 1$ (Lin et al. 2003), eq. (S16), $\lambda/a = c/(2ad) \ll 1$). In real life, the opposite is true (see parameters in Tab. 4.1). Thus, the leading front of the infected wave is, in fact, very steep. Also, finite population size is crucial, unless the population size is astronomically large (Rouzine et al. 2003). For the evolution speed (Lin et al. 2003) [eqs. (S3), (S12), and the equation before (S3)], obtained

$$c_{\text{Lin}} \approx \frac{2}{a}\sqrt{2U_b(R_0-1)}\left[1 - \frac{\pi^2}{2\ln(N)^2}\right]$$

which is an order of magnitude below the simulation results shown in Fig. 4.5. The predicted width of the infected and recovered distributions in genetic distance antigenic coordinate x is also completely off. For them, Lin et al. (2003) obtained the scaling ${U_b}^{1/2}$ and ${U_b}^{1/4}$, respectively (see their Fig. 3). In this section, the recovered distribution width does not depend on U_b, while the infected width does, but only logarithmically.

4.1.8 Multidimensional antigenic space

Analysis given in the previous sections assumed 1D antigenic space. Below this assumption is relaxed, and two generalizations of antigenic space are considered: a two-dimensional lattice and a tree with p neighbors per node. Results demonstrate that a quasi-1D snake-shaped wave is formed automatically in either case due to the random component of fitness landscape and the competition for resources between virus strains. Thus, the presence of multiple dimensions does not change the main conclusions.

4.1.8.1 Two dimensions with one antigenic coordinate

Consider a square lattice of integer pairs $\{i,j\}$ with one immunologically important (antigenic) coordinate $x_{ij} = i + \xi_{ij}$ and one neutral coordinate $y_{ij} = j + \eta_{ij}$, where ξ_{ij} and η_{ij} are random quantities uniformly distributed between $-V_x$ and V_x. The second coordinate of a strain chosen to be y– axis does not influence its antigenic properties. Turning the antigenic and neutral coordinates by 45 degrees does not cause substantial change in results (Rouzine and Rozhnova 2018). This, slightly rugged fitness landscape is equivalent to the random of variation of s.

As the generalization of the cross-immunity function, the 2D cross-immunity matrix between neighbor lattice points is defined as $K_{ij,mk} = K(x_{ij} - x_{mk})$, where $K(u>0) = 0$ is as in the 1D model (Tab. 4.1). For each vertex $\{i,j\}$, one writes the same dynamic equations as eqs. (4.1), except the integrals are replaced by the double sums over the vertices.

Mutation can occur in all four directions between adjacent vertices, with the same probability U_b.

Time snapshots demonstrate the spontaneous formation of a steady 1D-like wave starting from a flat front (Rouzine and Rozhnova 2018, fig. S5). The initial condition is the delta-function distribution for the infected and a uniform distribution for the recovered. In the course of time, the system develops several competing quasi-1D waves, "snakes," with an infected head and a recovered tail. If, due to the random component in the antigenic coordinate, the head one wave comes slightly forward, it will give it an advantage in the ability to infect the tails of the other waves (Rouzine and Rozhnova 2018, fig. S5). The competition between different "snakes" for the common resource, ultimately gets all but one "snakes" extinct. The typical length of a snake in the horizontal direction is determined by the parameters of the one-dimensional model, a and R_0, while the snake width in the neutral direction is proportional to V_x.

4.1.8.2 Two antigenic coordinates

To test the sensitivity of the assumption of a single antigenic coordinate, simulation was performed on a 2D lattice with two asymmetric antigenic coordinates (x,y). The cross-immunity matrix depends on the elliptical Euclidian distance, as follows

$$K_{ij,mk} = K\left(\sqrt{\left(x_{ij}-x_{mk}\right)^2 + \text{asym} * \left(y_{ij}-y_{mk}\right)^2}\right)$$

where $asym < 1$ is the asymmetry factor. Simulation, again, leads to the appearance of a single quasi-1D "snake" crawling in the direction of the easiest escape, although with some artifacts, such as the appearance of a backward wave (Section 4.1.9).

4.1.8.3 Many dimensions are equivalent to tree topology

As already mentioned in Section 4.1.5, in a high dimensional space, loops are very unlikely. The probability of returning back to the origin is proportional to the inverse number of dimensions and hence very small. Indeed, the number of nearest neighbors is equal to the number of amino acid positions in all epitopes combined, multiplied by 10. Hence, a tree with a large number of branches per node is a much better approximation to the real antigenic genetic space with than the 2D lattice.

Monte-Carlo simulation was carried on a tree in the same way as on the 2D lattice for different U_b, N, R_0, a and V_x. In this case, the antigenic coordinate of a strain was defined as $x = x_f + \xi$, where x_f is the antigenic coordinate of the parental strain, and ξ is a random contribution drawn uniformly from interval $[-V_x, V_x]$, $V_x \ll 1$. The antigenic distance between each two nodes i and j was calculated by adding all the branches connecting these two nodes. The cross-immunity function K_{ij} was asymmetric and depended on the antigenic distance as in Tab. 4.1.

An example of time-dependent phylogenetic tree with the quasi-1D "snake" is shown in (Rouzine and Rozhnova 2018, fig. S6). Model parameters are $U_b = 0.001$, $N = 10^8$,

$R_0 = 3, a = 10, V_X = 0.5$. Thus, a quasi-1D trajectory arises on both 2D and a tree topology. To conclude, the main result of the present work is not restricted by the 1D model.

4.1.9 Approximations

As discussed in Section 4.1.5, the analysis is based on three approximations, as follows:

Linear fitness landscape approximation: The fitness landscape was approximated with its linear Taylor expansion, eq. (4.9). The actual fitness $w(u)$ plateaus at large negative and positive u (Fig. 4.4). The plateau at large positive u may be critical at the high-fitness edge of the traveling wave of $i(u)$. The traveling wave theory (Rouzine et al. 2008) predicts the edge location, $u = u_0$, with respect to the peak, eq. (4.26). For the linear approximation to work, the edge must stay within the linear range. Comparing the (simulated) edge location to the fitness landscape (Fig. 4.4), the linear approximation turns to be accurate at reasonable mutation rates $U_b = 10^{-6} - 10^{-4}$. The deviation from linearity at much higher U_b creates an error. To correct for the nonlinearity, the derivation in Rouzine et al. (2008) was generalized for the arbitrary fitness landscape, $w(u)$. The final result has an approximate form of eq. (4.24), but with u_0 and σ given by expressions

$$w(u_0) = c \log \frac{c}{eU_b}$$

$$\sigma = \frac{[w(u_0)]^2}{2u_0 \langle w \rangle_{u_0}} \tag{4.32}$$

which replace eq. (4.26) and the second of eq. (4.9), respectively. Here $\langle w \rangle_{u_0}$ denotes the fitness $w(u)$ averaged over the interval $0 < u < u_0$. The new effective value of σ, eq. (4.32), accounts for the nonlinearity. The analytic results shown in Fig. 4.5 are corrected for this effect.

Asymmetry of cross-immunity matrix: The analysis assumed an asymmetric immunity matrix, $K(u > 0) = 0$, eqs. (2.1). Briefly, strains u were not allowed to infect forward in the antigenic coordinate, $u' > u$. This approximation is mostly self-consistent, because, in the end result, almost all recovered individuals lag behind the infected individuals anyway (Fig. 4.2). However, this approximation is important in the trailing tail of the wave, where it compensates for a nonbiological artifact of the single-memory approximation of the model. If the immunity function is symmetric, a second wave may emerge from the trailing tail of the main wave and start moving backwards (Fig. 4.3). Fortunately, in the relevant parameter range, either introducing finite population size N or prohibiting infection forward eliminates this problem (Fig. 4.3).

Single-memory approximation: A central assumption is that only memory cells from the last infection are taken into account. Otherwise, one would have to label each recovered by an infinite set of antigenic labels instead of one. In reality, the immune system of an individual “remembers” all the previous infections (Murphy 2011). The single-memory approximation could, theoretically, create a cycle of reinfection within a small group of strains, such as $x_1 \to x_2 \to x_1$. The backward wave in 1D is an example of such a reinfection loop, fortunately, easily eliminated by matrix asymmetry that does not affect anything else. Other than this caveat, the single-memory approximation is applicable at large a, because the consecutive strains infecting a typical individual are sufficiently far apart. Indeed, the average last infecting strain is at the average distance $\langle u \rangle = -a$ behind the current infection (Fig. 4.2A). Therefore, the earlier strains whose memories are neglected are located at average coordinates $\langle u \rangle = -2a, -3a, -4a, \ldots$. At these remote locations in the trailing tail of the wave (Fig. 4.2A), they are very few recovered individuals available or infection. Therefore, the approximation introduces a modest correction. The correction is especially small for the exponential form of $K(u)$.

The same consideration applies to the multidimensional genetic space even better, because the loops are very unlikely, return of the virus to the old memory is highly unlikely, and the genetic space is effectively a tree (Section 4.1.8). As before, the wave is effectively a one-dimensional snake, and the argument from the previous paragraph applies.

4.2 Evolution of CD8 T-cell epitopes of HIV

CD8 T cells, cytotoxic lymphocytes (CTL), which kill infected cells and decrease virus replication by secreting cytokines, are a major factor in the control of HIV replication. A viral infection is followed by the rise of CTL causing virus depletion after a peak. The virus escapes full clearance and establishes a chronic infection by the combination of two methods: by killing helper CD4 T cells and thus decreasing the sensitivity of the immune response (Letvin et al. 2006; Sun et al. 2006; Potter et al. 2007; Vingert et al. 2010; Rouzine 2022), as well as by the emergence of mutations in the antigenically important regions, epitopes, spreading across infected cells. As a result, the virus becomes partially resistant to the immune response.

The knowledge of the factors controlling the order of mutated sites is useful for predicting conserved epitopes and designing a better vaccination strategy, so far, absent in HIV due to rapid antigen evolution of that virus. Batorsky et al. (2014) developed a mathematical model of viral dynamics driven by the selection pressure from multiple CTL clones, which themselves change in time in response to antigen (virus). The model includes a cost of escape mutation to viral replication, Δf (Ganusov and De Boer 2006; Ganusov et al. 2011), as well as the variable benefit of escape due to the partial impairment of CTL recognition, Δr, a parameter not considered before (akin to the parameter $1/a$ from Section 4.1). The results reviewed below demonstrate that the process of antigenic escape is regulated by the tradeoff between the cost and benefit of escape mutations. A trajectory of escape is predicted to move from the epitope sites with a high recognition loss and a low fitness cost to the sites with a low recognition loss and a high fitness cost. The positive correlation between fitness costs and benefits of escape mutation sequence predicted by this work was observed in the polymerase gene of HIV.

The range of Δr inferred in this work presented from published experimental studies is $\Delta r = (0.01 - 0.86)$; the assumption of complete recognition loss, as in the previous studies, $\Delta r = 1$, leads to an overestimate of mutation cost. The commonly observed pattern of escape, in which escape mutations are observed transiently, is explained from the combined effect of time-dependence immune pressure and the partial recognition loss. The reviewed work concludes that the partial nature of recognition loss is as important for predicting the order of escapes and, ultimately, for predicting conserved epitopes that can be targeted by vaccines, as fitness loss.

Despite the vigorous immune response, HIV persists for years. The proof that CTL does control HIV/SIV replication comes from experiments on the artificial depletion of cytotoxic $CD8^+$ T cells in SIV infected animals causing a rapid surge in virus load (Jin et al. 1999; Schmitz et al. 1999). The rapid evolution of HIV in epitopes demonstrates that cytotoxic cells, which recognize specific RNA sequences, are functional and exert selection pressure on virus to change. This "antigenic escape" impedes vaccine design (Finlay and McFadden 2006) and is implicated in viral adaption to a host, as well as pathogenesis (Rouzine and Coffin 1999c; Rouzine 2020a).

The transmitted virus strain is targeted by many CTL clones (Karlsson et al. 2007; Turnbull et al. 2009; Liu et al. 2013), each recognizing a short sequence of 8 – 10 amino acids presented on the cell surface by MHC-I molecules. Escape mutations in CTL epitopes emerge rapidly within a month of infection and continue to appear slowly throughout chronic infection, sometimes decreasing the rate of virus replication in absence of immune response (intrinsic fitness cost) (Friedrich et al. 2004; Leslie et al. 2005; Troyer et al. 2009). Importantly, not all targeted epitopes escape. Among epitopes that escape, the rate of escape decelerates dramatically after a few months of infection. Which parameters decide the rate of escape in a given epitope, what is the order of escapes, which epitopes escape and which do not are the questions addressed below (Liu et al. 2011; Henn et al. 2012).

Mathematical modeling was undertaken to study late escape mutations (Althaus and De Boer 2008; Mostowy et al. 2011) and the influence of the distributed CTL pressure (Ganusov et al. 2011; van Deutekom et al. 2013). These works emphasized two parameters, the mutational cost, Δf, and the active epitope number, n. The cost of an escape mutation can be inferred from its reversion sometimes observed upon transmission into HLA-type-mismatched individuals, whose epitopes are located in different places (Friedrich et al. 2004; Leslie et al. 2005; Kearney et al. 2009; Fryer et al. 2010). Frequent fixation of compensatory mutations outside of the mutated epitope also indicates a cost to fitness (Kelleher et al. 2001; Crawford et al. 2011), which effect was invoked to explain the pattern of HIV evolution within patients (Rouzine and Coffin 1999c) (Section 1.1, Model 4). Experiments show a wide range of fitness costs (Troyer et al. 2009; Song et al. 2012; Boutwell et al. 2013).

However, and this is the main point of this section, Δf and n are not the only parameters important for escape dynamics. As it was shown in HIV-infected individuals and SIV-infected animals, an escape mutant does not completely prevent the recognition of an epitope, and the drop in recognition efficiency varies between mutations (Cale et al. 2011; Liu et al. 2011).

Below, a cost–benefit diagram of escape mutation is used to analyze the dynamics of antigenic escape. The benefit is a partial loss of recognition by CTL, and the cost is a decrease of intrinsic fitness. The basic model by Althaus and De Boer (2008) is generalized to include the new factor of partial escape and investigate how the two parameters together determine the rate and the order of antigenic escape (Ganusov and De Boer 2006; Ganusov et al. 2011). The predictions are then compared to the data on correlation between the two parameters in clinically important escape mutations from the polymerase gene (Mostowy et al. 2012), and are used to estimate the range of recognition losses from three published experimental studies (Schneidewind et al. 2008; Kawashima et al. 2009; Matthews et al. 2012). Notably, the partial nature of escape can reproduce the diverse temporal patterns observed in patients, such as the change in the dominant mutated epitope sequence over time (Goonetilleke et al. 2009; Fischer et al. 2010; Liu et al. 2011) and nonnested pattern due to the changing CTL pressure and partial CTL recognition loss.

4.2.1 Model of HIV dynamics in the presence of multiple epitopes

The model includes the dynamics of multiple viral strains in the presence of multiple CTL clones, as illustrated by the diagram in Fig. 4.7A. The model is formalized by a system of ODE, as follows:

$$\frac{dT}{dt} = \lambda - d_T\, T - \beta T \sum_i f_i I_i \tag{4.33}$$

$$\frac{dI_i}{dt} = \beta T\, f_i I_i - d_I I_i - \kappa I_i \sum_j r_{ij} E_j \tag{4.34}$$

$$\frac{dE_j}{dt} = \sigma + c\, E_j \sum_i \frac{(r_{ij} I_i)}{h_j + \sum_i r_{ij} I_i} - d_E E_j \tag{4.35}$$

where $f_i < 1$ and $r_{ij} < 1$ are the relative replication rate of strain i and its relative recognition by CTL clone j as compared to the initially transmitted sequence, respectively.

The model includes processes, as follows. Target cells, which number is denoted T, are replenished at a linear rate λ cells per day, die (or leave the highly infectable phase) at a rate d_T, and are infected at a rate proportional to viral fitness and the number of productively infected cells I_i in the system. Each genome includes n epitopes, each with m amino acid sites, which corresponds to 2^{mn} possible strains. An infected cell is labeled by integrated provirus i comprising n epitopes denoted

$g^i = \{e_1^i, e_2^i, \ldots, e_n^i\}$, where $e_j^i = \{a_{j1}^i, a_{j2}^i, \ldots, a_{jm}^i\}$, and binary variable $a_{jk}^i = 0, 1$ indicates the presence or absence of a mutation in epitope j at site k. A mutation in epitope j at site k confers cost s_{jk} to the logarithm of replication rate, as given by

$$f_i \equiv \exp\left(-\sum_j s_{jk} a_{jk}^i\right) \tag{4.36}$$

By the definition, the strain with all $s_{jk} \equiv 0$ is the transmitted strain with fitness 1. Equation (4.36) assumes that mutation has an additive effect in logarithm. In other words, it neglects epistasis discussed in Chapter 2.

Effector CTLs whose number is denoted, E_j, are assumed to be replenished at a rate σ, proliferate at a rate proportional to the number of infected cells expressing the cognate epitope j and to avidity $\frac{1}{h_j}$, with a maximal proliferation rate c. They die at rate d_E and kill infected cells at a maximal rate κ. The model assumes that CTL are the cause of CTL death rather than viral cytotoxicity, see the original work for discussion (Batorsky et al. 2014, *Methods*). Mutation in epitope j at site k decreases the logarithm of CTL recognition by a_{jk}, as given by

$$r_{ij} \equiv \exp\left(-\sum_k a_{jk} a_{jk}^i\right) \tag{4.37}$$

For the sake of simplicity, avidities of CTLs are assumed to be equal, $h_j \equiv h$ (Section 4.2.5). It is convenient to denote the relative loss in fitness and recognition as $\Delta r_{ij} \equiv 1 - r_{ij}$ and $\Delta f_i \equiv 1 - f_i$.

In simulations below, mutations occur randomly with the rate $\mu = 3 \cdot 10^{-5}$ per site per generation of infected cells (period $1/d_I$) (Mansky and Temin 1995). Because the total numbers of all cell types are very large, dynamics is simulated deterministically, with a low cutoff below which a strain is considered extinct. Model parameters and the estimated range are listed in Tab. 4.2.

4.2.1.1 Simplified model to study the order of escape mutations

In addition to the main model, it is convenient to introduce a simplified companion model, which does not explicitly consider dynamics, but predicts the trajectory of escape in the phase diagram (Δf, Δr). The values of the fitness cost Δf and the recognition loss Δr are randomly sampled from a uniform distribution $[0, 1]$ for $n = 10$ epitopes with $m = 10$ sites per epitope. The sites are ranked in order of $\Delta r - n\Delta f$. When CTLs decay in response to an escape in an epitope, the immune pressure on all other sites in that epitope is decreased. Hence, CTL decay can be introduced approximately, as follows: Δr is reduced for all sites in the epitope as given by

$$\Delta r(n+1) = \Delta r(n) \exp\left[-d'_E \sum_i \Delta r_i(0)\right]$$

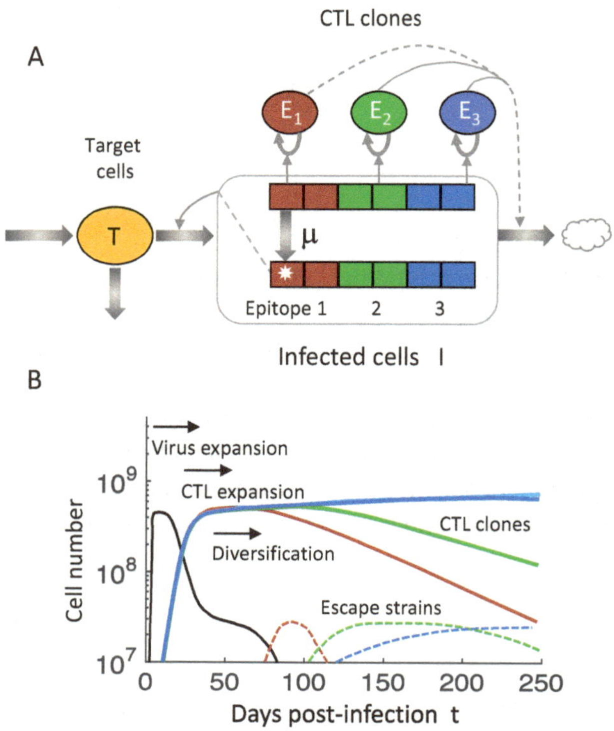

Fig. 4.7: Computational model of the interaction between HIV and multiple CTL clones.
(A) The model given by eqs. (4.33) to (4.37) comprises three interacting cell compartments: target cells (T), infected cells (I) and multiple CTL clones (E). Viral genomes contain multiple epitopes, which can mutate to partially abrogate CTL recognition. An escape mutation is denoted by a star symbol. Each CTL clone recognizes a single viral epitope and is stimulated to divide at a rate proportional to the number of infected cells with recognizable epitopes. The model is designed to study the rate of escape in epitopes when CTL pressure is distributed across multiple epitopes, as well as study intraepitope escape patterns when CTL respond dynamically to the infected cells that they recognize. Broad gray arrows: flux of cells from one compartment to another. Thin arrows: dependence of a flux on another compartment. Dashed lines represent attenuation of the interaction strength. (B) Simulation example showing three phases of HIV evolution. A single virus strain initiates the infection (transmitted strain, black curve). In response to the growing number of infected cells, multiple CTL clones are activated (solid color lines), and the system reaches a steady state. Finally, virus strains with escape mutations (dashed colored lines) replace the transmitted strain. In response to lowered activation signals, some CTL clones decline. The escape strains are colored to match the CTL clone against which an escape was most recently acquired. Model parameters: number of epitopes, $n = 6$; number of sites per epitope, $m = 1$. Epitopes 1 – 3 have parameters that allow escape $\Delta r_i = [0.1, 0.2, 0.3]$, $\Delta f = 0.01$, epitopes 4 – 6 have parameters that prohibit escape, $\Delta f_i = \Delta r_i = 0.1$. Other parameters are listed in Tab. 4.2. Based on Batorsky et al. (2014).

for all sites i in the epitope that have already escaped. Here parameter d'_E is defined as the decay rate per escape, in contrast to d_E in the main model, which is defined per day.

Tab. 4.2: Model parameters.

Parameter	Realistic value	Description	References
n	1 – 8	Number of epitopes recognized during the first 100 days	a
m	2 – 9	Number of sites per epitope important for recognition	a
d_T	$1.0/d$	Rate at which activated target cells leave the highly infectable phase	b
λ/d_T	5×10^8 cells	Activated target cell level	
d_I	$1.0/d$	Virus-induced infected cell death rate	c
$\kappa_j E_j^{ss}$	$4.0/d$	CTL-induced infected cell death rate	d
β	$1.1 \times 10^{-8} (\mathrm{d} \cdot \mathrm{cell})^{-1}$	Basic efficiency of target cell infection	d
s	$[0 - \infty]$	Intrinsic mutation cost	e
α	$[0 - \infty]$	Reduction of CTL recognition	e
Δf_i	$[0 - 1]$	Fractional reduction in intrinsic replication rate	Free parameter
Δr_{ij}	$[0 - 1]$	Fractional reduction of CTL recognition	Free parameter
κ	$1 \times 10^{-9} (\mathrm{d} \cdot \mathrm{cell})^{-1}$	CTL killing efficiency	f
σ/d_E	10^3 cells	Initial population of CTLs	g
c	1.0/d	Maximum growth rate of effector cells	g
d_E	0.1/d	Death rate of effector cells	d
h_j	2.5×10^8 cells	Number of recognized infected cells for half maximal proliferation of CTL (inverse avidity).	h

[a]Goonetilleke et al. (2009), Fischer et al. (2010), Liu et al. (2011), and Henn et al. (2012)
[b]Li et al. (2005), Sergeev, Batorsky, Coffin et al. (2010), and Sergeev, Batorsky and Rouzine (2010)
[c]Klatt et al. (2010)
[d]Kuroda et al. (1999), Ogg et al. (1999), and Sergeev, Batorsky and Rouzine (2010)
[e]Schneidewind et al. (2008), Kawashima et al. (2009), Troyer et al. (2009), Matthews et al. (2012), Song et al. (2012), and Boutwell et al. (2013)
[f]Ogg et al. (1998)
[g]Kuroda et al. (1999)
[h]Haase (2011)

4.2.2 Simulations of the dynamics of antigenic escape

4.2.2.1 Phases of HIV infection

The model in Section 4.2.1 was solved numerically in MATLAB™. The initial condition for eqs. (4.33)–(4.35) is that target cells T are at their normal steady-state level in an uninfected patient, there is a small initial number of effector CTL cells (naive cells), and a small amount of virus is introduced. The results show three phases of HIV infection (Fig. 4.7B), as follows:

Phase 1: The transmitted HIV strain expands and depletes target cells.

Phase 2: CTL clones that recognize the transmitted strain expand and decrease the number of infected cells, until a steady state is established (chronic infection) [Section 4.2.6, eqs. (4.43)–(4.45)].

Phase 3: Escape mutants emerge and expand. While the clonal composition of CTLs changes, their total number is weakly affected (Fig. 4.8).

The dynamics of virus and CTL during escape depends on the number of active epitopes, n, and the recognition loss per mutation, Δr (Tab. 4.2). In the simplest case, when there is only one CTL clone, which is in steady state, and a rising escape mutation completely abrogates recognition, $\Delta r = 1$, the clone contracts, and the mutant virus load grows until target cells are depleted enough to check this growth. In contrast, if the recognition loss due to mutation is partial, $\Delta r < 1$, the virus load increases only transiently, and the CTL clone expands until reaching a new steady state. When several CTL clones with similar avidity control the virus population (Fig. 4.7B), and an escape mutation emerges in an epitope, the cognate CTL clones contracts, because it proliferates at a smaller rate than other CTL clones and dies as fast, eq. (4.47). Due to the other CTL clones, the population of infected cells stays under control after the escape mutation. The predicted three-phase dynamics was observed in HIV-infected individuals including the decline of CTL clones to mutated epitopes (Goonetilleke et al. 2009; Fischer et al. 2010; Liu et al. 2011; Henn et al. 2012)

4.2.2.2 The determinants of the escape rate of a mutant strain

The presence of CTL creates positive selection for escape mutations (Phase 3 in Fig. 4.7B). Mutant clones in different epitopes grow with different rates, due to variation in the loss of both fitness and recognition. The exponential growth rates determine which mutant strain will be most-fit and thus dominate the virus population. Suppose, the initial state is the steady state when RHS of eqs. (4.33)–(4.35) are all equal to zero. Suppose also, an escape mutation occurs in Epitope 1 causing a fitness cost, Δf_i, and a loss of CTL recognition, Δr_i. As follows from eq. (4.34), the mutant strain grows initially as $I_i(t) = I_i(0)\exp(\epsilon_i t)$, with the expansion rate ϵ_i given by

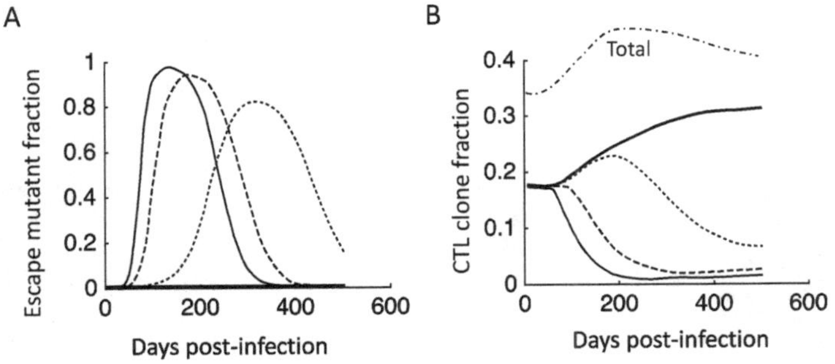

Fig. 4.8: Changing composition of CTL and infected cell populations.
Simulation of model given by eqs. (4.33)–(4.37) with the same parameters as in Fig. 4.7B. (A) Three escape variants in viral epitopes spread through the population of infected cells to abrogate CTL recognition, but then are cleared by a CTL clone cognate to another epitope. The other epitopes in this example cannot escape. (B) Dynamics of the first four clones of CTLs on time. The first three CTL clones decay due to escape mutations in cognate epitopes (A). Once an escape mutation has spread to the majority of infected cells, the corresponding CTL clone begins to decay with a rate proportional to the fraction of recognition lost. Note that the total CTL population given by the sum over the four curves remains roughly constant in a steady state (dash-dotted curve). Based on Batorsky et al. (2014).

$$\epsilon_i = \frac{1}{I_i}\frac{dI_i}{dt} = kE_{\text{tot}}\left(\frac{\Delta r_i}{n_1} - \Delta f_i\right) \tag{4.38}$$

where $1/n_1$ is the fraction of CTL population recognizing Epitope 1. Thus, the escape rate reflects the trade-off between recognition loss and fitness cost, which determines whether a mutant strain will grow, $\epsilon_i > 0$. Therefore, in order to infer the fitness cost of an escape mutation from the growth rate, as it is done routinely, one has to measure not only ϵ_i, but also the loss of recognition. In the data from HIV-infected individuals, escape slows down during the first 100 days post infection (Goonetilleke et al 2009; Ganusov et al. 2011; Liu et al. 2011; Henn et al. 2012). The model predictions agree with these findings (Fig. 4.9). The observed variation of escape rates over time and across sites can be obtained from the variation of Δr alone, assuming a minor fitness cost.

4.2.2.3 The trajectory of escape mutations in the cost-benefit plane

During chronic infection, HIV elicits a limited number (10 – 100) of detectable CTL responses against different epitopes (Turnbull et al. 2009), in which 5 – 30 escape mutations are fixed (Goonetilleke et al. 2009; Liu et al. 2011). The next task was to study the trajectory of escape mutations among epitopes and epitope sites in the cost–benefit plane. For this aim, the simplified model focusing only on the order of escape is easier to handle, leaving the details of model dynamics out of the picture. Parameters Δr and Δf were randomly und uniformly generated in the interval between 0 and 1 for a genome with $n = 10$ epitopes and $m = 10$ sites per epitope (Fig. 4.10). The sites are ranked in the descending order of the escape rate ϵ_i, eq. (4.38).

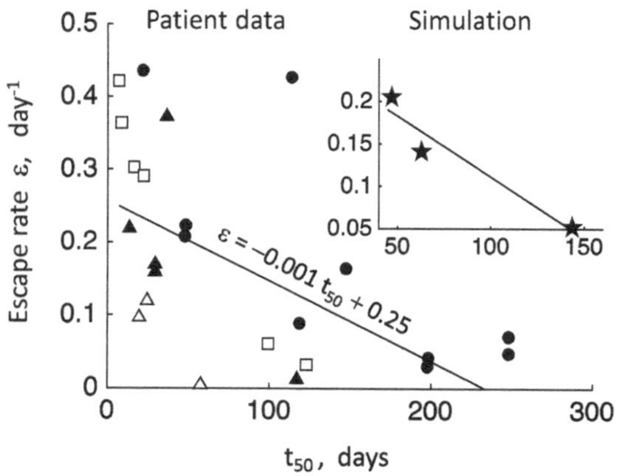

Fig. 4.9: Escape rate and t_{50} are negatively correlated in two experimental studies. The frequency of a mutated epitope for epitopes over time is fit to the curve $f(t) = \{1 + \exp[-\epsilon(t - t_{50})]\}^{-1}$, which describes deterministic selection on a single site with selection coefficient ϵ, in order to determine parameters ϵ and t_{50}. Symbols show data from a single patient studied in Liu et al. (2011) (filled circles) and multiple patients studied in Goonetilleke et al. (2009) CH40 (filled triangles), CH58 (open triangles), CH77 (open squares). Inset: Simulation example showing the correlation between escape rate, ϵ, and the time that the mutation spreads to 50% of the population of infected cells, denoted t_{50}. Parameters ϵ and t_{50} are found for the three escape mutations shown in Fig. 4.7B that occur in the first 200 days post infection. Based on Batorsky et al. (2014).

After a large number of random sets is generated, the values of Δr and Δf of a site of each rank are found to correlate negatively. A typical trajectory in the cost–benefit plane runs from a high recognition loss and a low fitness cost to a low recognition loss and a high fitness cost (Fig. 4.10A). The escape rate per epitope decreases with the number of escape (Fig. 4.10B). Consistent with observations (Goonetilleke et al. 2009; Henn et al. 2012), each epitope in simulation escapes at more than one site.

The interplay between CTL dynamics and partial escape determines the trajectory of HIV escape. As already mentioned, an escape mutation causes the decay of the cognate CTL clone. This CTL decay decreases the recognition loss due to other escape mutations in that epitope, because the immune pressure on the epitope is relaxed. Hence, the CTL decay has important consequences for the sequence of escape in the cost–benefit plane. In the absence of CTL decay, the trajectory stays straight until no more escapes are possible, because the cost exceeds the benefit (Fig. 4.10A). When CTLs are allowed to decay in response to an escape in an epitope, the positive selection pressure on all other sites in that epitope is decreased, and the average trajectory bends toward the X-axis (Fig. 4.10C). Due to CTL decay, the total number of escape mutations becomes less (from 55 in Fig. 4.10A to 20 in Fig. 4.10C), and escape mutations per epitope are fewer as well (Fig. 4.10B, D).

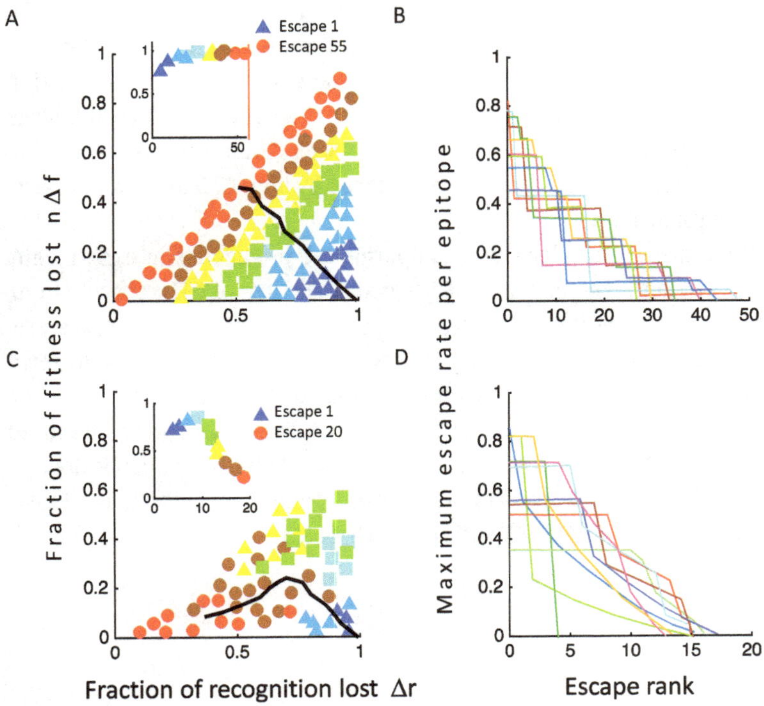

Fig. 4.10: The escape trajectory in the cost–benefit plane bends over time due to CTL decay. Fitness costs and recognition losses are randomly generated for 100 sites (10 epitopes with 10 sites per epitope) in order to study the sequence of escaped sites (black line) in the whole genome without CTL decay (A, B) or with CTL decay (C, D) for 1,000 simulation runs. (A) For each site that escapes, the fractional fitness cost, Δf, multiplied by the number of epitopes, $n = 10$, and fractional recognition loss, Δr, is shown [eq. (4.38) and Tab. 4.2]. Colors show the predicted rank of escape mutations, from early escape mutations (blue) to late escape mutations (red). The average trajectory over all runs (black curve) moves from high recognition loss, low fitness cost to low recognition loss, high fitness cost. Inset: The best-fit slope for each escape rank. (B) The maximum escape rate of any epitope site for all 10 epitopes for a representative simulation run. (C-D) As in (A-B), except including CTL decay. CTL decay is simulated by reducing recognition losses for all epitope sites in epitopes that have partially escaped according to $\Delta r(n+1) = \Delta r(n)\exp[-d_E' \sum \Delta r_i(0)]$, summing over all i sites in the epitope that have escaped with $d_E' = 0.1$ per escape event. Based on Batorsky et al. (2014).

4.2.3 Correlation between escape cost and benefit in Pol gene is explained

Simulation results in Fig. 4.10 help to understand the observations by Mostowy et al. (2012) that the fitness cost of escape mutations correlates positively with the HLA-binding loss in polymerase sequences sampled from HIV-infected individuals. The correlation is predicted by eq. (4.38), which implies that costly mutations will not be amplified unless they also offer a large benefit to virus. The weak strength of the observed correlation (slope = −0.12) is explained from the large number of targeted epitopes, $n \gg 1$. CTL decay caused by escape reduces the correlation further. Escape mutations sampled during the acute phase of HIV infection are predicted to have a much larger correlation slope than those observed during the chronic phase (Batorsky et al. 2014, fig. 3B). Since the majority of database sequences used by Mostowy et al. (2012) were sampled from chronic phase, the small slope of the correlation is caused by both CTL decay and a large epitope number in chronic phase.

An important nuance to consider when comparing the predictions to experiments in Mostowy et al. (2012) is the method of measurements. The loss of CTL recognition Δr is a complex parameter comprising the changes in antigen processing, cell-surface presentation, and the HLA-epitope-TCR binding (below "HLA binding"), while (Mostowy et al. 2012) use only the calculated loss of HLA binding, ΔB. To obtain the unit conversion, one has to compare data from publications in which both parameters were measured (Schneidewind et al. 2008; Kawashima et al. 2009; Matthews et al. 2012), which is done in Section 4.2.8. The result is a strong correlation between Δr and ΔB expressed as a linear relationship $\Delta B = 0.78\,\Delta r - 0.004$ [Fig. 4.11 and eqs. (4.52)–(4.55)].

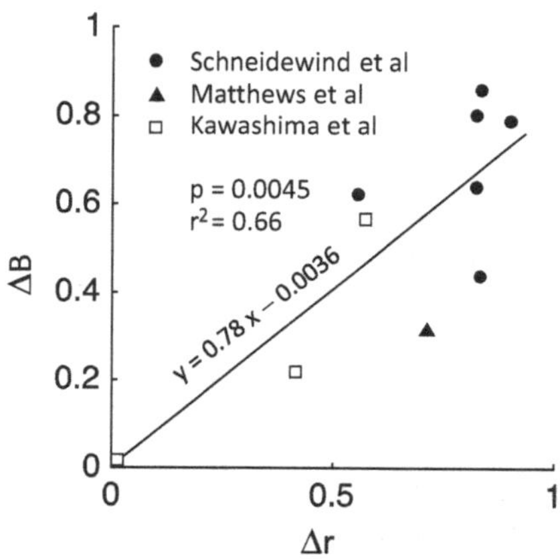

Fig. 4.11: Estimating the relationship between Δr and ΔB from three published experiments. (Schneidewind et al. 2008; Kawashima et al. 2009; Matthews et al. 2012). Based on Batorsky et al. (2014).

4.2.4 Three patterns of antigenic escape in an epitope with two sites

In response to CTL pressure, different combinations (haplotypes) of mutations can emerge in epitopes in the course of time. (Goonetilleke et al. 2009) observed that the order in which escape haplotypes appeared varied between epitopes. One can classify these orders into "simple," "nested," and "leapfrog" patterns of escape and illustrate each pattern using an epitope with two sites and four possible haplotypes. The escape mutation at a site is denoted by allele 1; otherwise, it is 0. The simple pattern (observed infrequently) is represented by a single escape haplotype, $00 \to 10$. The nested pattern adds a new mutation to the single mutant, $00 \to 10 \to 11$. This pattern is expected if both sites are under a constant, positive selection pressure, $\epsilon > 0$, throughout the course of infection. The leapfrog pattern is caused by a switch in the dominant single-site mutation: $00 \to 10 \to 01$.

In principle, at small population sizes, $N\mu \ll 1$, such a change could be possible due to clonal interference, when the first escape mutation is under weaker selection pressure than the second mutation that emerges later due to random delay and pushes the first clone out of the population (Gerrish and Lenski 1998; Rouzine 2020b). However, several groups (Rouzine and Coffin 1999b; Frost et al. 2000; Pennings et al. 2014) inferred from HIV genomics that the population size of HIV in untreated patients is very large, $N\mu \gg 1$, which implies the lack of such a random delay. In the present model, the leapfrog pattern of escape arises due to the time-dependent nature of selection pressure caused by the immune response. Below the region of leapfrog pattern is mapped in the plane of fitness and recognition loss.

Consider a genome composed of a large number $n \gg 1$ of simple two-site epitopes, $m = 2$. The results from this simplest case can be used to illustrate the general case $m > 2$. Note that, out of eight or nine amino acids total of a recognized peptide, some are extremely costly and can never escape. From eqs. (4.38), the escape rates for each of the three mutant haplotypes of epitope j are given by

$$\epsilon_{10} = kE_{\text{tot}} \left(\frac{\Delta r_1}{n_j(t)} - \Delta f_1 \right) \tag{4.39}$$

$$\epsilon_{01} = kE_{\text{tot}} \left(\frac{\Delta r_2}{n_j(t)} - \Delta f_2 \right) \tag{4.40}$$

$$\epsilon_{11} = kE_{\text{tot}} \left(\frac{\Delta r_1 + \Delta r_2 - \Delta r_1 \Delta r_2}{n_j(t)} - (\Delta f_2 + \Delta f_1 - \Delta f_1 \Delta f_2) \right) \tag{4.41}$$

where $1/n_j(t)$ is the time-dependent fraction of the total CTL population E_{tot} taken by clone E_j. Below the initial escape rate of haplotype 10 is assumed to be higher than that of 01, $\epsilon_{10} > \epsilon_{01}$.

Once the clone with the fastest escape haplotype becomes sufficiently large and partly replaces the wild type, the CTL clone E_j will decay in time. The total virus load will stay the same, because the other CTL clones will control virus. As the clone fraction $1/n_j(t)$ and, hence, the positive part of selection pressure on the epitope declines, the mutation costs, Δf_i, in eqs. (4.39)–(4.41) become more important, and the change in the best-fit haplotype occurs. In the long term, the decay of the CTL clone will cause the wild type (00) to become a dominant strain again, but the existence of compensatory mutations not considered in this work may stabilize the last escaped haplotype (Section 4.2.5). The sequence of dominant haplotypes can be one of three patterns, as follows (Fig. 4.12):

Simple pattern (Fig. 4.12A): This pattern of escape occurs if the second site cannot escape, because ϵ_{01} in eqs. (4.40) is negative, or if ϵ_{01} is positive but small, so that the double escape mutant 11 does not have time to grow to an observable level, before wild-

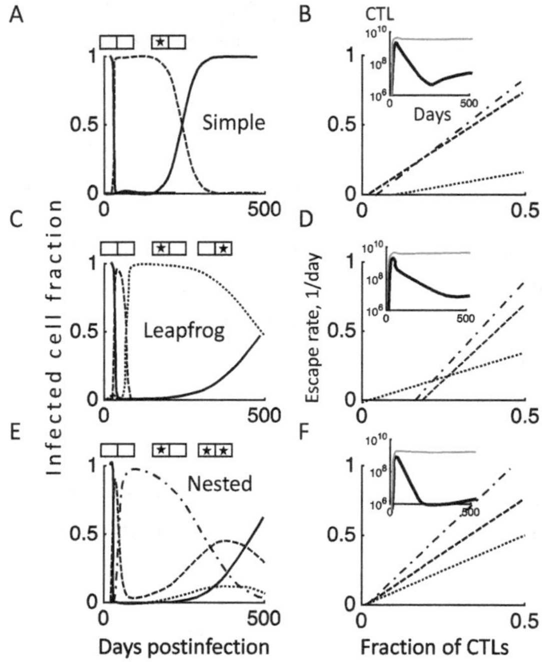

Fig. 4.12: Dynamical selection pressure from CTLs causes three possible patterns of intraepitope escape: example for an epitope with two sites.
Intraepitope escape in one epitope with two sites is studied for the model shown in Fig. 4.7 and eqs. (4.33) to (4.37). The sequence in which haplotypes are selected depends on the distribution of fitness and recognition losses within an epitope. The fraction of the infected cell population containing each of the four haplotypes in the escaping epitope is shown for "simple" (A), "leapfrog" (C) and "nested" pattern (E). (B, D, F) The dependence of the escape rate, eqs. (4.39) to (4.41), for each haplotype on the fraction of CTLs cognate to the epitope. The inset shows CTL dynamics: the size of the CTL clone to escaping epitope (thick black curve) and the total CTL number (thin gray curve). Parameters: (A,B) $\Delta r_{1k} = [0.4, 0.1]$, $\Delta f_{1k} = [0.01, 0.01]$, (C,D) $\Delta r_{1k} = [0.6, 0.2]$, $\Delta f_{1k} = [0.1, 0.003]$, (E,F) $\Delta r_{1k} = [0.4, 0.25]$, $\Delta f_{1k} = [0.003, 0.003]$ with $n = 2, m = 2$ for all panels; other parameters are given in Tab. 4.2. Based on Batorsky et al. (2014).

type 00 regains the fitness advantage. Observation of this pattern requires a large difference in the values of fitness costs or recognition loss between the two sites.

Nested pattern (Fig. 4.12E): When both single-mutant haplotypes 01, 10 preserve a positive growth rate during a long period of time, the double-mutant haplotype has the highest escape rate. The nested pattern requires that the parameters of the two single mutants are similar.

Leapfrog pattern (Fig. 4.13C): This pattern of escape occurs if, due to CTL decay, haplotype 01 gains the advantage over initially dominant 10 before 10 loses to the wildtype. For this to occur, site 1 must have both a larger fitness and a larger recognition loss than site 2

$$\Delta r_1 > \Delta r_2, \quad \Delta f_1 > \Delta f_2 \tag{4.42}$$

Note that the cognate CTL clone decays faster rate during the initial time period when variant 10 is dominant and then slows down (inset in Fig. 4.12D). The relatively high recognition loss of the mutation at the first site is the reason. The full conditions required for this pattern to be observed are studied in Section 4.2.7.

To determine which pattern of escape takes place for each set of recognition and fitness losses at each of the two epitope sites, numerical simulation was used. The epitope number n was fixed. The rate of escape of the first haplotype ϵ_{10} and the ratio of the fitness costs between two epitope sites $\Delta f_1/\Delta f_2$ were chosen at the values representative of either acute or chronic infection, and for either fast decaying (Fig. 4.13) or slow decaying CTLs (Batorsky et al. 2014, fig. S3). The ranges of escape rates observed in HIV-infected patients at different times are shown in Fig. 4.9.

At large escape rates ϵ_{10}, the leapfrog is observed at large Δr_1 and a broad range Δr_2 (Batorsky et al. 2014, fig. S3). In some cases, haplotype 11 is observed as a short intermediate between haplotypes 10 and 01, which is labeled "nested leapfrog" in Fig. 4.13. The time period within which a given haplotype dominates the population depends on values Δf_k and Δr_k of the sites. The less costly haplotype is the longer it dominates the population (time t_{01} in the insets of Fig. 4.13). Cognate CTL decay the fastest when variant 11 is the largest (inset in Fig. 4.12F).

4.2.5 Approximations

Using a model with multiple CTL epitopes, the rate of viral escape is predicted to depend critically on the degree of recognition loss, as well its fitness costs and the total number of acting CTL clones. Furthermore, the fact that CTL populations change their relative sizes in response to escape has observable consequences for the order of escape mutations over time. Declining CTL levels relax selection pressure on an escaped causing the dominant epitope haplotype to switch in a nonnested fashion.

The model has several built-in assumptions that deserve a comment, as follows.

Preexisting mutations: The population of infected cells is assumed to be large, so that all single escape mutants are assumed to preexist in the population before the rise of the immune response (Althaus and De Boer 2008; Mostowy et al. 2011). This assumption follows from the studies demonstrating a large effective population size of HIV in untreated patients (Rouzine and Coffin 1999b; Kouyos et al. 2006; Pennings et al. 2014; Rouzine, Coffin et al. 2014). The preexistence of CTL escape mutations can also be inferred from the observed preexistence of mutations resistant to replication inhibitors (Richman et al. 1994; Lech et al. 1996). Furthermore, because the average number of escape mutations per epitope is larger than that per drug, the smallest average mutation cost of an escape

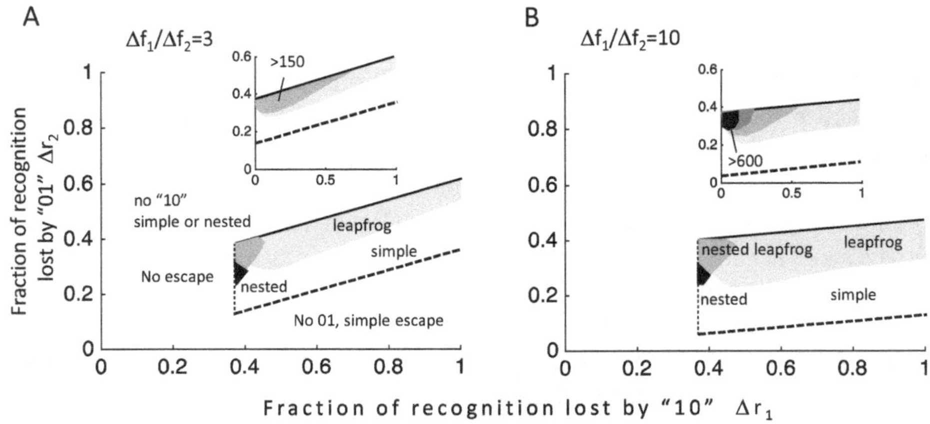

Fig. 4.13: The pattern of emergence of escape variants in a single epitope contains information about the fraction of recognition and fitness lost by single-site mutations in the epitope.
The pattern of escape (Fig. 4.12) is obtained from simulation of the model (Fig. 4.7), eqs. (4.33) to (4.37) with two sites per epitope, $m = 2$ for a range of recognition and fitness losses. The pattern that is obtained is plotted as a function of the parameters of recognition loss at the first and second site, Δr_2 and Δr_2, respectively. Inequalities (4.48) (straight solid line) and (4.51) (straight dashed line) determine the region where the leapfrog pattern can be observed. Regions that require $\Delta f_1 < 0$ are not allowed (dotted line). The shaded regions between these three lines correspond to regions of parameter space where both sites escape. The corresponding patterns are: "leapfrog" 10 $\longrightarrow$ 01, Fig. 4.12C), "nested" 10 $\longrightarrow$ 11, Fig. 4.12E), "nested leapfrog" 10 $\longrightarrow$ 01. Observation of the leapfrog pattern in an epitope tightly constrains the fraction of CTL recognition loss conferred by sites in an epitope. The inset shows the length of time during which haplotype 01 is dominant in the escaping epitope. Fixed parameters are: the escape rate of the first haplotype $\epsilon_{10} = 0.5$ and the number of targeted epitopes $n = 3$, values which correspond to escape mutations that occur in acute infection [see original work (Batorsky et al. 2014), Fig. S3 legend, for parameters that correspond to chronic infection]. Fitness costs are chosen such that the second site is less costly than the first: $\Delta f_1 / \Delta f_2 = 3$ (A) or much less costly than the first, $\Delta f_1 / \Delta f_2 = 10$ (B). Other parameters are given in Tab. 4.2. Based on Batorsky et al. (2014).

mutation is only lower than for drug resistance. These estimates cast doubts on the interpretation of very late first-in-epitope mutations as mutation *de novo* (Liu et al. 2006).

Reversion is prevented by compensatory mutations: Mutations compensating for fitness losses for drug-resistance variants (Quinones-Mateu and Arts 2002; Bonhoeffer et al. 2004; Hinkley et al. 2011) and for immune-escape variants (Kelleher et al. 2001; Troyer et al. 2009; Crawford et al. 2011) are well-documented for HIV [see Poon et al. (2010) for review]. Mutations compensating for early escape mutants have been proposed as the main reason for a large number of diverse sites in HIV genome (Rouzine and Coffin 1999c) and for the emergence of "swine flu" pandemics of 2009 (Pedruzzi and Rouzine, 2021). However, the evolutionary dynamics in the presence of epistasis in a multi-locus system is a very complex matter (Chapter 2). Therefore, this simple model is not able to account for compensatory mutations and predicts the eventual reversion of any escape strain with a finite fitness cost to the transmitted variant. As the next best thing, this work calculates the period of maintenance of an escape haplotype in the population before the reversion, when compensation is necessary to prevent it (inset of Fig. 4.13). When the leap-frog patter is observed in patients, the escape haplotype is usually short-lived compared to the next escape variant, and the transmitted strain is not observed (Goonetilleke et al. 2009; Fischer et al. 2010; Liu et al. 2011). The model predicts that the first escape is immunologically effective, but also costly compared to the second variant, and that the second variant is compensated gradually.

Multiple CTL clones of equal avidity: This is a reasonable assumption for the first year of infection. At this time, a large number of CTL clones are activated and are maintained for many months at similar levels (Karlsson et al. 2007; Turnbull et al. 2009). Therefore, they must have similar avidity. The steady-state viral load is predicted to be proportional to the inverse avidity of the best CTL clone h, eq. (4.44) (Section 4.2.6). Kadolsky and Asquith (2010) estimated that the average viral load increases by only 0.051 decimal log copy/mL per one CTL escape (while the average log of HIV load is around 4.5 – 5). The present model interprets this spacing as the average avidity spacing, which justifies the above assumption. Given closely spaced avidities, recognition and fitness losses control the order and timing of escape mutations, rather than avidity variation.

Other assumptions: Additional simplifications are, as follows.

(i) *The proliferation rate of CTL clones does not depend on the total CTL level*: In the original model (Althaus and De Boer 2008), the total number of CTL limits the growth of individual CTL clones competing for targets. While this ceiling is likely important in the CTL peak (Rouzine 2022), its importance in chronic infection, when only a few percent of CD8 T cells are HIV-specific (Schmitz et al. 1999), is doubtful. Also, the target availability is hardly important, because most CTL are not attached their targets even at the peak of infection (Haase 2011).

(ii) *CTL are short-lived in the absence of antigen*: The average lifetime of 10 days is consistent with the modeling results (Rouzine et al. 2006; Sergeev, Batorsky, Coffin et al. 2010; Sergeev, Batorsky and Rouzine 2010; Rouzine et al. 2015) informed by the experimental infection of rhesus macaques (Ogg et al. 1998; Kuroda et al. 1999).

(iii) *Recombination is ignored*: In the long term, recombination may accelerate antigenic escape by collecting mutations from different epitopes, as suggested by experiment (Mostowy et al. 2011) and modeling of HIV data (Neher et al. 2010; Batorsky et al. 2011; Neher et al. 2013). For the intraepitope dynamics, recombination is negligible.

(iv) *New CTL clones against escape mutants are neglected*: The reason for this assumption is the time delay required to produce new clones, and that the new clone has a smaller avidity, otherwise he would have already emerged earlier. Another reason it that an escaped epitope is still under partial control of an old clone and the other clones control virus load. Combined, these factors either delay or make impossible the emergence of clones cognate to escaped epitopes.

(v) *Fitness costs are small*: Based on eq. (4.38), a costly escape with $\Delta f_i \approx 1$ cannot occur, because recognition losses are partial, and the number of CTL clones is large, $n \gg 1$. The justification is that such conserved sites are excluded from analysis, which is why the effective number of sites per epitope is less than its total length 8 – 9.

(vi) *Fitness costs are positive*: Some mutations transmitted to an individual can revert slowly, that is, they may have a negative fitness cost. However, they are observed outside of the individual's epitopes (Kearney et al. 2009).

(vii) *Epitopes evolve separately*: Typically, escapes in epitopes do not overlap much in time, except for a few early escape mutations during the acute viremia peak (Goonetilleke et al. 2009). In general, however, linkage effects may complicate the picture and need to be kept in mind (Rouzine 2020b, chapter II).

To summarize Section 4.2, the incomplete loss of CTL recognition during escape affects the rate and the order of escape mutations not less than the fitness costs to mutation and the diversity of CTL clones. The simple model also naturally explains the positive correlation between recognition loss and fitness cost in escape mutations observed in HIV (Mostowy et al. 2012). The results emphasize the need for direct measurements of recognition losses of different escape mutations, which connects this work to Section 4.1, where this parameter was described by the cross-immunity function $K(u)$. The simple trajectory approach (Fig. 4.10) can serve for the prediction of conserved epitopes of use in vaccines.

4.2.6 Derivation of the steady state, escape rate, and clone contraction

The remaining Sections 4.2.6–4.2.8 serve as a mathematical appendix.

Steady state: Before antigenic escape, the virus population is uniform in the transmitted strain I_{trans} and expanding, which causes all n CTL clones to expand and decrease the virus load. Eventually, the expansion of CTL is balanced by their death (the first and second phases in Fig. 4.7B) and a steady state is approached. By setting all time derivatives in eqs. (4.33)–(4.35) to zero, with all $r_{ij} \equiv 1$ and $f_i \equiv 1$, one gets

$$T^{\text{ss}} = \frac{\lambda}{d_T + \beta I^{\text{ss}}} \tag{4.43}$$

$$I^{\text{ss}} \approx \frac{d_E h}{c}, \quad d_E \ll c \tag{4.44}$$

$$E_{\text{tot}} \equiv \frac{\beta T^{\text{ss}} - d_I}{\kappa}, \quad \sum_j E_j^{\text{ss}} = E_{\text{tot}} \tag{4.45}$$

The infected cell number in the steady-state I^{ss} is proportional to the inverse avidity of the CTL clones, h, and n is the number CTL clones, assumed to have equal avidities and precursor frequencies (Section 4.2.5). The steady state represents an indifferent equilibrium, in which each CTL clone takes $\sim 1/n$ of the total CTL number, E_{tot}.

The escape rate of the first escape mutant, eq. (4.38), can obtained from eq. (4.34), as follows. In the initially uniform virus population with $r_{ij} \equiv 1$ and $f_i \equiv 1$, a mutated strain with a single mutation in epitope 1 has fitness and recognition reduced by $\Delta f_i = 1 - e^{-s}$ and $\Delta r_i = 1 - e^{-r_i}$, respectively. From eq. (4.34), one gets

$$\frac{dI_i}{dt} = \left[\beta T^{ss}(1 - \Delta f_i) - d_I - \kappa \left(\sum_j E_j^{ss} - E_1^{ss} \Delta r_i \right) \right] I_i \tag{4.46}$$

Assuming cell populations have reached steady state, eqs. (4.43)–(4.45), we can substitute eq. (4.45) and $E_1 = \frac{kE_{tot}}{n_1}$, where $1/n_1$ is the fraction of CTL population occupied by CTL clone 1, into eq. (4.46). Upon mutual cancelation of the two main terms in (4.46), one arrives at eq. (4.38).

4.2.6.1 Contraction of CTL clones after escape

Suppose, an escape mutation in an epitope has occurred and is rising in a population. Let I_j^{th} be the level of infected cells that keeps the CTL clone recognizing epitope j in a steady state (the threshold). Before escape mutations rise to high frequencies in the population, all CTL clones have the same threshold, $I_j^{th} = I^{ss}$. When the escape mutation has taken over the majority of the virus population, CTLs to the escaping epitope recognize infected cells less efficiently, and the number of infected cells necessary to maintain the clone (threshold) in steady state is elevated. If escape mutation reduced recognition by Δr_j, the new plateau is given by

$$I_j^{th} = \frac{I^{ss}}{1 - \Delta r_j}$$

CTL clones that target unescaped epitopes maintain their original, lower threshold I^{ss}. Therefore, as long as at least one epitope has not escaped yet, the level of infected cells will be pinned by the CTL populations targeting conserved epitopes (Fig. 4.7B). CTLs to escaped epitopes are below their thresholds and decline as

$$E_j(t) = \frac{E_{tot}}{n} e^{-d_E \Delta r_j t} \tag{4.47}$$

where eq. (4.35) and the strong inequality $\frac{d_E}{c}(1 - \Delta r_j) \ll 1$ (Tab. 4.2) are used. Thus, CTL clones recognizing escaped epitopes decline due to the presence of the CTL clones recognizing conserved epitopes (Fig. 4.7B).

The size of the total CTL population, E_{tot}, is determined by the change in the replication rate of the virus βT^{ss}, eq. (4.45). At small fitness costs assumed in this work,

$\Delta f_j \ll 1$, the steady-state number of target cells, eqs. (4.43), and the total CTL population, E_{tot}, change little, which is consistent with the experimental observation that the overall levels of CTL and target cells change slowly during chronic infection, while CTL and virus composition changes a lot.

4.2.7 Leapfrog pattern of escape

4.2.7.1 Necessary conditions

The leapfrog pattern is characterized by a switch 00 → 10 → 01, as opposed to the simple escape pattern 00 → 10 or the nested pattern 00 → 10 → 11. The latter is predicted in a deterministically large population if both sites are advantageous, $\epsilon > 0$, and selection pressure is constant. In the absence of clonal interference due to a large population size, the leapfrog pattern implies time-dependent CTL selection pressure. The conditions on the fitness and recognition loss of both epitope sites necessary for the leapfrog pattern to occur are derived below.

Without the loss of generality, consider two haplotypes with escape rates, $\epsilon_{10} > \epsilon_{01}$, given by eqs. (4.39) and (4.40):

$$\frac{\Delta r_1}{n} - \Delta f_1 > \frac{\Delta r_2}{n} - \Delta f_2 \tag{4.48}$$

Equations (4.39)–(4.41) can be used to determine conditions under which the haplotype with the largest escape rate switches from 10 to 01.

Suppose 11 does not reach observable frequency before 01 becomes dominant, and the target cells stay constant at their steady state. Both approximations are valid as long as all fitness costs are sufficiently small, as given by Δf_1 and $\Delta f_2 \sim 1/n \ll 1$, as assumed throughout this work. For the leapfrog pattern to be observed, haplotype 01 must become better fit than both 10 and the transmitted haplotype 00, before 00 becomes the best-fit again. The CTL number at which 00 overcomes 10 should be compared with the CTL number at which 01 becomes better-fit than 10. To find the condition of switch from haplotype 10 to 01, note the two have the same expansion rate, when the right-hand sides of eqs. (4.39) and (4.40) are equal. The fraction of the total CTL population taken by cognate CTL clone of size E_j denoted $1/n_{switch}$, at this switch point is given by

$$\frac{1}{n_{switch}} = \frac{\Delta f_1 - \Delta f_2}{\Delta r_1 - \Delta r_2} \tag{4.49}$$

Next, the transmitted strain 00 has a higher intrinsic fitness than 10, and will begin to grow once $\varepsilon_{10} = 0$. Reversion of haplotype 10 will begin once $E_j = 1/n_{rev}$, when RHS of eq. (4.39) is equal to zero, which is equivalent to

$$\frac{1}{n_{\mathrm{rev}}} = \frac{\Delta f_1}{\Delta r_1} \tag{4.50}$$

The leapfrog pattern can be observed if the switch from 10 to 01 occurs before the reversion of haplotype 10

$$n_{\mathrm{switch}} > \frac{1}{n_{\mathrm{rev}}} \tag{4.51}$$

Together, eqs. (4.48)–(4.51) represent the necessary conditions for the leapfrog pattern to occur. They can be met only if site 1 has both a greater fitness and recognition loss than site 1, as given by eq. (4.42).

4.2.7.2 The phase region of leapfrog pattern

Thus, the partial recognition loss is a major determinant of the interepitope pattern of escape (simple, leapfrog, nested). To determine the observed pattern as a function of the recognition losses at the two sites, simulation based on the main model introduced in Section 4.2.1 was performed. Each of the three pattern was mapped onto the plane $(\Delta r_1, \Delta r_2)$ (Fig. 4.13). The number of epitopes, n, the escape rate for the first haplotype, ϵ_{10}, and the ratio of the fitness costs in the two epitope sites, $\Delta f_1/\Delta f_2$ were fixed. Two values of ϵ_{10} were taken from the opposite ends of the observed range (Fig. 4.9). At large escape rates, the leapfrog pattern is observed at large values of Δr_1 and in a broad range of Δr_2. At smaller Δr_1, leapfrog is observed in a narrow range of Δr_2. There also a region with a fourth pattern where haplotype 11 is observed for a short time between haplotypes 10 and 01, which is labeled "nested leapfrog" (Fig. 4.12).

The time interval in which a haplotype dominates the population depends on Δf_k and Δr_k. (Goonetilleke et al. 2009) found out that, in 7 out of 18 epitopes with the leapfrog pattern, the first haplotype is short-lived compared to the second, and the transmitted epitope persists in a population at a low frequency after the initial decline. Reversion to the transmitted epitope is not always observed, either because the fitness costs are very small, or because they are compensated before reversion. These data provide the estimates of the time interval of compensation. The absolute lifetime of the second haplotype t_{01} represents the amount of time during which the compensatory mutations evolve and can be compared to the predictions (inset of Fig. 4.13). For a small fitness cost of the second haplotype, reversion to the wild-type can take years (Fig. 4.13C and D).

Equations (4.48)–(4.51) and the condition that the fitness cost of the first site is positive, $\Delta f_1 > 0$, bound the region of the leapfrog pattern on the outside. However, whether the second escape haplotype has sufficient time to grow to dominate the population before starting to revert depends on CTL decay rate. If CTL are very long-lived, the escape haplotypes live longer as well, and the leapfrog phase region expands (Batorsky et al. 2014, fig. S3).

4.2.8 Relationship between Δr and HLA binding loss from three experiments

4.2.8.1 Finding Δr from virus dynamics in the presence of CTL

Schneidewind et al. (2008) analyzed the virus growth rate in the presence and absence of CTL for several escape mutations in the KK10 epitopes of the Gag protein (Fig. 4C in that work). The CTL killing rate, kE, can be found from the comparison of the growth of infected cells (labeled with green fluorescent protein, GFP) between the cases when CTLs are present and absent. Next, by comparing the values of kE between the wild-type and a mutant, one infers the loss of CTL recognition in the mutant, Δr, as follows. The dynamics of GFP_+ cells is given by

$$\mathrm{GFP}_+(t) = \mathrm{GFP}_+(0)e^{(\beta - kE)t} \tag{4.52}$$

where β denotes the virus replication rate in the absence of the immune response in units per day. For each strain, we use the experimental data on GFP_+ at day 7, with and without CTL, to obtain

$$kE = \frac{1}{7}\left[\log\frac{\mathrm{GFP}_+\mathrm{CTL}_-(7)}{\mathrm{GFP}_+(0)} - \log\frac{\mathrm{GFP}_+\mathrm{CTL}_+(7)}{\mathrm{GFP}_-+(0)}\right] =$$

$$\frac{1}{7}\log\left[\frac{\mathrm{GFP}_+\mathrm{CTL}_-(7)}{\mathrm{GFP}_+\mathrm{CTL}_+(7)}\right]$$

Note that the initial value $\mathrm{GFP}_+(0)$ cancels out. Comparing the CTL killing rate of the wild type, denoted kE_w, to the CTL killing rate of the mutant, denoted kE_m, yields

$$\Delta r = 1 - \frac{(kE)_m}{(kE)_w} \tag{4.53}$$

4.2.8.2 HLA-binding loss ΔB from competition binding assay

According to the “absolute criterion” (Mostowy et al. 2012), the binding impairment of the mutant peptide, ΔB, is given by

$$\Delta B = \frac{1}{5 \cdot 10^4}\log\left(\frac{IC50_m}{IC50_w}\right) \tag{4.54}$$

where we use the values from Schneidewind et al. (2008) (Tab. 1 in that work).

4.2.8.3 ΔB and Δr from the measurement of CTL activity

In the cases when viral dynamics over time is unknown, a different method of estimation of Δr is needed. The method described below is based on data from (Kawashima et al. 2009; Matthews et al. 2012) who measured the loss of HLA binding and the loss of CTL function (specific lysis or INF-γ secretion) for the same epitopes.

(Tomiyama et al. 2005) measured the ability of HLA-B*5101 restricted CTLs to recognize epitopes in several genes. The idea of the method is to find a relationship between the concentration for 50% maximal binding, BL50, and the concentration for 50% maximal lysis, LL50. In order to calculate a peptide concentration relevant to both binding and lysis assays, all units are converted to nM, and the best linear fit is obtained. Then, BL50 is calculated by fitting the relative mean fluorescence intensity (MFI) versus the peptide concentration in the binding assay in Kawashima et al. (2009) (their fig. 1 G) and Matthews et al. (2012) (their fig. 6A). The following interpolation function is used:

$$y = \frac{\text{bottom} - (\text{top} - \text{bottom})}{\left[1 + \frac{\log(\text{peptide concentration})}{\text{BL50}}\right]^{\text{slope}}} \tag{4.55}$$

The function has four fitting parameters: *top* and *bottom* set the maximum and minimum values, respectively, BL50 sets the characteristic value of the logarithm of the peptide concentration, and *slope* sets the shape of the curve.

The best-fit parameters are found by fitting eq. (4.55) to the specific lysis % vs. peptide concentration in Kawashima et al. (2009) (their fig. 1 F) and the INF-γ SFC count in Matthews et al. (2012), see their Fig. 5B. Then, the lysis value or INF-γ SFC count is determined at the peptide concentration corresponding to BL50 (termed LL50 at BL50). This is done in order to ensure that the peptide concentration considered is relevant for both binding and lysis assays. Finally, Δr is found from eq. (4.53), where kE for these two experiments is found for the wild type and mutant strains by either the percent lysis or the INF-γ SFC count measured at the peptide concentration LL50 at BL50 found from the fitting above. The results are listed in Tabs. 4.3 and 4.4. Combining the estimates from the three experiments, a strong correlation between Δr and ΔB is obtained, $\Delta B = 0.78\Delta r - 0.004$ (Fig. 4.11).

Tab. 4.3: Estimating Δr and ΔB from Kawashima et al. (2009) for the three mutations in HLA-B*51 restricted epitope TAFTIPSI (RT 128135) that showed detectable binding to HLA-B*51.

Quantity	**I135**	**I135V**	**I135L**	**I135T**
Log BL50(nM)	4.07	4.15	5.06	6.64
Log LL50 (nM)	0.11	0.37	0.82	0.94
Log LL50atBL50 (nM)	0.61	0.64	1.03	1.72
kE = Lysis at LL50atBL50 for I135	65.48	64.13	18.6	8.99
Δr		0.01	0.42	0.58
ΔB		0.02	0.21	0.54

Tab. 4.4: Estimating Δr and ΔB from Matthews et al. (2012) for HLA-B*3501 restricted epitope NPPIPVGDIY (Gag 253 – 262) for one mutation.

Empty	**N260D**	**N260E**
Log BL50(nM)	0.89	2.01
Log LL50 (nM)	– 0.3	1.63
Log LL50atBL50	– 0.3	1.45
kE = Lysis at LL50atBL50 for N260D	1427.33	73.94
Δr		0.72
ΔB		0.30

4.3 Stability of HIV in the presence of defective interference particles in a cell and a host

If a virus genome undergoes extensive deletions, it will lack elements essential for its replication. To replicate, it must be complemented by wild-type virus infecting the same cell. Such conditionally replicating virus variants are termed “defective interfering particles” (DIP) (Huang and Baltimore 1970; Holland 1990). Replicating at the expense of the wild-type, DIPs represent secondary parasites interfering with its replication by depleting its products. DIP was proposed many times as a therapy for various diseases including HIV. The main problem with this, apparently straightforward idea is the questionable evolutionary stability of such an interaction. The evolution of resistance to DIP seems especially probable for HIV, because natural DIPs for HIV, unlike DIPs for many other viruses, have not been detected.

To understand the reason, Rouzine and Weinberger (2013b) tested if the lack of natural lentiviral DIPs may be due to natural selection within a host. The aim was to predict the parameters of DIP construct needed to maintain the suppression of HIV load that would be stable both dynamically and evolutionary. Upgrading the existing attempts, a simple but realistic mathematical model of virus dynamics was developed for two coupled biological levels, a cell and a host. The fitness effect of a DIP-resistant mutation in HIV genome was derived analytically in terms of the model parameters at the cellular level.

The analysis by Rouzine and Weinberger (2013b) followed in this section makes key predictions for two types of DIP interference: (i) Interference by codimerization between DIP and HIV genomes is evolutionary unstable. (ii) Interference by stealing virus capsids is evolutionarily stable provided the wild-type virus produces an excess of capsids over its genomes. The study offers an idea of an experimental set-up that could shed light on the lack of naturally occurring lentiviral DIPs. Hypothetical therapeutic approaches and their evolutionarily robustness are discussed as well.

4.3.1 Missing lentiviral DIP and evolutionary stability

Viral genomes express *cis*- and *trans*-acting elements that both can be stolen by DIP. *Trans*-elements are simply translated proteins, such as capsid and envelope proteins, transcription factors, proteases. *Cis*-elements are posttranscription (but pretranslation) products, such as various spliced mRNA species and genomic RNA. They also include the service regions of viral genome that interact with viral proteins and cellular molecules to ensure all phases of viral replication in a cell, such as promoters and packaging signals. If the wild-type virus undergoes a mutation or mutations that result in the loss of one or more important protein, but retain all necessary elements in its genome required for its replication (provirus transcription, in the case of a retrovirus), it becomes a DIP.

Several human and animal pathogens have naturally occurring DIPs, including vesicular stomatitis virus (Holland and Villarreal 1975), murine leukemia virus (Chattopadhyay et al. 1989), influenza virus (McLain et al. 1988), Rous sarcoma virus (Voynow and Coffin 1985), and Dengue fever virus (Aaskov et al. 2006; Li et al. 2011). The notoriously error-prone replication of RNA viruses generates many defective mutants (Huang and Baltimore 1970; Holland 1990). In murine leukemia virus, mutations within DIPs result in expression of new proteins, which enhance viral cytotoxicity or the immune response to an infected cell (Chattopadhyay et al. 1989; Kubo et al. 1996; Cook et al. 2003; Paun et al. 2005). Other DIPs interfere with the wild-type replication or attenuate virulence in animal models (Cave et al. 1985; Dimmock 1985; Barrett and Dimmock 1986; Levine et al. 2006; Marriott and Dimmock 2010). No wonder that DIPs have been proposed for use as therapeutic and gene transfer agents (Weinberger et al. 2003; Levine et al. 2006; D'Costa et al. 2009; Marriott and Dimmock 2010; Vignuzzi and Lopez 2019; Chaturvedi et al. 2021; Rezelj et al. 2021; Sharov et al. 2021). DIPs can arise spontaneously and be constructed artificially (Voynow and Coffin 1985; McLain et al. 1988; Chattopadhyay et al. 1989; Li et al. 2011; Chaturvedi et al. 2021; Shirogane et al. 2021; Xiao et al. 2021). They can also serve as a live vaccine eliciting broad neutralizing responses against a virus (Xiao et al. 2021).

Below, a broad class of lentiviral DIP designs with mutations or deletions in various proteins is considered. Possible designs range from minimal DIPs completely stripped of all proteins, to DIPs expressing some, including transcriptional transactivating proteins (Tat) and genomic export proteins (Rev). Such HIV mutants can replicate by adding the missing viral proteins (Dull et al. 1998), or by coinfecting with a homologous replication-competent virus that supplies the missing proteins (Turner et al. 2009). Because DIPs replicate at the expense of the wild-type virus, they represent secondary parasites.

As already mentioned, despite large volumes of sequencing data for HIV and SIV variants, and in contrast to many other virus families, natural lentiviral DIPs have not been discovered. Although many HIV-infected cells harbor inactive variants of HIV provirus, the vast majority of infected cells in an average host contain a single transcriptionally active provirus (Kearney et al. 2009; Neher and Leitner 2010; Batorsky et al. 2011). Cells coinfected with a replication-competent and a replication-deficient provirus replicating at the expense of the first have not been found. There are three possible reasons for this absence: (i) insensitive assays; (ii) a molecular block to DIP formation, or (iii) natural selection against DIP or against the wild-type virus that can sustain the parasite.

While the ability to form and spread DIPs varies between viruses (O'Hara et al. 1984; DePolo and Holland 1986; Giachetti and Holland 1988), an inherent block to the generation of large deletions within the HIV genome does not exist (Li et al. 1991). Nor is there any known obstacle to mobilization of sub-genomic HIV, which are routinely produced during the generation of retrovirus vectors for gene therapy from packaging cell lines (Dull et al. 1998). Many engineered HIV variants lacking proteins conditionally replicate in the presence of full-length HIV in tissue culture (Chen et al. 1992; Paik et al. 1997; Bukovsky et al. 1999; Finzi et al. 1999; Evans and Garcia 2000; Klimatcheva et al. 2001). An artificial HIV-based construct, which can be viewed as a DIP encoding a gene-therapy element, was shown to replicate within a humanized-mouse model of HIV (Mukherjee et al. 2010) and in HIV-infected patients (Levine et al. 2006). Thus, there is no indication for a molecular block to DIP formation, at least, in artificial constructs.

In contrast, the explanation that HIV develops the resistance to DIP by divergent evolution seems to be plausible. While the dynamics and dynamic stability of various DIPs has been addressed in large body of work (Kirkwood and Bangham 1994; Nelson and Perelson 1995; Weinberger et al. 2003; Thompson and Yin 2010; Metzger et al. 2011), the evolutionary stability of lentiviral DIPs was not sufficiently studied before the reviewed work. The matter is far from being trivial. For example, "resource stealing" by a DIP of vesicular stomatitis virus exerts strong selective pressures on its helper virus. In long-term cultures over a period of 5 years and several hundred passages, coevolution of both wild-type virus and DIP was observed (Perrault and Holland 1972; O'Hara et al. 1984). In the course of this Red Queen chase, the wild-type virus was able to gain resistance to the DIP (Horodyski and Holland 1980; DePolo et al. 1987; Steinhauer et al. 1989). For a recent review of virus-DIP coevolution for other viruses, see Vignuzzi and Lopez (2019).

Similarly, HIV is expected to be under the selection pressure to mutate to reduce its deleterious interaction with a DIP. Because HIV is the fastest-evolving virus in nature (Rouzine and Coffin 1999c and reference therein), it is quite likely to escape. On the other hand, the genetic plasticity of the DIPs would be as high and under a high selective pressure to survive by staying in step with HIV. This, again would lead to an evolutionary chase between the two virus species, the Red Queen effect. Will DIP be able to keep up, or HIV will escape eventually? The outcome is far from being obvious, especially in the presence of multiple scales of biological organization. In this section, this question is addressed at a cell and a host scale. The host population (epidemiological) scale is reviewed in the next section, Section 4.4.

The field of population genetics offers appropriate tools to analyze such questions related to evolutionary and coevolutionary processes (Rouzine 2020b). Models of population genetics, introduced into virology in the end of the twentieth century, proved to be powerful tools for the analysis of HIV sequence data and led to predictive mathematical theories on the HIV adaptation, resulting in an explosion of theoretical work (Coffin 1995; Leigh-Brown 1997; Rouzine and Coffin 1999c; Frost et al. 2001; Rouzine et al. 2003;

Rouzine and Coffin 2005; Desai et al. 2007; Neher et al. 2010; Rouzine and Coffin 2010; Batorsky et al. 2011; Mostowy et al. 2011; Good et al. 2012; Walczak et al. 2012; Neher et al. 2013; Batorsky et al. 2014; Kryazhimskiy et al. 2014; Neher et al. 2014; Good and Desai 2015; Jerison and Desai 2015; Rouzine et al. 2015; Neher et al. 2016; Pedruzzi et al. 2018; Rouzine 2020b, a; Barlukova and Rouzine 2021). Application of similar models to time-series sequence data or time-variation of virus diversity allowed the estimation of the selection coefficients of mutations, including those responsible for virus adaptation to untreated patients and mutations conferring resistance to antiretrovirals and immune escape (Rouzine and Coffin 1999c; Frost et al. 2000; Neher and Leitner 2010; Batorsky et al. 2011; Ganusov et al. 2011; Hinkley et al. 2011; Wang et al. 2011; Messer and Neher 2012; Mostowy et al. 2012; Barlukova and Rouzine 2021). Combined with the basic models of HIV dynamics (Perelson et al. 1996; Nowak and May 2000; Perelson and Ribeiro 2013), this field provide methods to answer the question of whether the interaction between HIV and DIP would be evolutionarily stable.

It is important to determine the direction of HIV and DIP evolution within an infected individual, to find out whether a specific mechanism of interference is evolutionarily stable. The biomedical aim is to inform the future design of therapeutic DIP. Testing virus stability in a cell culture is possible but difficult, because the total number of cells is too small, and the condition in culture differs from that in vivo. Before attempting expensive animal tests (Chaturvedi et al. 2021; Xiao et al. 2021), it is important to test it *in silico*. Recently, computational and experimental studies tested the effect of DIP for various viruses including SARS CoV-2 (Chaturvedi et al. 2021; Rezelj et al. 2021; Shirogane et al. 2021; Xiao et al. 2021).

Following Rouzine and Weinberger (2013b), this section analyzes a model of HIV and hypothetical DIP dynamics at two bioscales: the single-cell scale describing the formation of HIV and DIP particles, for each type of DIP interference, and, the individual-host scale, where previous models (Mohri et al. 1998; Nowak and May 2000) are generalized to include the presence of multiple copies of the hypothetical DIP. The fitness effect of mutation on the host scale is derived analytically directly from the single-cell model, and its sign is tested, to find out whether it can decrease the stealing of HIV products by DIP.

4.3.2 DIP interference with HIV by competition for genomic RNA leads to divergent evolution

In principle, DIP can interfere with HIV by binding its genomic RNA (gRNA) ("genome stealing"). Lentiviral genomes are packaged into virions as RNA dimers. The pairing is initiated at a six-nucleotide palindrome, the dimerization initiation signal (DIS), located within stem loop 1 (SL1) of the HIV genome. It has consensus sequence GCGCGC (Moore et al. 2007).

Suppose that DIP lacks the transactivator protein of HIV and, hence, can be expressed only in cells coinfected by HIV. A minimal model, which includes only nondimerized and

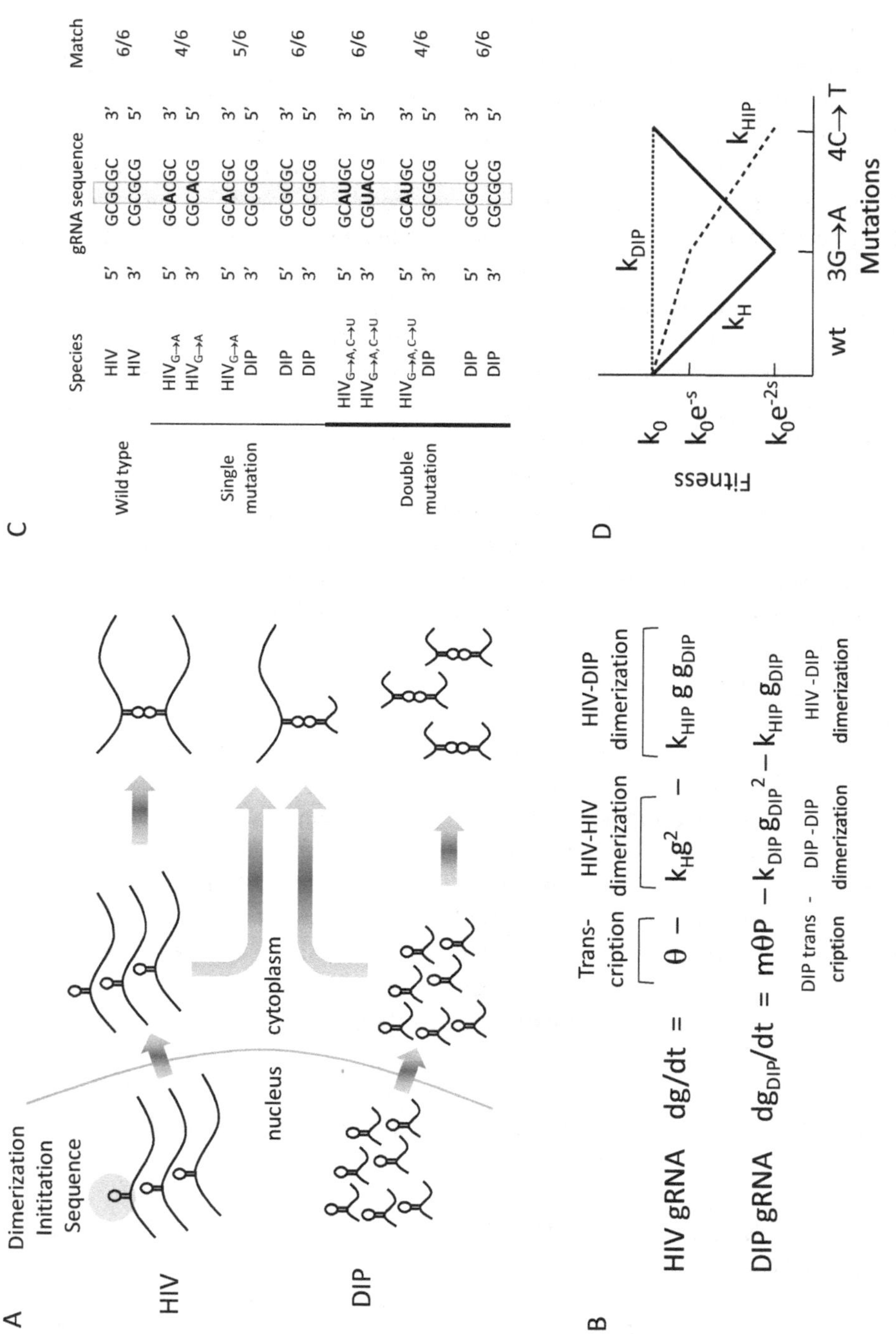

Fig. 4.14: DIP interference by "genome-stealing" is evolutionary unstable due to the divergent evolution of the dimerization initiation sequences.

dimerized genomes can predict the evolution of the genome dimerization between DIP and HIV (Fig. 4.14A, B). This model (and the models in the following sections) incorporate virological facts, as follows:

(i) Experimental evidence demonstrates that sub-genomic RNAs preserving consensus DIS palindrome can bind wild-type genomes (Moore et al. 2007; Chen et al. 2009). Heterozygous virions containing such RNA heterodymers cannot replicate and serve only as a suppressor of HIV replication.
(ii) The majority of HIV-infected cells contain a single integrated HIV provirus (Batorsky et al. 2011; Josefsson et al. 2011), due to the short lifetime of infected cells (Ho et al. 1995) and *Nef*-induced downregulation of virus receptor CD4 (serving to enhance the release of virions from a cell). In contrast, DIP integration is not restricted to a single copy, because it causes neither effect. Multiple copies of integrated DIP provirus increase the production of the genomic RNA of DIP by the factor of m, where $m = 1, 2, 3, \ldots$, is the number of copies, which varies among cells randomly, and whose average is proportional to the virus concentration in the body.
(iii) Shorter RNA genomes of DIP without splicing sites (D'Costa et al. 2009; Koldej and Anson 2009) are transcribed faster than those of HIV, so that DIP RNA genomes are expressed faster than HIV RNA genomes (An et al. 1999; Bukovsky et al. 1999; D'Costa et al. 2009; Koldej and Anson 2009). The expression asymmetry will be denoted P, where $P > 1$. Thus, the net expression asymmetry between DIP and HIV is mP. Due to the DIP overabundance, most HIV genomic RNAs are stolen by DIP to produce nonviable virions (Fig. 4.14A), leaving fewer to be used to form HIV-HIV RNA dimers. This is a form of interference.

To analyze the evolutionary stability, consider a single mutation in HIV DIS of type GCG<u>C</u>GC → GCG<u>A</u>GC. It causes a mismatch and hence decreases in the probability of

Fig. 4.14 (continued)
(A) Mechanism of interference. Genomic RNA (gRNA) monomers of HIV and DIP form three types of dimeric complexes (HIV-HIV, HIV-DIP, and DIP-DIP) due to a so-called kissing loop between the dimerization initiation sequences, which include a palindromic sequence with consensus sequence GCGCGC. With a faster transcription rate and multiple copies of provirus, DIP gRNAs are more abundant, so that HIV-DIP heterodimers outcompete HIV gRNA monodimers. (B) A simplified model describing the abundance of gRNA monomers for HIV and DIP in the infected cell, $g(t)$ and $g_{DIP}(t)$, respectively; θ is a composite parameter proportional to the linear rate of gRNA production, and P is the asymmetry of gene expression between HIV and DIP. Parameters k_H, k_{IP}, and k_{HIP} are dimerization coefficients for HIV-HIV, DIP-DIP, and HIV-DIP, respectively. (C) Two mutations in the kissing loop result in the divergent evolution of HIV and DIP. Top row: Exact match for any gRNA pair (HIV-HIV, HIV-DIP, and DIP-DIP). Middle rows: If a single mutation arises within HIV (green rectangle), an HIV-HIV homodimer has two mismatches, a HIV-DIP heterodimer has a single mismatch, and a DIP-DIP homodimers has none. Bottom rows: The second compensatory mutation occurs in HIV, and heterodimerization is disfavored. (D) Evolutionary "fitness" of homodimers and heterodimers estimated from B and C. Although a single mutation favors heterodimerization, the second mutation favors HIV homodimerization. Based on Rouzine and Weinberger (2013b).

DIP-HIV heterodimerization, but it also results in two mismatches in HIV-HIV homodimerization, which has a stronger effect on the dimerization probability (Fig. 4.14C). However, a compensatory symmetric mutation restores a new perfect palindrome in HIV and makes DIP-HIV dimerization less favorable than HIV-HIV. Thus, assuming that dimerization coefficient is a function of the number of mismatches, one predicts the divergent evolution in DIS sequence between HIV and DIP (Fig. 4.14D).

Of course, this simple argument does not take into account the cost of the palindrome DIS being different from the consensus GCGCGC sequence (Moore et al. 2007) and considers only one two generations of virus. However, a more realistic model also predicts the progressive decrease in heterodimerization relative to homodimerization (Fig. 4.15).

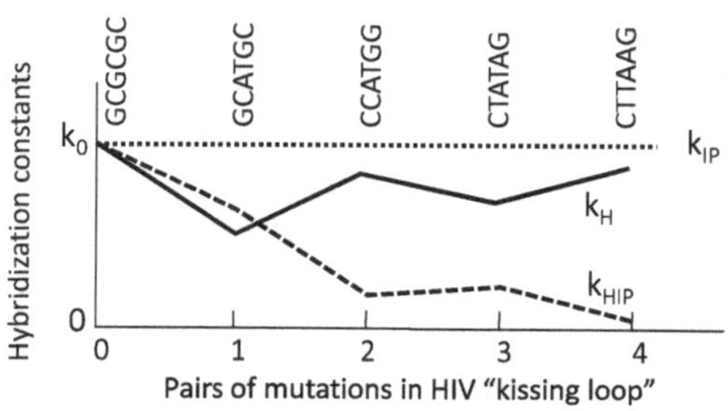

Fig. 4.15: Sequence divergence between HIV and DIP due to evolution in the dimerization initiation sequence.
An example of an evolving palindromic DIS sequence is shown on the top. Initial sequence for DIP and HIV is the same, GCGCGC, and DIP sequence remains unchanged. Dimerization coefficients for HIV-HIV, HIV-DIP and DIP-DIP pairs, as defined in Fig. 4.14B, are shown qualitatively versus the number of mutation pairs. A pair includes a mutation causing a DIS palindrome mismatch, and a compensatory mutation, which restores the match. Each dimerization constant k_H, k_{IP}, k_{HIP} has a component determined by the number of mismatches (Fig. 4.14C) and a fluctuating component, depending on palindrome sequence. HIV-DIP cross-dimerization coefficient decreases as the loop sequence diverges from GCGCGC, with HIV-HIV dimerization coefficient fluctuating around a constant level. Based on Rouzine and Weinberger (2013b).

4.3.3 An alternative mechanism of DIP interference: protein stealing

With the knowledge that "genome stealing" is unstable (Section 4.3.2), the next attempt is interference by competition for proteins. DIPs depend on the proteins of the wild-type to package and leave the infected cell, thus competing for proteins with HIV. To be specific, below stealing of capsids is discussed, although the analysis is sufficiently general to apply to the theft of other proteins, such as envelope, protease, or reverse transcriptase. Depending on its design, DIP can also steal several proteins at once. The only difference is in the point of virus replication cycle where the stolen protein(s) are important.

The key to the success is to integrate two bio levels: a single-cell model describing the competition for capsids and a model of virus dynamics in a host. Unlike the previous multi-scale modeling of virus infections (Krakauer and Komarova 2003; Haseltine et al. 2008; Guedj and Neumann 2010), this two-scale model is designed specifically for HIV in the presence of a hypothetical DIP and, unlike the previous attempt in this direction (Metzger et al. 2011), it is tuned into the specific virology of HIV. The model has to be constructed to cover possible DIP designs, from (i) a minimal DIP that does not code for any proteins at all, to (ii) a DIP that codes for the Tat and Rev proteins thus allowing for the expression and export of genomes to cytoplasm in the absence of HIV, but does not express capsid proteins or the Nef protein, which mediate superinfection protection, to (iii) a DIP that codes for everything but one essential protein, such as capsid.

In the following subsections, both biological levels of models and their integration are discussed. Selection coefficient for the resistance mutants of HIV is derived explicitly from the model equations. Results are discussed in Section 4.3.5. Analytical derivations are found in Sections 4.3.6 and 4.3.7. Numerical calculations were performed in software MATLAB™ (version R2011*a*).

4.3.3.1 The single-cell model with capsid stealing

To make analysis transparent, the single-cell model (Fig. 4.16A right) includes only dynamics of dimerized HIV RNA genomes, whose number in a cell is denoted G, encapsidation-competent capsids, C, and dimerized DIP RNA genomes, G_{DIP}. Model parameters are defined in Tab. 4.5. The DIP and HIV number of virions per cell predicted by the model is used as the input for the individual-patient model described below [Fig. 4.16A left, eqs. (4.59)–(4.64), Tab. 4.6].

The model equations have the form

$$\frac{dG}{dt} = \theta - k_{\text{pck}}GC - \alpha G \tag{4.56}$$

$$\frac{dC}{dt} = \eta\theta - k_{\text{pck}}(G + G_{\text{DIP}})C - \beta C \tag{4.57}$$

$$\frac{dG_{\text{DIP}}}{dt} = mP\theta - k_{\text{pck}}G_{\text{DIP}}C - \alpha G_{\text{DIP}} \tag{4.58}$$

The processes include the production and decay of HIV genomes with linear rate θ, the packaging of these genomes into capsids with coefficient k_{pck}, assumed to be the same for HIV and DIP (Section 4.3.5), the capsid production with linear rate $\eta\theta$, and the competition for encapsidation between DIP and HIV genomes. All genomes are homodimers, either HIV-HIV or DIP-DIP. The model neglects heterozygous genomes due to the divergence in the dimerization initiation sequence (Section 4.3.2). A DIP provirus expresses more RNA copies than an HIV provirus by a fixed factor of $P > 1$ (An et al. 1999; Bukovsky et al. 1999; D'Costa et al. 2009; Koldej and Anson 2009) (Section 4.3.1). The net asymmetry between DIP and HIV expression in a dually infected cell is product mP.

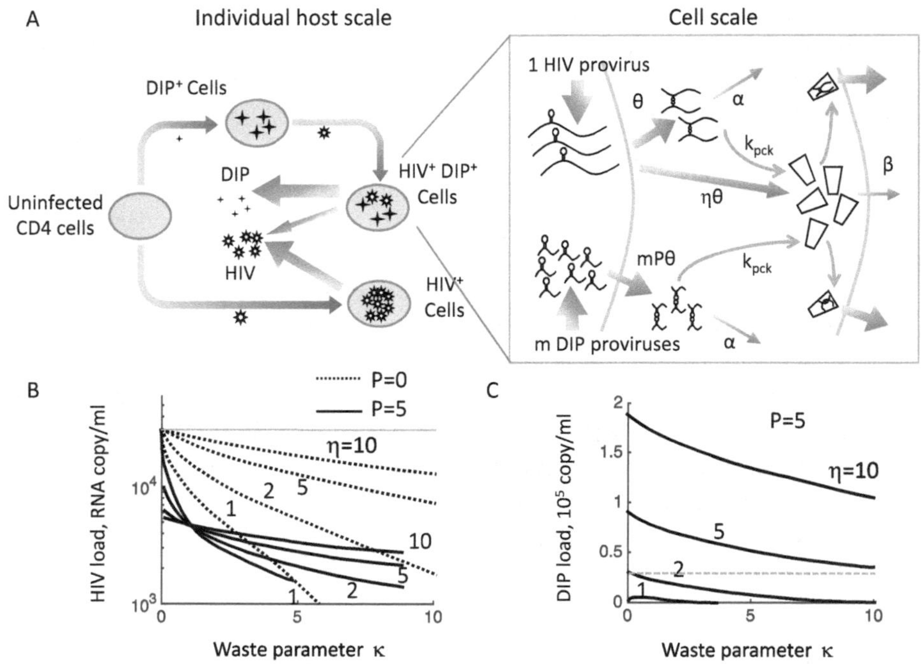

Fig. 4.16: DIPs that steal capsid stably suppress HIV load across a broad range of parameters. (A) Schematic of the model comprising two scales of biological organization. The individual host scale is the standard model of HIV replication expanded to include DIPs, eqs. (4.59)–(4.64). Uninfected cells can be infected with either HIV or DIP; DIP$^+$ cells can be superinfected with HIV to become dually infected. The single-cell model is described by eqs. (4.56)–(4.58). A dually infected cell has one integrated HIV provirus and m copies of DIP provirus. A fraction of HIV RNA is translated into proteins that form empty capsids. DIP does not express proteins. Broad arrows represent multi-stage processes. A fraction of stable dimer genomes and full capsids is also lost. Remaining genomes, HIV or DIP, are packaged with capsids and released as infectious particles. (B) Steady-state HIV load and (C) steady-state DIP load at different values of two single-cell parameters: the "capsid waste parameter" κ and the capsid-to-genome production ratio η (see Tab. 4.5). The horizontal gray line in (B) shows HIV viral load in the absence of capsid waste and DIP ($\kappa = P = 0$), which is assumed to be the average load in untreated humans ($3 . 10^4$ RNA copies/mL blood). Parameters in (B) and (C): DIP-to-HIV expression ratio $P = 5$ and the basic reproduction ratio $R_0 = 10$ (Tab. 4.5). The decrease in HIV load in the presence of capsid waste ($\kappa > 0$, red lines), as compared to the untreated HIV steady-state level (gray line), is partly due to the loss of HIV products (black dotted lines calculated at $P = 0$) and partly due to DIP, which competes with HIV for available target cells and steals HIV capsid in dually infected cells. The first effect is more important at $\eta \sim 1$, and the DIP suppression factor is stronger at large $\eta \gg 1$ [see Fig. S2 in Rouzine and Weinberger (2013b)]. Based on Rouzine and Weinberger (2013b).

Dually infected cells are classified based on DIP copy number m. Genomes and capsids are degraded with exponential rates α and β, respectively. Due to the time-scale separation between processes in a cell (minutes to hours) and a host (days to months), all three state variables, G_{DIP}, and C are assumed to reach a steady state rapidly, and their values are derived analytically below [Section 4.3.6, eqs. (4.71)–(4.75)].

Tab. 4.5: Model parameters and state variables for a single-cell capsid-stealing model, Fig. 4.16A and eqs. (4.56)–(4.58).

State variables				
Notation	**Definition**	**Units**		
G	Concentration of full-length dimerized HIV genomic mRNAs	$1/\mu L$		
G_{DIP}	Concentration of full-length dimerized DIP genomic mRNAs	$1/\mu L$		
C	Concentration of encapsidation-competent capsids in cytoplasm	$1/\mu L$		
m	DIP (integrated) provirus copy # (MOI)	*Dimensionless*		
Model parameters				
Notation	**Definition**	**Units**	**Value**	**References**
θ	Accumulation rate of HIV genomes	$1/\mu L/day$		Free
k_{pck}	Packaging constant	$\mu L/day$	$K = \frac{\alpha\beta}{\theta k_{pck}}$ (for θ, k_{pck}, α)	
α	Loss of genome rate	1/day		
β	Loss of capsid rate	1/day		
η	Capsid-to-genome accumulation ratio	1	$1.2 - 5$	Chen et al. (2009) and Sergeev, Batorsky and Rouzine (2010)
P	Expression asymmetry between DIP and HIV	1	$8 - 10$	D'Costa et al. (2009)

4.3.3.2 The individual-host model

The single-cell model introduced in the previous subsection is used to express HIV and DIP numbers in a cell in terms of cell-model parameters. Then, these values are used as input for the model of viral dynamics in an individual host described below. That model is upgraded from the basic model of virus dynamics (Mohri et al. 1998; Ribeiro et al. 2002a, b) to include DIP, the coinfection of cells with DIP and HIV, and the fact that the dually infected cells produce less HIV than cells infected with HIV only. Based on the very small frequency of cells coinfected with two active proviruses of HIV (Batorsky et al. 2011), a single HIV provirus per (singly or dually) infected cell is assumed. However, multiple DIP provirus copies per cell are allowed, in which the model differs from that in Metzger et al. (2011).

The model comprises uninfected, but infectable, $CD4^+$ T cells, T; cells actively infected with HIV, I; $CD4^+$ T cells harboring m copies of DIP provirus but not infected with HIV, $T_{\mathrm{DIP}\,m}$; dually infected cells harboring a copy of HIV and m copies of DIP, $I_{\mathrm{D}\,m}$,; free virus concentration in peripheral blood, V; and DIP particle concentration V_{DIP}. By the definition, $T_{\mathrm{DIP}\,0} = T$. The equations have the form

$$\frac{dT}{dt} = b - (d + kV + kV_{\mathrm{DIP}})T \tag{4.59}$$

$$\frac{dI}{dt} = kVT - \delta I \tag{4.60}$$

$$\frac{dT_{\mathrm{DIP}\,m}}{dt} = kV_{\mathrm{DIP}}T_{\mathrm{DIP}\,m-1} - (d + kV + kV_{\mathrm{DIP}})T_{\mathrm{DIP}\,m}, \quad m = 1, 2, 3, \ldots \tag{4.61}$$

$$\frac{dI_{\mathrm{D}\,m}}{dt} = kVT_{\mathrm{DIP}\,m} - \delta I_{\mathrm{D}\,m}, m = 1, 2, 3, \ldots \tag{4.62}$$

$$\frac{dV}{dt} = n\delta I + n\delta \sum_{m=1}^{\infty} \psi_m I_{\mathrm{D}\,m} - cV \tag{4.63}$$

$$\frac{dV_{\mathrm{DIP}}}{dt} = n\delta \sum_{m=1}^{\infty} \rho_m \psi_m I_{\mathrm{D}\,m} - cV_{\mathrm{DIP}} \tag{4.64}$$

The model parameters (Tab. 4.6) are the inverse lifespan of uninfected cells, d, the linear generation rate of uninfected cells, b, the infectivity coefficient, k, the death rate of singly and dually infected cells, which is assumed to be the same because it is caused by the cytotoxic CD8 T cells (Kuroda et al. 1999; Schmitz et al. 1999), δ, and the HIV burst size from a singly infected cell, n.

Two extra parameters due to the presence of DIP are the ratio of HIV burst size between a singly infected cell and a dually infected cell with m copies of DIP provirus, ψ_m, and the ratio of DIP to HIV virion number coming from a dually infected cell with m copies of DIP provirus, ρ_m.

The processes included in eqs. (4.59)–(4.64) are, as follows. Uninfected cells permissive for viral replication, T, are replenished from a linear source (Mohri et al. 1998) with the possible mechanism discussed by Rouzine and Coffin (1999d) and Grossman et al. (1999). This compartment is depleted due to their natural death, infection by HIV particles, or infection by DIP, eq. (4.59). Infected cells, I, produce n virions and die at average rate $\delta \sim 1$/day. A cell can also be infected with m copies of DIP provirus, $T_{\mathrm{DIP}\,m}$, eq. (4.61). These cells do not express HIV proteins and die at the same rate as uninfected cells. Upon coinfection with HIV, such a cell becomes "dually infected," $I_{\mathrm{D}\,m}$, eq. (4.62). These HIV + DIP + cells die as rapidly as HIV + DIP − cells, I, after making $\psi_m n$ particles of HIV and $\rho_m \psi_m n$ particles of DIP, eqs. (4.63) and (4.64). Finally, free virus particles V and V_{DIP} are cleared at rate c, which is the fastest rate in the host model (Tab. 4.6).

Extra parameters ρ_m, ψ_m represent the output of the single-cell model, eqs. (4.56)–(4.58). They can be expressed in terms of single-cell model parameters and the steady-state numbers of genomes and capsids (Section 4.3.7):

$$n = \frac{k_{\text{pck}}[GC]_{P=0}}{\delta} \tag{4.65}$$

$$\psi_m n = \frac{k_{\text{pck}} GC}{\delta} \tag{4.66}$$

$$\rho_m \psi_m n = \frac{k_{\text{pck}} G_{\text{DIP}} C}{\delta} \tag{4.67}$$

Because the focus is on long-term asymptomatic HIV infection, the system defined by eqs. (4.59)–(4.63) is assumed to be in a steady state. The steady-state values of the state variables are derived in Section 4.3.6, eqs. (4.84)–(4.93).

4.3.3.3 Parameter values

Tables 4.5 and 4.6 summarize model parameters at both biological levels. Luckily, we can reduce the total number of parameters to a smaller number of composite parameters by changing units (Section 4.3.6). All results can be expressed in terms of four dimensionless parameters

(i) The composite waste parameter

$$\kappa = \frac{\alpha\beta}{\theta k_{\text{pck}}}$$

(i) The ratio of encapsidation-competent capsids to dimerized HIV genomes produced per unit time, η, referred to below as the capsid-to-genome production ratio.

(ii) Basic reproduction ratio, R_0

(iii) DIP-to-HIV genome expression ratio P.

Lentiviral DIPs with $P = 8 - 10$ have been engineered (D'Costa et al. 2009). The focus below is on the conservative value $P = 5$, but interval $P = 2 - 30$ was also investigated in the original study (Rouzine and Weinberger 2013b).

4.3.3.4 Dynamically stable suppression of HIV at the host level

The normalized steady-state values of HIV and DIP at the individual-patient scale can be expressed as a function of the four dimensionless parameters: κ, η, P, R_0, three of which belong to the single-cell model. Analytical solution presented in Section 4.3.6 demonstrates that the presence of DIP decreases viral load of HIV in a broad range of parameters (Fig. 4.16B). The degree of HIV suppression and DIP load are sensitive to the DIP-to-HIV expression asymmetry, $P > 1$ (Rouzine and Weinberger 2013b, fig. 2D,E).

Tab. 4.6: State variables and parameters for the individual-host model, eqs. (4.59)–(4.64). Two bottom rows: G, G_{DIP},and C are steady-state values of state variables of the intracell model [Section 4.3.6, eqs. (4.75)–(4.78)].

State variables				
Notation	**Definition**		**Units**	
T	Uninfected $CD4^{+}$ T cells permissive for viral replication		Cell/μL	
I	$CD4^{+}$ T cells infected with HIV only		Cell/μL	
$T_{\mathrm{DIP}\ m}$	$CD4^{+}$ T cells infected with m copies of DIP provirus but not infected with HIV, $T_{\mathrm{DIP}\ 0} = T$		Cell/μL	
$I_{\mathrm{D}\ m}$	Dually infected cells with an HIV and m copies of DIP provirus		Cell/μL	
V	HIV viral load		RNA copy / mL	
V_{DIP}	DIP viral load		RNA copy / mL	
Model parameters				
Notation	**Definition**	**Units**	**Value**	**References**
b	Linear production rate of uninfected cells	Cells/μL/day	$R_0 = \dfrac{bkn}{cd} \sim 10$	Nowak et al. (1997)
d	Death rate of uninfected cells	1/day		
k	Infectivity factor	mL/day/RNA copy		
n	HIV burst size from a singly infected cell	RNA copy/cell		
c	Virion clearance rate	1/day		
δ	Death rate of HIV-infected cells	1/day	1.0/day	Markowitz et al. (2003)
$n\psi_m$	HIV burst size from a dually infected cell with m copies of DIP provirus	1	$\dfrac{k_{\mathrm{pck}} GC}{\delta}$	From single-cell model
$np_m\psi_m$	DIP burst size from a dually infected cell with m copies of DIP provirus	1	$\dfrac{k_{\mathrm{pck}} G_{\mathrm{DIP}} C}{\delta}$	

The dynamic stability condition of DIP in a population is given by eq. (4.96), which, under a small waste parameter $\kappa \ll 1$, has the form

$$\eta > \eta_c = \frac{1+P}{P}\frac{R_0}{R_0 - 1}$$

The biological meaning of this condition obtained in Section 4.3.6, eq. (4.97), is that, for DIP to be stable, HIV must generate extra capsids as compared to its genomes, for

DIP to parasitize on. If capsids are limiting, no DIP will arise. For moderately wasteful process $\kappa \sim 1$, the condition is relaxed, and η can be a bit smaller than unity (see Fig. 4.16C at $\eta = 1$).

The decrease in HIV load as compared to its value at $=0$, $R_0 = 10$ (Fig. 4.16B). is only partly due to the presence of DIP. The remainder of the decrease is due to increased waste, κ, which decreases the HIV burst size n, as one can show from eqs. (4.75) and (4.76) (Section 4.3.6 below). As a control, HIV load at zero DIP load (i.e., at $P = 0$) is shown in Fig. 4.16B (dotted lines).

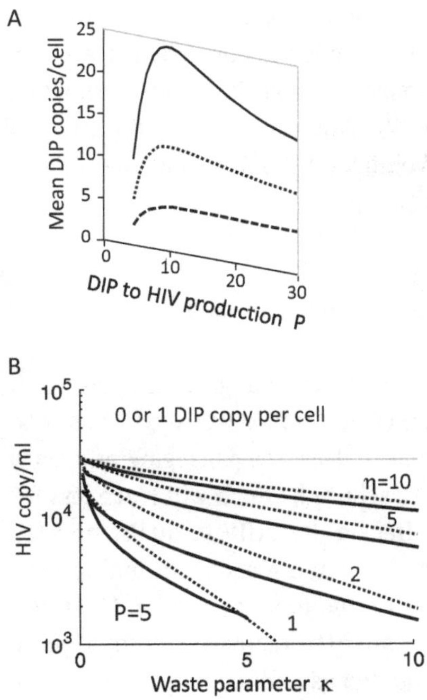

Fig. 4.17: High average multiplicity of DIP infection enhances dynamic stability of DIP and suppression of HIV.
(A) The average number of DIP copies per cell, $E[m]$, as a function of expression asymmetry P at zero waste parameter, $\kappa = 0$, and fixed values of η equal to 2 (dashed curve), 5 dotted), and 10 (solid). Y-axis: $1/(1-q)$, where q is the ratio of the cell number with $m+1$ copies to the cell number with m copies (Section 4.3.6), so that $q = 0$ corresponds to a DIP-free population. DIP population is unstable if $\eta \sim 1$ or less. (B) Steady-state HIV load as a function of waste parameter κ, when multiplicity of DIP infection, m, is artificially restricted to 1. Suppression of HIV is primarily due to the loss of HIV products at large κ (cf. Fig. 4.16B). Parameters: $P = 5$, $R_0 = 10$ are as in Fig. 4.16. Based on Rouzine and Weinberger (2013b).

4.3.3.5 HIV suppression is due to high multiplicity of DIP infection

One reason for the strong suppression of HIV is that the multiple infection of cells by DIP amplifies its effect on HIV. The average multiplicity of DIP infection denoted $E[m]$ is rather large even at modest values of η and P (see Fig. 4.17A). Hence, the results essentially depend on the multiple integration of DIP per cell. The number of DIP copies, m, varies among dually infected cells. Using the individual-patient model, eqs. (4.59)–(4.64), the average multiplicity of DIP infection, $E[m]$, was derived analytically, eqs. (4.88), (4.89), (4.92) in Section 4.3.6 below, as a function of capsid waste, κ, and expression asymmetry, P, and capsid-to-genome ratio, η (Fig. 4.17A).

The average multiplicity of integrated DIP genomes, $E[m]$, increases with both P and η (Fig. 4.17A). The increase with either parameter leads to more DIP virions, which results in greater multiplicity of DIP infection. Thus, the ratio of DIP to HIV genomic mRNA within a cell is determined by P and η in two ways: directly, through the molecular architecture of the DIP, and indirectly, through the amplification of this factor by m.

To test the contribution of the multiplicity of DIP to HIV interference, a variant of the model was considered, where DIP was limited to a single copy per cell ($m = 1$ or $m = 0$). Then, the HIV viral loads was recalculated (Fig. 4.17B). This "control" demonstrated that, when $E[m] \leq 1$, the suppression by DIP is modest and contributes little to the decrease in HIV viral load. Furthermore, if $E[m] \leq 1$,DIP loses stability at low η as κ increases. Even modest suppression of HIV load requires a high excess of capsids.

In summary, the multiplicity of integrated DIP genomes, m, is critical for DIP-mediated suppression of HIV. From the other hand, multiple infections by DIP amplify even modest expression asymmetry P and create relatively high DIP loads even when η is just slightly larger than 1 (Figs. 4.16C and 4.17). This conclusion hold across a broad range of $P > 1$ and η values (Rouzine and Weinberger 2013b, fig. 3d,e), provided the dynamic stability condition is met.

4.3.4 Testing evolutionary stability: the effective selection coefficient

The next question is whether HIV can diverge from DIP by changing its genome. In principle, HIV could resist DIP by mutating to increase the capsid waste parameter, κ. For example, HIV could mutate its packaging signal to decrease packaging efficiency k_{pck} allowing more capsids be degraded instead of packaged, which will leave less for DIP to use. The idea is that DIP would be more affected than HIV in dually infected cells due to the higher DIP expression asymmetry, $P > 1$, and integration multiplicity, $m > 1$. Thus, increased capsid waste could benefit HIV due to a decrease in DIP interference. However, this mutation has also a cost, because the increased waste of products would lower the virion output. The HIV load plotted as a function of κ shows the benefit and the cost as the two components of HIV suppression: one due to decrease in HIV burst, and another due to DIP interference (Fig. 4.16B). As κ is increased, one component becomes larger and another smaller. It is not clear *a priori* which effect dominates.

To determine if HIV is under selection pressure to increase or decrease waste, the selection coefficient, ∂s, defined as a small increase in fitness of a mutant HIV strain due to a small change in waste parameter κ was derived analytically [eqs. (4.98)–(4.108) in Section 4.3.7]. Positive values of ∂s would imply that a mutation is selected for.

Because the effect of mutation on κ, denoted $\partial\kappa$, is unknown and may vary among bases, the resulting selection coefficient is expressed in a normalized form $\partial s/(\partial\kappa/\kappa)$. It is worth stressing that the selection coefficient here is derived from a molecular model. Most studies use selection coefficient as input parameter and determine it by fitting data (Rouzine and Coffin 1999c; Frost et al. 2000; Neher and Leitner 2010; Batorsky et al. 2011;

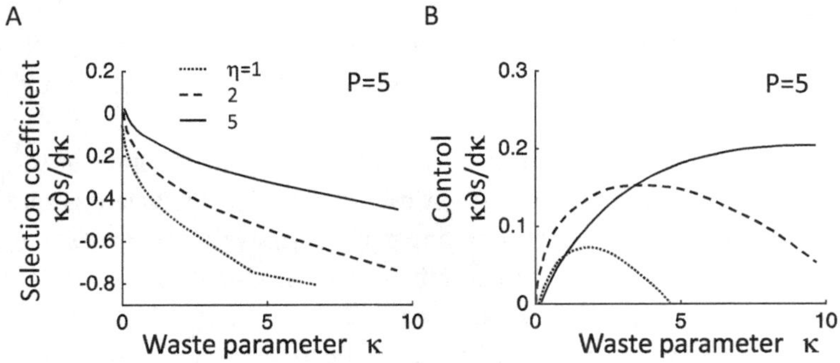

Fig. 4.18: HIV cannot escape DIP by increasing the waste of resources.
(A) Normalized selection coefficient $\partial s/(\partial\kappa/\kappa)$ as a function of κ, for various capsid-to-genome production ratios η. B) Negative control for $\partial s/(\partial\kappa/\kappa)$ within HIV^+ DIP^+ dually infected cells when the burst-size changes due to increased capsid waste are neglected, the 1st term in eq. (4.105). The derivation is given in Section 4.3.7. Parameters $R_0 = 10, P = 5$. Based on Rouzine and Weinberger (2013b).

Ganusov et al. 2011; Hinkley et al. 2011; Wang et al. 2011; Messer and Neher 2012; Mostowy et al. 2012).

Even a beneficial allele is likely to become extinct due to the combination of random drift and linkage effects (Rouzine et al. 2001). To become fixed in a population, mutation must occur within a high-fitness strain (Neher et al. 2010; Good et al. 2012). In the present work, this complexity is neglected, because the focus is on the general direction of evolution rather than its exact speed, and in the sign of ∂s as the pointer. The results demonstrate that HIV mutants with increased capsid waste are selected against since $\partial s/(\partial\kappa/\kappa) < 0$ for a range of η and κ (Fig. 4.18A), despite the fact that DIP interference is decreased with κ (see Fig. 4.16B for low values of η). Hence, the cost of mutation to HIV burst size in a singly infected cell is higher than the gain from the escape in dually infected cell. To make it completely obvious, the effective selection was recalculated by dropping the term in eq. (4.105) corresponding to the change in HIV burst size. As expected, in this case, $\partial s/(\partial\kappa/\kappa) > 0$, and HIV evolves toward higher capsid waste and less DIP interference (Fig. 4.18B). The direction of evolutionary selection is robust across a broad range of $P > 2$ (Rouzine and Weinberger 2013b, fig. 3d,e). The intuitive reason for this result is that most of HIV comes from singly infected cells. Hence, the HIV burst size in these cells and not DIP interference within dually infected cells where HIV is strongly suppressed dominates the direction of HIV evolution in a patient.

Another possible route of HIV escape from DIP interference is to reduce the excess of capsid material DIP thrives on, that is, decrease parameter η. Yet, the analysis in Section 4.3.7 shows that HIV always evolves toward high values of η; that is, HIV mutants that produce more capsids are selected for in a large range of η and κ (Fig. 4.19A). The absolute value of the selection coefficient depends on η and the presence of DIP. At $\eta < 1$, when DIP is absent, HIV benefits from the increase in η due to the increases in the virion progeny [first term in eq. (4.105); 2nd term is zero]. At $\eta_c < \eta < P+1$, when DIP is present, an increase in capsid production is also beneficial, due to the presence of DIP, which steals most capsids in dually infected cells [second term in eq. (4.105); the first term in eq. (4.105) is zero].

Interestingly, in a narrow interval of η, such that $1 < \eta < \eta_c$, in the limit of low waste $\kappa \to 0$, and below DIP's dynamic stability threshold, selection coefficient is zero [in eq. (4.105), $\partial s = \partial_n = q = 0$]. Intuitively, HIV has more capsids than it needs, and does not have to share them with DIP. Only a finite rate of product loss (finite $\kappa > 0$) results in a small, positive selection coefficient: in eq. (4.105), $\partial s > 0, \partial_n > 0, q = 0$.

As a control, the procedure was repeated with HIV burst size kept constant [putting $\partial_n = 0$ in eq. (4.105)]. The resulting value of the selection coefficient is positive everywhere where DIP is dynamically stable (Fig. 4.19B). Please note the discontinuity at $\kappa = 0$ and $\eta = 1+P$ caused by the fact that DIP stops competing with HIV for capsids at $\eta > 1+P$, when there are enough capsids for both of them. The selection coefficient depends weakly on P (Rouzine and Weinberger 2013b, fig. 4D-E).

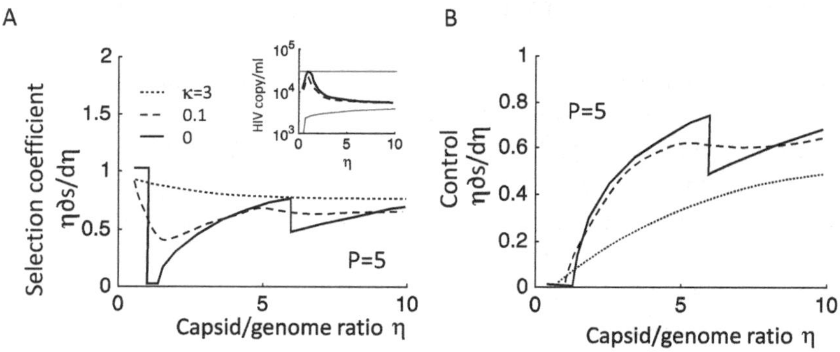

Fig. 4.19: HIV cannot escape DIP by decreasing the capsid-to-genome ratio.
(A) Normalized effective selection coefficient $\partial s/(\partial\eta/\eta)$, as a function of η, for three values of the waste parameter κ. Inset: HIV load as a function of η at three values of waste parameter κ. The positive values of $\partial s/(\partial\eta/\eta)$ imply that mutation is selected for, and HIV evolves toward increasing η. (B) A negative control neglecting HIV burst-size changes due to mutation [$\partial_n = 0$ in eq. (4.105)]. Parameters are as in Fig. 4.16: $R_0 = 10, P = 5$. Based on Rouzine and Weinberger (2013b).

4.3.5 Discussion

The present work addressed the question of whether an engineered DIP would stably interfere with HIV. It identified a class of designs based on genome stealing as unstable due to compensatory mutations causing the divergence of HIV away from DIP. This finding was in contrast to the results of Ke and Lloyd-Smith (2012), who found the evolutionary stability for DIPs based on genome stealing due to various hypothetical assumptions divergent from HIV virology (Rouzine and Weinberger 2013c). Different HIV subtypes can, indeed, display diverse DIS sequences (Paillart et al. 2004).

The model has a number of assumptions and limitations. The main limitation is that the model does not explain why, unlike simple "gamma" retroviruses, lentiviruses do not exhibit spontaneous formation of DIP (Marriott and Dimmock 2010). Due to qualitative differences in the replication strategy between these two groups of retroviruses, the present host model cannot be applied to the gamma-retroviral lifecycle. Firstly, HIV-infected cells short-lived, a day or less, while in the gamma retrovirus setting, infected cells are very long-lived (Rainey and Coffin 2006). Secondly, HIV replicates only by the infection of new uninfected cells, as assumed here, while simple retroviruses replicate in a host, mostly, by the division of infected cells; horizontal transmission is rare and is used for interhost transmission. To determine whether these obvious differences account for the different outcome of DIP observation, or whether some hidden factors are at play, would require additional analysis.

Also, all process times are assumed to be Poisson-distributed, neglecting various time delays, which would require a more complex time-structured approach (Rouzine and McKenzie 2003). Also, in a cell and a host, all dynamic variables are assumed to be constant in time and in steady state. A more complex model would include deterministic chaos. For example, Kirkwood and Bangham (1994) introduced age-structured dynamics of infected cells under the additional assumption that wild-type virions are produced by singly infected cells, because wild type is completely suppressed. They predicted chaotic oscillations, as is often the case with age-structured dynamics (Rouzine and McKenzie 2003). Engineered lentiviral DIPs (Chen et al. 1992; Bukovsky et al. 1999; Klimatcheva et al. 2001) do not exhibit such perfect interference, nor the present model predicts it. Therefore, whether age-structured dynamics with time delays will lead to chaotic dynamics for DIP and HIV, or even if it does, whether it will alter the main conclusions of our analysis, is unclear.

The interference by protein stealing is predicted to be dynamically and evolutionarily stable, provided that the ratio of capsid-to-dimerized-genome production, η, is larger than a threshold value, which is larger and close to 1. The main factor of the dynamic stability is multiple number of DIP proviruses per cell, in contrast to HIV, which has one active integrated genome per cell. The model assumes that DIP is not expressed in the absence of HIV and, hence, does not cause cell death by viral toxicity or the immune response. In contrast, HIV-infected cells have an average lifespan of ~1 day (Perelson et al. 1996; Markowitz et al. 2003). Also, cells infected with DIPs that lack HIV protein Nef would not downregulate the CD4 entry receptor to prevent superinfection (Michel et al. 2005).

The analysis also predicts evolutionary stability for capsid stealing, because the most promising route of escape, starving DIP by increasing capsid waste, κ, by decreasing packaging efficiency, k_{pck}, has a larger cost to HIV replication than the benefit from curbing DIP production.

Note that packaging constant k_{pck} is assumed to be the same in the single-cell model. The basis for this assumption is that hypothetical DIP does not express capsid, and the two particles use the same capsid expressed by HIV provirus. Any genetic changes in the packaging domain of capsid equally affects packaging of virus variants that have the same stem SL3 sequence in their genomic RNA. In principle, a double mutation, one in HIV capsid decreasing k_{pck} and a compensatory mutation in HIV SL3 could decrease k_{pck} for DIP strongly while keeping it high HIV, similar to what happens with the genome-stealing model (Section 4.3.2). However, in this case, the situation is different. Firstly, the compensatory mutation can occur in DIP SL3 as well. Secondly, DIP can follow HIV easily with a single mutation, when HIV is escaping with a double mutation is a single mutation. Because mutation probability is small, 3 10^{-5} per nucleotide per generation, the first process is much faster. Therefore, it is likely that the equality of the two packaging constants will be preserved. The same consideration applies, of course, to any type of protein stealing: HIV has to adjust more sites to escape than DIP to follow. Nevertheless, this fact is not entirely trivial, because HIV can use preexisting double mutants with mutual compensation (Chapter 2). A Red Queen chase taking into account epistatic effects deserves a separate study.

Another possible escape route considered is starving DIP by decreasing the excess of HIV capsids over its genomes, η. However, the sign of selection coefficient indicates that, in the presence of DIP, HIV evolves toward larger η, to decrease competition for capsids (Fig. 4.19A). If $\eta < 1$, where DIP is unstable, the analysis above implies rapid evolution toward the increase of η to $\eta = 1$. Indeed, each genome requires a capsid, so this evolution is expected without calculations. At $\eta > 1$, in the absence of DIP, HIV is still predicted to increase η, but at a much slower rate (at $\kappa \ll 1$), and only to compensate for the capsid decay. But once η reaches the DIP stability threshold slightly above $\eta = 1$, HIV evolution toward larger η accelerates again due to decrease of competition for capsids and hence DIP interference, until η reaches a molecular limit (Fig. 4.19A).

The central findings of this analysis, the evolutionary instability of genome stealing and the evolutionary stability of protein stealing in broad range of parameters conform to the conventional wisdom that a parasite (DIP) that harms its host (HIV) without benefit to itself will not be evolutionary stable in the long term. Conversely, in the capsid-stealing scenario, a DIP that depletes a surplus of HIV capsids is expected to survive.

The value of η critical for dynamic DIP stability was not measured directly. However, the fraction of empty capsids is known and can provide an estimate (Fig. 4.20). The fraction of filled capsids in primary SIVmac251 infection is ~20% (Kuroda et al. 1999; Letvin et al. 2006), which implies $\eta \sim 5$ (Sergeev, Batorsky and Rouzine 2010). These data differ from the *in vitro* measurements in a tissue culture, where the fraction of filled HIV capsids was 90%, which sets an upper bound of $\eta \leq 1.1$. Still, measurements in vivo are generally reliable, and it is possible that that viral production in this 293T tissue-culture setting was optimized compared to the in vivo setting. Thus, direct e measurements of η should be carried out across a large sample of patients. In addition, it would be useful to introduce the host population level, to predict the value of η that could evolve in the natural host. Partly, this is done in the next section, Section 4.4. Systematic population-level theoretical studies will be required to determine when DIPs may arise and transmit in the population, as has occurred for Dengue virus (Aaskov et al. 2006).

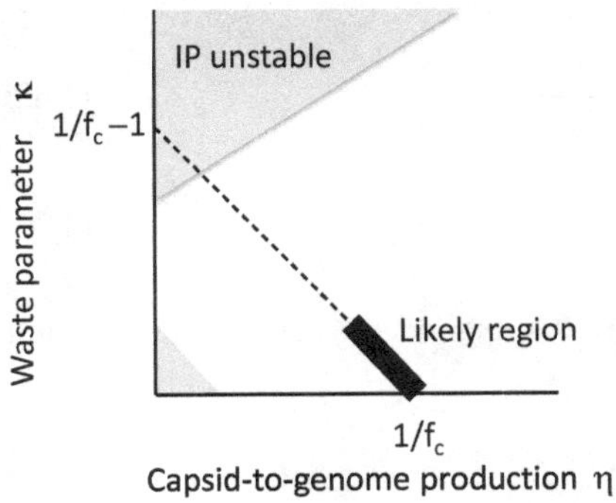

Fig. 4.20: Estimates for waste parameter κ and capsid-to-genome production η are inversely related. The fraction of nonempty virions is $f_C \sim 0.2$ for HIV (Sergeev, Batorsky and Rouzine 2010). The likely region of actual parameters (black rectangle) is far from regions of DIP instability (shaded, compare with Fig. 4.16C). Based on Rouzine and Weinberger (2013b).

Concerns have been expressed over the safety of DIPs as an antiviral strategy. An old concern is the possibility of generating a virulent phenotype by recombination between DIPs and the full-length HIV. However, the recombination of two retroviral strains occurs during the phase of reverse transcription (unique for these viruses and HBV), and is conditioned on heterodimerization and copackaging. This process would not be possible, because sequences of HIV DIS and DIP DIS would diverge (Section 4.3.2). Production of a virulent recombinant is much faster by recombination between different replication-competent HIV strains, given that HIV is extremely polymorphous. DIP-HIV recombinants will tend to have lower fitness and be selected out.

To conclude, DIPs are predicted to be evolutionary stable within an individual provided that the capsid to genome production ratio is sufficiently large. In Section 4.4, the third bioscale, a host population is analyzed to determine the direction of HIV evolution with respect to DIP interference.

4.3.6 Derivation of HIV and DIP loads

The remaining subsections of Section 4.3 serve as a mathematical appendix. Two bioscales are considered: (i) a single infected cell and (ii) an individual infected person.

4.3.6.1 Single cell

Dynamics of subpopulations of genomes and capsids in a cell coinfected with an integrated HIV provirus and m copies of a DIP provirus is described by eqs. (4.56)–(4.58). The packaging coefficients for HIV and DIP are assumed to be the same, k_{pck} (Section 4.3.5).

After an initial delay and several hours of transcription and translation, the system approaches a steady-state (Razooky and Weinberger 2011; Razooky et al. 2015). Setting all three time derivatives in eqs. (4.56)–(4.58) to zero, the steady-state amounts of genomes and capsids, for a HIV$^+$ DIP$^+$ cell, are given by the equations

$$G = \frac{\theta}{\alpha + k_{\text{pck}} C} \tag{4.68}$$

$$C = \frac{\eta\theta}{\beta + k_{\text{pck}}(G + G_{\text{DIP}})} \tag{4.69}$$

$$G_{\text{DIP}} = \frac{mP\theta}{\alpha + k_{\text{pck}} C} \tag{4.70}$$

For a HIV$^+$ DIP$^-$ cell, the same equations with $m = 0$ apply. A new convenient notation is the normalized capsid number

$$y = k_{\text{pck}} C/\alpha \tag{4.71}$$

In this notation, eqs. (4.68) and (4.70) take the form

$$G = \frac{\theta}{\alpha(1+y)} \tag{4.72}$$

$$G_{\text{DIP}} = \frac{mP\theta}{\alpha(1+y)} \tag{4.73}$$

where y is found by solving a quadratic equation

$$\kappa y^2 + (mP + 1 - \eta + \kappa)y - \eta = 0 \tag{4.74}$$

and κ is the waste parameter combining four raw parameters (Tab. 4.5). The solution of eq. (4.74) has the form

$$y = \frac{1}{2\kappa}\left[-(mP + 1 - \eta + \kappa) + \sqrt{(mP + 1 - \eta + \kappa)^2 + 4\eta\kappa}\right] \tag{4.75}$$

The output for the single-cell model is the input for the next-level model of an individual host and represents the total number of particles of each type produced by a single and dually infected cell. The total numbers of virus particles per cell are given by eqs. (4.65)–(4.67). Substituting the steady-state values of G, G_{DIP}, and C from eqs. (4.71)–(4.73) into eqs. (4.65)–(4.67), we obtain

$$n = \frac{\theta}{\delta}\left[\frac{y}{y+1}\right]_{P=0} \tag{4.76}$$

$$\psi_m n = \frac{\theta}{\delta} \frac{y}{y+1} \tag{4.77}$$

$$\rho_m = \frac{G_{\text{DIP}}}{G} = m \tag{4.78}$$

where the normalized HIV capsid concentration, y, is given by eq. (4.75), and the number of DIP proviruses, m, is any positive integer.

4.3.6.2 Small waste parameter, $\kappa \ll 1$

As shown below, HIV evolves toward the decrease of κ (Section 4.3.7). Consider the case of small κ, $\kappa \ll 1$. Then eq. (4.75) for y simplifies to

$$y = \begin{cases} \dfrac{\eta}{1+mP-\eta} & 1+mP > \eta \quad (4.79) \\[2ex] \dfrac{\eta-1-mP}{\kappa} & 1+mP < \eta \quad (4.80) \end{cases}$$

Substituting y from eqs. (4.79) and (4.80) into eqs. (4.76) and (4.77) and evaluating the limit $\kappa \to 0$, one obtains

$$n = \frac{\theta}{\delta} \min(\eta, 1) \tag{4.81}$$

$$\psi_m n = \frac{\theta}{\delta} \min\left(\frac{\eta}{1+mP}, 1\right) \tag{4.82}$$

respectively. Combining eqs. (4.81) and (4.82), HIV virion output from a dually infected cell is suppressed by the factor of

$$\psi_m = \frac{\min\left(\frac{\eta}{1+mP}, 1\right)}{\min(\eta, 1)} \tag{4.83}$$

As easy to see, function $\psi_m(\eta)$ is a nondecreasing function of η. It has a low plateau value $1/(1+mP)$ at $\eta < 1$, a high plateau value 1 at $\eta > 1+mP$, and increases linearly with η in between the two plateaus.

4.3.6.3 Individual host

Dynamic equations have the form of deterministic ODE, eqs. (4.59)–(4.64), with model parameters defined in Tab. 4.6. The entire work is focused on a chronic HIV infection, which represents an approximate steady state. Setting the time derivatives to zero, one gets

$$T = \frac{b}{d(1 + \nu + \nu_{\mathrm{DIP}})} \tag{4.84}$$

$$I = \frac{b\nu}{d(1 + \nu + \nu_{\mathrm{DIP}})} \tag{4.85}$$

$$T_{\mathrm{DIP}\,m} = Tq^m \tag{4.86}$$

$$I_{\mathrm{D}\,m} = Iq^m \tag{4.87}$$

$$1 + \nu + \nu_{\mathrm{DIP}} = R_0\left(1 + \sum_{m=1}^{\infty} \psi_m q^m\right) \text{ or } \nu = 0 \tag{4.88}$$

$$(1 + \nu + \nu_{\mathrm{DIP}})^2 = R_0\nu\left(1 + \sum_{m=1}^{\infty} \rho_m \psi_m q^{m-1}\right) \text{ or } \nu_{\mathrm{DIP}} = 0 \tag{4.89}$$

where a new convenient notation is introduced

$$R_0 \equiv \frac{nkb}{cd} \tag{4.90}$$

$$\nu \equiv \frac{kV}{d}, \nu_{\mathrm{DIP}} \equiv \frac{kV_{\mathrm{DIP}}}{d} \tag{4.91}$$

$$q = \frac{V_{\mathrm{DIP}}}{d + kV + kV_{\mathrm{DIP}}} = \frac{\nu_{\mathrm{DIP}}}{1 + \nu + \nu_{\mathrm{DIP}}} \tag{4.92}$$

Here, R_0 is the reproduction ratio in the beginning of infection before target cell depletion (Tab. 4.6), ν and ν_{DIP} are dimensionless HIV and DIP loads, and $1/(1-q)$ represents the average number of DIP provirus copies in a dually infected cell $E[m]$, as given by

$$E[m] = \frac{\sum_{m=1}^{\infty} mq^m}{\sum_{m=1}^{\infty} q^m} = \frac{q\frac{d}{dq}\sum_{m=1}^{\infty} q^m}{\sum_{m=1}^{\infty} q^m} = \frac{1}{1-q} \tag{4.93}$$

HIV load ν and DIP load ν_{DIP} can be obtained by solving eqs. (4.88) and (4.89) with respect to ν and ν_{DIP}.

Note that q entering these equations depends on ν and ν_{DIP} as given by eq. (4.92) and must be calculated self-consistently. MATLAB™ (version R2011a) was used to perform this calculation through numerical iteration. In certain important cases, such as the case of small κ and large P, this calculation can also be performed analytically in the general form, with asymptotic accuracy. The two parameters of the in vivo model reflecting the effect of DIP, ρ_m, ψ_m, can be expressed in terms of intracellular parameters κ, η and mP, as given by eqs. (4.75)–(4.78). Therefore, the total rescaled HIV load and the total DIP load, as well as other important properties of the steady state in an individual, depend on four dimensionless parameters: R_0, P, κ, η. Results for HIV and DIP loads are shown in Fig. 4.6.

4.3.6.4 Dynamic stability of DIP in a host

One approach would be to determine the parameter range in which the HIV steady state with DIP is stable once established. A more stringent criterion is to test whether DIP can autonomously spread between HIV-infected individuals, that is, whether a small amount of DIP added to a DIP-free steady-state virus population will expand and result in a new steady state.

Let us start from the DIP-free state, $V_{\mathrm{DIP}} = I_{\mathrm{D}\,m} = T_{\mathrm{DIP}\,m} = 0$, when eqs. (4.84)–(4.89) reduce to a well-known result (Nowak and May 2000)

$$T^{\mathrm{ss}} = \frac{b}{dR_0}, \quad V^{\mathrm{ss}} = \frac{d}{k}(R_0 - 1), \quad I^{\mathrm{ss}} = \frac{b(R_0 - 1)}{dR_0} \tag{4.94}$$

At time $t = 0$, a small amount of DIP is introduced, $V_{\mathrm{DIP}}(0)$. The goal is to determine whether state variables $V_{\mathrm{DIP}}(t), I_{\mathrm{D}\,m}(t), T_{\mathrm{DIP}\,m}(t)$ will expand or contract in time. There is no need to solve the entire set of eqs. (4.84)–(4.89), because DIP load is initially low so that that variables $T, I,$and V can be approximated with their previous steady-state levels, eqs. (4.94). The DIP-related state variables $V_{\mathrm{DIP}}(t), I_{\mathrm{D}\,m}(t), T_{\mathrm{DIP}\,m}(t)$ are found by solving the linearized versions of eqs. (4.61), (4.62), and (4.64)

$$\begin{aligned}
\frac{dT_{\mathrm{DIP}1}}{dt} &= kV_{\mathrm{DIP}}T^{\mathrm{ss}} - (d + kV^{\mathrm{ss}})T_{\mathrm{DIP}1} \\
\frac{dI_{\mathrm{D}1}}{dt} &= kV^{\mathrm{ss}}T_{\mathrm{DIP}1} - \delta I_{\mathrm{D}1} \\
\frac{dV_{\mathrm{DIP}}}{dt} &= n\delta\rho_1\psi_1 I_{\mathrm{D}1} - cV_{\mathrm{DIP}}
\end{aligned} \tag{4.95}$$

Multiple-infected cells, $I_{\mathrm{D}\,m}$ and $T_{\mathrm{DIP}\,m}$ for $m \geq 2$, are neglected, because they are higher-order terms in the small variable, V_{DIP}.

After a short initial period, the three variables in eqs. (4.95) will be proportional to $\exp[\lambda_{\max}(t)]$, where $\lambda_{\max}$ is the largest eigenvalue of the dynamic matrix in eq. (4.95), denoted $\boldsymbol{D}$. DIP expands if $\lambda_{\max} > 0$. From the eigenvalue equation $\det[\boldsymbol{D} - \lambda 1] = 0$ and substituting V^{ss} and T^{ss} from eqs. (4.94) and R_0 from (4.90), we obtain the DIP expansion condition

$$\rho_1\psi_1 > \frac{R_0}{R_0 - 1} \tag{4.96}$$

An equivalent condition was previously obtained for the model version that assumed a single copy of DIP provirus in dually infected cells (Metzger et al. 2011). This coincidence is natural, because, when DIP load is still low, multiple infection with DIP is negligible.

At a small waste parameter, $\kappa \ll 1$, using eqs. (4.78), (4.83), and (4.96), the dynamic stability condition, eq. (4.96), takes the form

$$\eta > \eta_c = \frac{1+P}{P}\frac{R_0}{R_0-1} \tag{4.97}$$

which is more than 1. For example, for $P=5$ and $R_0=10$, the stability interval is $\eta > 1.3$. In Section 4.4, it is shown that the stability interval is expanded tab 4.8 if the DIP-positive but HIV-negative cells are allowed to divide.

4.3.7 Selection coefficient of HIV in terms of intracellular parameters

Suppose, a mutation event causes a small change in one of its parameters. The goal of this subsection is to find out the fitness effect on virus ∂s. The starting point is a steady-state population, with state variables given by eqs. (4.84)–(4.89). We postulate that mutation changes the numbers of virions for HIV and DIP for two cell types, as given by

$$n \to n(1+\partial_n), \psi_m \to \psi_m(1+\partial_{\psi m}), \rho_m \to \rho_m(1+\partial_{\rho m}) \tag{4.98}$$

where small increments $(\partial_n, \partial_{\psi m}, \partial_{\rho m})$ are considered given input parameters. (Please note that virus infectivity parameter k can also change, but it enters only in combination with n and does not need to be considered separately.)

The central idea of calculating ∂s is to inject a small amount of the mutant virus strain, $V^{\text{mut}}(0)$ and derive the exponential rate of its expansion. The dynamics of the related state variables, $V^{\text{mut}}(t)$, $I^{\text{mut}}(t)$, $I^{\text{mut}}_{\text{D}m}(t)$, is calculated from eqs. (4.60), (4.62), (4.63), as follows. While the mutant strain is still at a small level, the rest of population is perturbed weakly and can be approximated with their steady-state values, eqs. (4.84)–(4.89). It is convenient to introduce the normalized subpopulations, $x(t)$, $y_m(t)$, and $z(t)$, as given by

$$\begin{aligned}
I^{\text{mut}}(t) &= I^{\text{ss}}x(t) \\
I^{\text{mut}}_{\text{D}m}(t) &= I^{\text{ss}}_{\text{D}m}y_m(t) \\
V^{\text{mut}}(t) &= V^{\text{ss}}z(t)
\end{aligned} \tag{4.99}$$

In the new notation, eqs. (4.60), (4.62), (4.63) become

$$\frac{dx}{dt} = (z-x)\delta \tag{4.100}$$

$$\frac{dy_m}{dt} = (z-y_m)\delta, \quad m = 1, 2, 3, \ldots \tag{4.101}$$

$$z + \frac{1}{c}\frac{dz}{dt} = \frac{R_0(1+\partial_n)}{1+\nu+\nu_{\text{DIP}}}\left(x + \sum_{m=1}^{\infty}\psi_m(1+\partial_{\psi m})q^m y_m\right) \tag{4.102}$$

Mutational change in the DIP burst size, ∂_{pm}, is absent from these equations. Because HIV virions are short-lived compared to infected cells, $c \gg \delta$ (Tab. 4.6), one can neglect the time derivative on the left-hand side of eq. (4.102). Substituting $R_0/(1+\nu+\nu_{\text{DIP}})$ from eq. (4.88) into eq. (4.102), one gets

$$z = \frac{1+\partial_n}{1+\sum_{m=1}^{\infty}\psi_m q^m}\left(x + \sum_{m=1}^{\infty}\psi_m(1+\partial_{\psi m})q^m y_m\right) \tag{4.103}$$

Taking eq. (4.103) into account, eqs. (4.100) and (4.101) represent a system of two linear ODEs. The system has a long-term asymptotic solution for $x(t), y_m(t)$, and $z(t)$ that increases exponentially in time proportionally to $\exp[t\partial s]$. The constant prefactors are related to each other, as given by

$$z = (1+\partial s)x = (1+\partial s)y_m \tag{4.104}$$

Substituting eq. (4.104) into (4.103) and neglecting the second-order terms in ∂s, one has

$$\partial s = \partial_n + \frac{\sum_{m=1}^{\infty}\psi_m q^m \partial_{\psi m}}{1+\sum_{m=1}^{\infty}\psi_m q^m} \tag{4.105}$$

where ψ_m is determined by eqs. (4.75) to (4.77) [or, for small κ, by eq. (4.83)]. The value of q is obtained from eqs. (4.88), (4.89), and (4.92) solved together numerically.

Thus, the selection coefficient eq. (4.105), has two contributions: (i) the relative change in the virion progeny of an HIV-infected cell in the absence of DIP, and (ii) the relative change in HIV suppression by DIP. The intuitive meaning of the coefficient given by eq. (4.105) is clear from the fact that it is exactly equal to the relative change in the virion progeny of an HIV-infected cell averaged over the TIP provirus number, including DIP-negative cells, as given by

$$\partial s = \left[\frac{\partial n_{\text{av}}}{n_{\text{av}}}\right]_{q=\text{const}}$$

$$n_{\text{av}} = n\frac{1+\sum_{m=1}^{\infty}\psi_m q^m}{1+\sum_{m=1}^{\infty}q^m}$$

Indeed, the average cell progeny number n_{av} is proportional to the Maltusian fitness of HIV in the presence of DIP in a steady state, and the selection coefficient ∂s is equal to the relative change in fitness due to mutation.

4.3.7.1 Change in waste parameter

In the previous subsection, the selection coefficient, ∂s, was expressed in terms of the host-model parameters. The aim of this subsection is to express it in terms of parameters of the single-cell model. Equations (4.75)–(4.78) give the virion progeny number of single and dually infected cells as functions of η, κ, P, m, and θ/δ. The expression asymmetry, P, is assumed to be chosen by the molecular engineer of the DIP and fixed. Let us start with mutation changing κ and consider change in η later on. The waste parameter can be changed by modifying packaging parameter k_{pck} depending on binding energy of a domain in HIV capsid protein *gag* and the cognate sequence in SL3 loop of genomic RNA of DIP and HIV.

Let us denote the change in parameter κ due to mutation as ∂_κ. Because ∂s is proportional to ∂_κ, which can be smaller or larger and is determined by molecular details and the choice of the mutating genomic site. The aim is to calculate the semi-log derivative representing the relative change in fitness over the relative change in waste

$$\kappa \frac{\partial s}{\partial_\kappa} = \frac{\partial s}{\partial_{\log \kappa}}$$

Mutations that decrease packaging have $\partial_\kappa > 0$. In a singly infected cell, such a mutation is deleterious to HIV. However, singly infected cells are rare, because their fraction is on the order of $1/E[m] \approx 1 - q$, from eq. (4.87), where $1 - q$ is small (Fig. 4.17A). In dually infected cells, the mutation may be beneficial, because decrease in packaging also decreases the capsid stealing by DIP. In other words, the two terms in the numerator of eq. (4.105) have different signs, $\partial_n < 0, \psi_m > 0$. Below both effects are shown to be of the same order, but the first effect wins and results in natural selection against the increase in κ.

The relative changes in the HIV virion progeny in singly and dually infected cells, n and $\psi_m n$, are obtained from eqs. (4.76) and (4.77) and are given by

$$\partial_n = \partial_\kappa \frac{\partial}{\partial \kappa} \log \left[\frac{y}{1+y} \right]_{P=0}$$

$$\partial_n + \partial_{\psi m} = \partial_\kappa \frac{\partial}{\partial \kappa} \log \frac{y}{1+y} \tag{4.106}$$

respectively, where y as a function of κ is given by eq. (4.75). Substituting $\partial_n, \partial_{\psi m}$ from eqs. (4.106) into (4.105) and calculating q from eqs. (4.88), (4.89), and (4.92) numerically, one arrives at the desired value of ∂s (Fig. 4.18A). As expected, the factor of HIV suppression by DIP favors mutations increasing waste parameter κ. However, the overall decrease in the HIV burst size dominates evolution, and HIV evolves toward smaller waste parameters.

4.3.7.2 Change in capsid-to-genome production ratio

Now consider mutation altering the capsid-to-genome production ratio η by ∂_η. Replacing the derivatives in κ with derivatives in η in eq. (4.106), we get

$$\partial_n = \partial_\eta \frac{\partial}{\partial \eta} \log \left[\frac{y}{1+y}\right]_{P=0}$$

$$\partial_n + \partial_{\psi m} = \partial_\eta \frac{\partial}{\partial \eta} \log \frac{y}{1+y}$$

Here y as a function of η is given by eq. (4.75).

As it was shown above, HIV evolves toward small waste parameters. In the limit $\kappa \ll 1$, eqs. (4.81) and (4.83) can be used for the burst size n and the suppression factor ψ_m, whence

$$\partial_n = \frac{\partial_\eta}{\eta} \Theta\,(\eta < 1) \tag{4.107}$$

$$\partial_{\psi m} = \frac{\partial_\eta}{\eta} \Theta\,(1 < \eta < 1 + mP) \tag{4.108}$$

Here step-function $\Theta\,(X) = 1$ if argument X is true and 0 if it is false. The selection coefficient ∂s is obtained by substituting eqs. (4.107) and (4.108) into eq. (4.105) and using ψ_m from eq. (4.83). As usual, parameter q is obtained by solving together eqs. (4.88), (4.89), and (4.92) numerically. DIP must also be stable in a host dynamically, which adds an additional condition $\eta > \eta_c$, eq. (4.97). The final result is shown in Fig. 4.19A.

4.3.8 Estimate of parameters κ and η in infected individuals

All the results obtained depend on four parameters: R_0, P, η and κ (Figs. 4.16–4.19), of which R_0 is a host-level parameter, and the other three are cell-level parameters. Let us estimate their values in a typical HIV infection. Parameter R_0 was measured in patients with an average value $R_0 \approx 8$ (the references in Tab. 4.6), parameter P is determined by the molecular design of DIP. Parameters η and κ are unknown. Direct estimates may be difficult, because they depend on processes comprising of many consecutive stages. However, one can estimate them indirectly, relating to the "successful" fractions of genomes and capsids, denoted f_G and f_C, respectively, referring to the genomes that are packaged within released virions, as opposed to being decomposed in a cell, and the capsids released with a dimerized HIV genome inside, as opposed to be wasted (decomposed in a cell or released empty or with irrelevant RNA, which processes are all included in the decay rate β). Both f_G and f_C can be derived from the present model in the absence of DIP and compared to the data in the literature, which results in a relation between η and κ.

In a singly infected cell, the steady-state conditions for eqs. (4.56) and (4.57) have the form

$$\theta = k_{\text{pck}}GC + \alpha G$$
$$\eta\theta = k_{\text{pck}}GC + \beta C \tag{4.109}$$

The fractions of capsids and genomes that end up to be a part of virions are

$$f_C \equiv \frac{k_{\text{pck}}GC}{\eta\theta},\ f_G \equiv \frac{k_{\text{pck}}GC}{\theta} \tag{4.110}$$

whence

$$\eta = \frac{f_G}{f_C} \tag{4.111}$$

Parameter κ (Tab. 4.5) can be written as

$$\kappa = \frac{\alpha\beta}{\theta k_{\text{pck}}} = \frac{\alpha G}{\theta}\frac{\beta C}{k_{\text{pck}}GC} = (1-f_G)\frac{1-f_C}{f_C} \tag{4.112}$$

The second equality is based on eqs. (4.109) and (4.110). Excluding f_G from eqs. (4.111) and (4.112), a linear relationship between η and κ follows

$$\eta + \frac{\kappa}{1-f_C} = \frac{1}{f_C} \tag{4.113}$$

One can estimate f_C as the fraction of nonempty virions estimated in simians as $f_C \sim 0.2$ (Sergeev, Batorsky and Rouzine 2010, appendix D). To obtain the estimate, the cited authors compared two measurements of the height of the average viremia peak in MAMU A*01 rhesus macaques infected with SIVmac251: (i) by p24 Ab assay (Kuroda et al. 1999) and (ii) by sensitive branching DNA assay (Letvin et al. 2006; Sun et al. 2006). These two methods counted HIV capsids and HIV genomes, respectfully. A study in cell culture, using a two-RNA labeling technique for an engineered HIV strain infecting a cell line, predicted a much higher value, $f_C \approx 0.9$ (Chen et al. 2009). Both estimates of f_C have to be taken cautiously, due to the limited fidelity of the assays. The first estimate is preferable, because it is obtained in vivo.

The relationship between η and κ given by eq. (4.113) and the region of dynamic instability of DIP are shown in Fig. 4.20. Thus, the increase in estimate of η causes the decrease in κ, and neither parameter can exceed $1/f_C$ (or larger). According to the above estimate, $1/f_C \approx 5$. Because HIV tends to evolve toward small κ and large η (Figs. 4.18 and 4.19; Section 4.3.7), it is reasonable to conjecture that η is close to $1/f_C \approx 5$ and, thus, far from the instability region of DIP, $\eta < \eta_c$, eq. (4.97). Including the division of DIP+HIV- cells neglected above further expands the DIP stability region to η far below 1 (Section 4.4).

4.4 Stability of HIV in the presence of defective interference particles in a host population: evolutionary conflicts and the "tragedy of the commons"

As discussed in the previous sections, the efficiency of the natural immune response, vaccines, and anti-viral therapeutics against RNA viruses is impeded by their rapid evolution (Sections 4.1 and 4.2) and, for some viruses, viral latency (Section 3.2). To prevent the emergence of resistance, engineering of defective interfering particles (DIP) that can potentially suppress HIV levels and yet not cause the evolution of resistance was proposed. A simple model neglecting epistasis demonstrated that HIV could be evolutionary stable in an individual host in the presence of suppressing DIP (Section 4.3). Whether the virus would also be resistance-proof at the epidemiological scale is a different question. In this section based on Rast et al. (2016), the multi-scale model of HIV-DIP dynamics and evolution is expanded to the epidemiological scale, in order to test for the possible propagation of DIP-resistant HIV mutants along the transmission chain. The existence of conflicting selection forces acting on HIV mutants at different scales is discovered. HIV mutants beneficial to the virus at the population scale may be deleterious to the virus within an individual host, and the other way around. Furthermore, DIP-resistant HIV mutants that can propagate within the DIP-treated population are found to revert to DIP-sensitive virus in DIP-untreated population, which limits their spread. The overall consequence of the evolutionary conflicts is the possibility of evolutionary stability of HIV in the presence of DIP at every biological scale. However, these findings put constraints on the design of DIP therapies that would be resistant to HIV escape.

As discussed in Section 4.3, DIP cannot replicate on their own and depends on the replication-competent virus, which provides the missing components to enable DIP replication (Huang and Baltimore 1970; Holland 1990). In this way, DIPs represent "cheaters" among virus variants emerging naturally during evolution. DIP can emerge spontaneously or be engineered (Voynow and Coffin 1985; McLain et al. 1988; Chattopadhyay et al. 1989; Li et al. 2011; Chaturvedi et al. 2021; Shirogane et al. 2021; Xiao et al. 2021). Their use as a therapeutic and gene transfer agent has been proposed and demonstrated experimentally, as a proof of principle, by many groups (Weinberger et al. 2003; Levine et al. 2006; D'Costa et al. 2009; Marriott and Dimmock 2010; Vignuzzi and Lopez 2019; Chaturvedi et al. 2021; Rezelj et al. 2021; Sharov et al. 2021). Some DIPs are engineered to work as a live vaccine, for example, eliciting broad neutralizing responses against SARS CoV-2 (Xiao et al. 2021). DIPs can suppress the replication of the wild-type virus by stealing its resources (Cave et al. 1985; Dimmock 1985; Barrett and Dimmock 1986; Levine et al. 2006; Marriott and Dimmock 2010). DIPs engineered to suppress HIV replication have the potential to impact the HIV/AIDS pandemic by self-propagation, without the use of expensive anti-retroviral drugs (Weinberger et al. 2003; Metzger et al. 2011; Ke and Lloyd-Smith 2012; Rouzine and Weinberger 2013b;

Notton et al. 2014). Although DIPs have not been observed in natural HIV infection, they have been engineered (Bukovsky et al. 1999; Evans and Garcia 2000; Levine et al. 2006; Mukherjee et al. 2010). While a DIP cannot eradicate the wild-type HIV in principle, because it depends on its products, it can suppress its levels significantly, as demonstrated experimentally (An et al. 1999; Klimatcheva et al. 2001) and using multiscale modeling (Section 4.3).

Such a decrease in virus load can potentially represent a "functional cure," slowing down progression to AIDS (Section 3.1) and lowering HIV transmission (Fraser et al. 2007; Shirreff et al. 2011) and, thus, having both therapeutic and protective effect. On the other hand, the decrease in both virus load in a host and virus transmission rate between individuals by DIP creates selection pressures for the evolution of DIP-resistant HIV mutants that could take over an individual patient and spread through a host population. In Section 4.3, it was demonstrated that HIV mutants that starve DIP by either decreasing the production or increasing the waste of proteins critical for DIP replication will be selected against within individual patients, at the moment of their nucleation: they would hurt the wild-type virus in singly infected cells more than help it in DIP-coinfected cells (Rouzine and Weinberger 2013b). The epistatic effects or the insertion of large amounts of resistant mutants were not considered in that analysis.

Even though a DIP-resistant mutant of HIV might be hard to nucleate in a host, once it is established, for example, by a large fluctuation, it could then propagate in DIP-treated individuals. Due to a higher virus load, it has a higher transmission rate (Fraser et al. 2007) than the wild-type and may eventually replace it. In this section, this possibility is analyzed, with the focus on the demographic groups at a high risk for HIV infection (Woodhouse et al. 1994; Lloyd-Smith et al. 2005; Baggaley et al. 2006). Previous modeling has suggested that DIPs have the potential to automatically concentrate and lower HIV prevalence within the high-risk groups (Metzger et al. 2011). Hence, these high-risk groups represent the key population where DIP-resistant mutants would arise. In this section following Rast et al. (2016), the two-scale model of DIP intervention (Section 4.3) is extended to a third scale, a host population, in order to understand how these DIP-resistant HIV mutants would spread in a population.

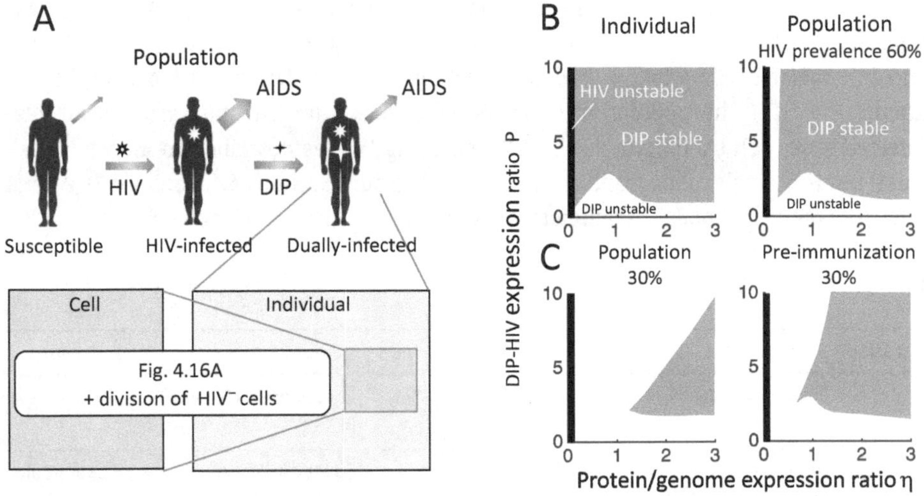

Fig. 4.21: DIP and HIV are dynamically costable across multiple biological scales.
(A) Schematic of the multi-scale mathematical model that tracks HIV and DIP levels across three biological scales: the single-cell scale, individual patient scale, and the population (epidemiological) scale. The cell-scale model quantifies DIP and HIV production and competition in individual cells (Fig. 4.16A, Section 4.3). The outputs of the single-cell model (DIP and HIV burst sizes) are used as inputs for the host-scale model. The host-scale model is based on Fig. 4.16A in Section 4.3, but includes division of $CD4^+$ T cells, both uninfected and infected with DIP (but not HIV). The output of the host-scale model (DIP and HIV viral loads) are used as input for the population-scale model. The population scale model is Susceptible-Infected model (Anderson et al. 1992) generalized to include DIP, eqs. (4.114) and Tab. 4.7. Death and transmission rates are calculated from HIV and DIP viral loads in hosts, eqs. (4.119)–(4.124), following Fraser et al. (2007). (B) DIP dynamic stability at the individual-host scale (left column) and the epidemiological scale (right column) quantified in a high-risk population of 60% starting HIV prevalence prior to DIP introduction. (C) Comparing DIP stability at the epidemiological scale in a high-risk population (30% starting HIV prevalence prior to DIP introduction) allowing DIPs to "preimmunize" hosts prior to HIV infection or under the maximally conservative assumption where DIPs can only infect hosts after a stable HIV infection has been initiated. DIPs engineered with $P > 3$ are stable except at low η values (near region of HIV extinction); as the initial HIV prevalence increases, the DIP-stability regime at the population-scale approaches the DIP-stability regime within hosts. Based on Rast et al. (2016).

4.4.1 Three-scale model of HIV and DIP dynamics

The model comprises three scales of biological organization (Fig. 4.21A). Each scale is represented by a system of deterministic ordinary differential equations describing replication and transmission of HIV and DIP. Equations are deterministic, because the the infected cells in a host and the individuals in a population number in millions. The aim here is not to describe the process of evolution subject to complex interplay between deterministic and stochastic effects (Rouzine 2020b), but only to find out its overall direction by calculating selection coefficients of DIP-resistant mutants, which can be done in a deterministic approximation (Section 4.3).

The individual-host and single-cell models are considered in Section 4.3. In this section, the host-level model is slightly modified to account for the division of uninfected and DIP-infected cells, which expands the stability region of DIP toward low capsid production by HIV. The population scale model is based on the standard susceptible-infected approach (Anderson et al. 1992) including DIP, as described in Metzger et al. (2011) but in a single demographics, the high-risk group (Sections 4.4.6 and 4.4.7). All the relevant composite model parameters are summarized in Tabs. 4.7 and 4.8.

Tab. 4.7: Population-level composite parameters.

Notation	Description	Units	Value	Reference
R_0^{pop}	Basic reproduction ratio in a population	1[a]	Equations (4.130) and (4.132)	
μ	HIV transmission rate in HIV^+ TIP^+ hosts relative to HIV^+ TIP^- hosts	1	Equations (4.136), (4.119), and (4.120)	Fraser et al. (2007)
ϕ	TIP transmission rate relative to HIV^+ TIP^- hosts	1	Equation (4.137), (4.119), (4.121)	Fraser et al. (2007)
B	High-risk life span decrease from an unsuppressed HIV infection	1	3.5, eq. (4.131)	Fraser et al. (2007)
τ	Decrease in HIV death rate due to superinfection with TIP	1	Equations (4.138), (4.123), and (4.124)	Shirreff et al. (2011)
δ_I	Death rate for HIV^+ TIP^- individuals	1/year	0.1, eq. (4.124)	Fraser et al. (2007)

[a]1 stands for "dimensionless."

Let us recap briefly the biology of DIP interference. The single-cell dynamics is described by eqs. (4.56)–(4.58). In the absence of HIV, DIP can infect $CD4^+$ T cells in many copies and integrate its genetic material into the cellular genome as a gene (provirus), but no DIP gene expression occurs assuming DIP does not encode trans-activating factors (Weinberger et al. 2005). After the cell is coinfected by HIV, the viral proteins required for DIP expression and virus particle production are expressed by HIV provirus. As a result, DIP genomic RNAs (gRNAs) are expressed as well, dimerize, and can compete with HIV gRNAs for HIV proteins. HIV and DIP heterodimerization is neglected, because it was predicted to be evolutionarily unstable (Rouzine and Weinberger 2013b, c). The key parameters at the single-cell scale are the production ratio of RNA genomes of DIP to HIV, P, and the ratio of HIV-capsid to HIV-genome production, η (Tab. 4.8). Below (η, P) is referred to as the "parameter plane."

In an individual host, DIP compete with HIV for available $CD4^+$ T cells to infect. Cells coinfected with both HIV and DIP produce considerably fewer virions compared to cells infected only with HIV, but instead produce plenty of DIP virions diverting

Tab. 4.8: Cell-level and host-level dimensionless parameters.

Notation	Description	Value	References
Cell-level parameters			
κ	Capsid and genome waste parameter	0.01	Table 4.5
P	TIP to HIV transcription rate ratio	>1	Table 4.5
η	Capsid-to-genome production rate ratio for HIV	>0	Table 4.5
Host-level parameters			
$R_{00} = [R_{0\,host}]_{\eta=1,\kappa=0}$	Basic reproduction ratio in an individual host	10	Table 4.6; Nowak et al. (1997)
δ/d	Cell life span decrease due to HIV infection	10	Mohri et al. (1998)
c/d	Viral to infected cell death tare	10	Perelson et al. (1996)
h_0	Maximum division number of a target cell per lifespan	3.3	Mohri et al. (1998)

Listed are composite dimensionless parameters relevant for the final results at the population level. For the definitions of κ and R_{00} in terms of raw model parameters, see Tabs. 4.5 and 4.6 in Section 4.3. All parameters are introduced in Section 4.3, except for h_0 , which is a new additional parameter.

HIV protein resources. As a result, HIV viral loads in DIP-treated individuals are lower than HIV-only infected individuals, but the corresponding DIP viral load may be very large due to a high multiplicity of DIP infection in a cell (Section 4.3). DIP and HIV are asymmetric in three respects, as follows:

(i) Within an individual cell, DIP requires the presence of HIV to replicate.
(ii) In an HIV-infected cell, expression of Vpr protein by HIV gene stops the division of infected cells (Re et al. 1995; Frankel and Young 1998), and expression of Nef protein suppresses superinfection (Bamdres et al. 1994; Frankel and Young 1998). In contrast, integrated DIP gene does nothing of a sort, because it is not expressed. Hence, DIP can divide with the cell.
(iii) HIV-infected cells live one day, while DIP-infected cells live as long as uninfected cells, for many months (Ho et al. 1995; Perelson et al. 1996). As a result, although DIP cannot replicate without HIV, multiple copies of the DIP provirus can accumulate in a cell genome prior to HIV infecting that cell (Section 4.3).

Equations of the individual-host scale have the form of eqs. (4.59)–(4.64) from Section 4.3, with one correction: the uninfected CD4 T cells T and DIP-infected cells T_{DIP} m can divide at a rate hd, here d is their natural death rate, and h is the maximum division number per cell lifespan. The division rate is homeostatic: it depends on the total number of all cells and decreases when it approaches the normal level of target cells in an uninfected host (Rast et al. 2016, S1 text, eqs. (17), (18), (51), 52)). Parameters are given in Tab. 4.8.

Now we add, to these two scales, the third scale: a population of hosts. The epidemiological model considers a well-mixed population of individuals at a high-risk of exposure to HIV with three compartments (Fig. 4.21, top) described by the system of ODE, as follows

$$\frac{1}{\delta_I}\frac{dS}{dt} = \frac{1}{B}(1-S) - R_0^{\text{pop}}(S\,I + \mu\,S\,I_D)$$

$$\frac{1}{\delta_I}\frac{dI}{dt} = R_0^{\text{pop}}(S\,I + \mu\,S\,I_D - \phi\,I\,I_D) - I$$

$$\frac{1}{\delta_I}\frac{dI_D}{dt} = R_0^{\text{pop}}\phi\,I\,{}_D - \tau\,I_D \tag{4.114}$$

Here the three compartment sizes are the number of individuals susceptible to HIV, $S(t)$, HIV-infected individuals, $I(t)$, and dually infected HIV^+ TIP^+ individuals, $I_D(t)$. All three compartment sizes are normalized to the size of the susceptible uninfected population. For the initial equations in absolute units, see eqs. (4.116)–(4.118) in Section 4.4.6.

Equation (4.114) include the following epidemiological processes. The susceptible individuals, S, enter the high-risk group at a constant linear rate and leave it at a fixed exponential rate. They can be infected with HIV coming from singly or dually infected individuals (at different rates) and become the singly infected, I. The singly infected can be superinfected with DIP from dually infected individuals and become the dually infected themselves, I_D. HIV infection progresses to AIDS as a function of HIV viral load in the patient, which is modeled as removal from the high-risk population. Superinfection with DIP slows progression to AIDS (and reduces transmission of HIV from that individual) by reducing the HIV viral load. DIPs only replicate within HIV-infected individuals, and we use a simplifying conservative assumption that DIPs only infect already HIV-infected patients; when this assumption is relaxed, DIPs have a broader region of dynamical stability (Fig. 4.21B, C; Section 4.4.8).

The model parameters are defined as follows: δ_I is the average AIDS progression rate for HIV-infected individuals, with δ_I/B being the removal rate from the high-risk population; R_0^{pop} is the basic HIV reproduction number in a population, eq. (4.130); μ and ϕ are the respective ratios of HIV and DIP transmission rates from HIV^+ DIP^+ individuals relative to HIV transmission from singly infected individuals, eqs. (4.136) and (4.137), respectively; τ is the decrease in the progression rate to AIDS due to HIV suppression by DIP, eqs. (4.138) and (4.123). The estimates of parameter values and the literature sources are given in Tab. 4.7.

The three biological scales (a cell, a host, and a host population) are connected together by assuming a steady-state at each lower scale and using the steady-state values of the state variables as input for the higher-scale model parameters. The steady-state assumption is justified, because the processes at lower scales are much faster than those in upper scales (minutes/hours, days/months, and years/decades, respectively). Given the input parameters at a cellular scale, η, κ, and P, the single-cell model predicts the total number of HIV and DIP virions produced by an infected cell (Tab. 4.5 in Section 4.3). Taking the viral progeny as input, the host-level model predicts the steady-state viral loads for HIV and DIP (Tab. 4.6 in Section 4.3). Finally, the viral loads are used to calculate the epidemiological parameters in eq. (4.114), such as the progression time to AIDS and the transmission rates of HIV and DIP based on previous work by Fraser et al. (2007); Shirreff et al. (2011) [eqs. (4.119)–(4.124) in Section 4.4.6]. The upper-level output is the total frequency of singly and dually infected individuals in a population calculated as a function of η and P, which can serve to evaluate the therapeutic value of DIP as a self-propagating agent. Model assumptions and the robustness to model variations will be discussed in Sections 4.4.5 and 4.4.8, respectively.

4.4.2 Conditions of HIV and DIP coexistence and HIV suppression in high-risk populations

The updated individual-host model reproduces qualitatively the results of Section 4.3. HIV and DIP achieve a dynamically stable steady-state in a host, where HIV is stably suppressed. The method of its calculation represents a variation on the method detailed in Section 4.3, with the extra cell division parameter, h (Rast et al. 2016, S1 text, section B).

Including the division of $DIP^+ HIV^-$ cells into the host model expands the region of DIP stability toward $\eta < 1$, compared to the old stability condition, which requires $\eta > 1$, eq. (4.97). If the number of cell divisions per its lifespan h is larger than $1 + R_{00}/P$, where R_{00} is the host reproduction ratio at $\eta = 1, \kappa = 0$ (Tab. 4.8), DIP can be dynamically stable at almost any value η (Rast et al. 2016, S1 text, eq. (50)). More exactly, DIP can be stable at $\eta > 0.1$, given $R_{00} = 10$. At smaller values of η, HIV cannot expand in a host anyway, so this region is of no interest (Fig. 4.21B, C, vertical black strip). However, stability requires a sufficiently large expression asymmetry, P. At small values of P, HIV can propagate in a host, but DIP cannot (Fig. 4.21B, left). In the remainder of parameter plane, HIV and DIP stably coexist. The maximum value of P at which DIP is unstable in the host is $P_{crit} \approx 3$, which corresponds to $\eta \approx 1$. The condition $P > P_{crit}$ represents an important restriction on DIP design. Below, the region $P > P_{crit}$ is considered.

On the population scale, DIP is stable within a broad parameter region (Fig. 4.21B, right). The exact boundaries of the stability region depend on HIV prevalence within the high-risk population before DIP has been introduced into it, and whether DIPs can preinfect individuals before they have been infected with HIV (Fig. 4.21C) (Metzger et al. 2011).

If one makes a conservative assumption that DIP cannot preinfect, the subpopulation available for DIP infection are only HIV-infected individuals. Because HIV prevalence before DIP depends on basic reproduction ratio in a population, R_0^{pop} (Nowak and May 2000), the effective reproduction ratio of DIP denoted R_{eff} also depends on it. In this case, the condition for the propagation and stability of DIP has the form

$$R_{\text{eff}} = R_0^{\text{pop}} - \frac{B\tau}{\phi} > 1 \tag{4.115}$$

(Tab. 4.7; Section 4.4.6). When the prevalence of HIV is increased due to the increase in R_0^{pop}, the second term in eq. (4.115) becomes relatively small, and DIP stability region in a population expands and approaches the within-host stability region (Fig. 4.21B). For example, if 60% of a high-risk population is HIV positive, stability of DIP in an individual usually implies its stability in the population.

The stability condition, eq. (4.115), assumes that DIP can only super-infect individuals previously infected with HIV. If DIPs can preinfect individuals and remain latent until a subsequent infection with HIV, as assumed by Metzger et al. (2011) and supported by the studies of HIV latency (Rouzine et al. 2015) (Chapter 2), HIV prevalence affects the DIP stability region much less (Fig. 4.21C). With preinfection or without, whenever DIP stably coexists with HIV, the incidence and prevalence of HIV infection in a population will be strongly suppressed (Rast et al. 2016, fig S2). The evolutionary stability of this suppression is analyzed in the next section.

4.4.3 HIV escape mutants that are resistant to DIP face conflicting selection pressures

To analyze the evolution of HIV mutants resistant to DIP, let us introduce a mutant HIV strain into the host population infected with HIV and DIP and analyze its competition with the present HIV strain. The two HIV strains are assumed to differ by the value of parameter η. The engineered DIP's parameter, P, as in Section 4.3, is assumed to be fixed. As in the dynamic stability analysis above, we take advantage of the time-scale separation: the processes on the host scale are much faster than on the population scale.

Each biological scale can develop DIP-resistant mutants of HIV: HIV mutants that increase HIV load in an individual, and HIV mutants that increase HIV prevalence. Because HIV and DIP are dynamically coupled at two scales, it is not *a priori* obvious whether a resistant mutant spreads in an individual or a population. In Section 4.3, the selection coefficient was obtained for an individual host. Below, selection pressures are compared between the two scales.

HIV load in an $\text{HIV}^+ \text{DIP}^+$ individual reaches a maximum at a value of $\eta = \eta_c$ slightly less than $\eta = 1$ (Fig. 4.22A, left). HIV mutants with $\eta = \eta_c$ are the best candidates for DIP resistance, because any mutation in η toward η_c increases the HIV load in a dually infected host. At the same time, a mutant that reduces η will increase HIV

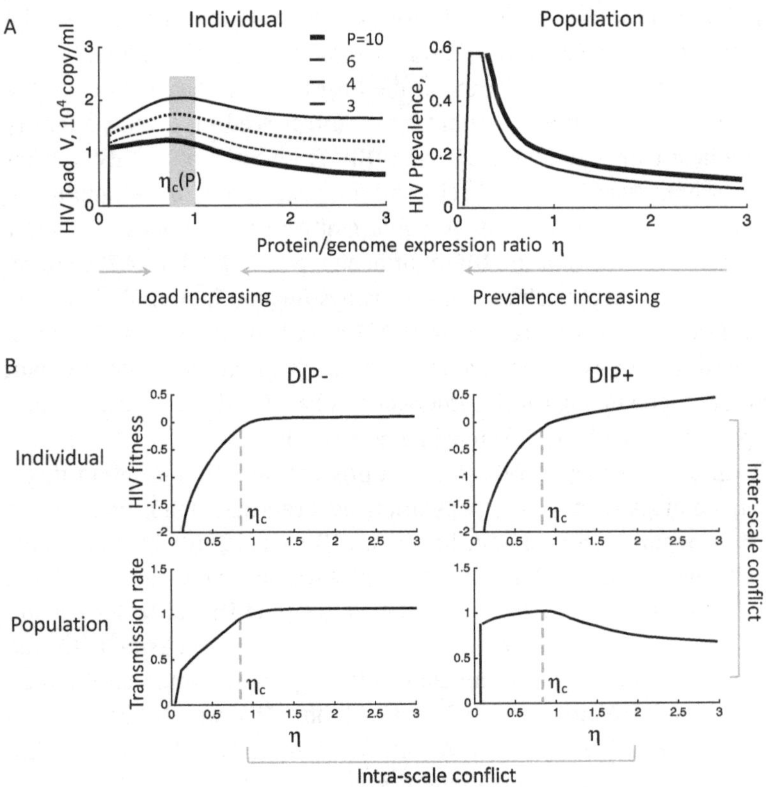

Fig. 4.22: DIP-resistant HIV mutants face conflicting selection pressures.
(A) Steady-state HIV load within a host normalized to the maximum value (left) and the steady-state prevalence of HIV in a high-risk population (right), as functions of P and η. For $P > 3$, HIV load is maximal at a value $\eta = \eta_c \approx 1$ (translucent gray plane). For $P < 3$, see Fig S2 in Rast et al. (2016). (B) Different fitness landscape in η at the individual-host scale (top panels) and the host population scale (bottom panels), for DIP- cells or individuals (left panels) or DIP$^+$ cells or individuals (right panels). Individual host-level fitness is calculated based on the rate of expansion or contraction of an HIV mutant, with the slope (selection coefficient) given by eq. (4.105) in Section 4.3. The average HIV transmission rate in a population is calculated from the steady-state viral load, as given by eqs. (4.119)–(4.122). Both fitness and transmission rate are normalized to their values at η_c. Expression asymmetry $P = 6$, and the plot is representative for the interval $P > 3$ (Rast et al. 2016, fig S3). Thus, the presence of DIP$^+$ and DIP$^-$ individuals exerts opposing selection pressures on the value of η in a population. Based on Rast et al. (2016).

prevalence in a dually infected population (Fig. 4.22A right). Both mutants, the viral-load increasing mutant and the prevalence-increasing mutant, can lead to a decrease in HIV suppression. Moreover, they can drive the system to full resistance at each scale, to the region of P and η where DIPs are unstable (Fig. 4.21B-C). If $P > 3$, the full resistance to DIP treatment in a host only occurs at low η values, near the HIV extinction threshold. Resistance in a population can occur, in principle, in a broader range.

After mapping the parameter regions where escape mutants could emerge, the next step is to calculate the selective benefit or disadvantage of these mutants. Section 4.3 described how the fitness landscape at the host level can be found from the rate of expansion or contraction of a mutant strain introduced in a small quantity into a steady-state host coinfected with HIV and DIP. The fitness of an HIV variant depends on viral outputs from individual cells and the distribution of DIP multiplicities among cells in the infected individual. A higher η always corresponds to a larger viral progeny per cell, regardless of the DIP multiplicity, eqs. (4.76) and (4.75). Consequently, HIV mutants with larger values of η are always favored in an individual, regardless of the presence or preexisting level of DIP [Fig. 4.22B, top; see also (Rast et al. 2016), Fig. S3]. This result represents an example of "tragedy of the commons" (Rankin et al. 2007): a mutation enhancing capsid production is beneficial, but it also enables the increased exploitation of all HIV variants in a host by DIP.

Interestingly, in the presence of DIP, selection pressure acts in a different direction depending on a biological scale. The population-level reproduction ratio of HIV is determined by its transmission rate and, hence, on HIV viral loads in individuals (Fraser et al. 2007). The HIV and DIP transmission rates introduced in eqs. (4.114) and Tab. 4.7 are calculated from the viral load, as specified in eqs. (4.119)–(4.124) in Section 4.4.6. Connecting virus load to single-cell parameters as described in Section 4.3, the transmission rates of HIV variants are functions of their η and P. Note that the individuals infected with only HIV and those infected with both HIV and DIP have different transmission rates (Fig. 4.22B bottom). Within DIP^+ individuals, the transmission rate and the viral load both have a maximum at $\eta = \eta_c$. In contrast, in DIP^- individuals, transmission always increases with η. Indeed, in DIP^+ hosts, HIV making too many resources means its suppression by DIP.

These conflicting selection pressures push the evolution in different directions along the η-axis. Overall, the model predicts two evolutionary conflicts: the interscale conflict between the host-level and the population-level fitness landscapes, and the intrascale conflict between transmission from individuals with a discordant DIP status (Fig 4.22B).

4.4.4 Evolutionary conflicts prevent the establishment of DIP-resistant HIV mutants

The next task is to determine which HIV mutants could spread through a population under these conflicts. For this end, the population-level model is generalized to include two strains of HIV: the initial strain a steady state, and a potentially resistant strain [Fig. 4.23A; Section 4.4.7, eqs. (4.144)]. For each strain, transmission rates are calculated from eqs. (4.114) and Tab. 4.7. Host fitness becomes important, when the two strains coinfect the same host, resulting in a rapid fixation of the more fit strain (vertical arrows in Fig. 4.23A). Such coinfection is frequently observed in a high-risk population [see Rouzine and Coffin (1999c) and references therein]. The competitive exclusion within a host occurs very rapidly relative to the dynamics of virus prevalence in a population. At each moment of time, individuals transiently coinfected with two HIV strains are few and can be neglected.

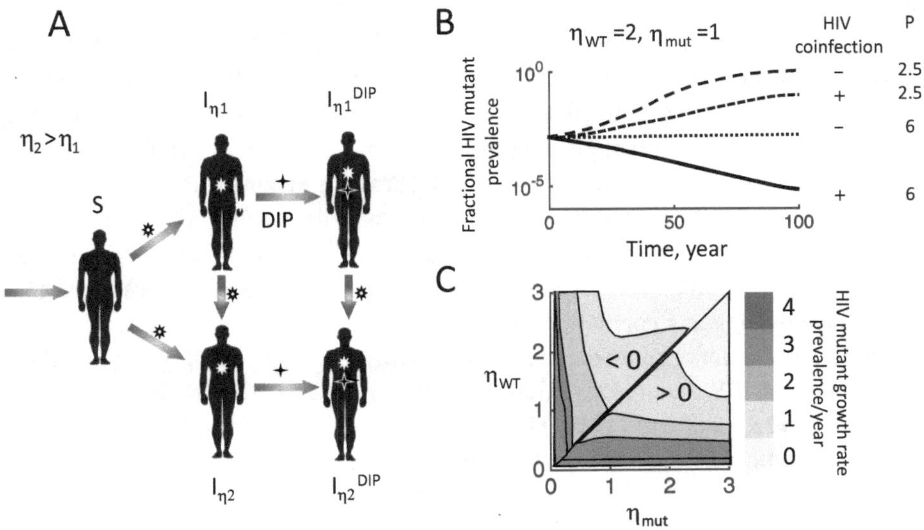

Fig. 4.23: Facing conflicting selection pressures, DIP-resistant HIV mutants cannot propagate in a host population.
(A) Schematics of the evolutionary model with a conflict. Two HIV mutants with different η values compete for transmission across a DIP-treated population, eqs. (4.144). When two HIV strains coinfect the same individual, the strain with the higher η replaces the other strain due to its higher fitness (Fig. 4.22B). (B) Prevalence of an HIV mutant with reduced η as a function of time ($\eta_{wt} = 2$, $\eta_{mut} = 1$) with the initial mutant fraction 0.1% and two values of P (shown). Two cases, with and without HIV strain coinfection are shown. Coinfection of a host with two HIV strains selects for HIV strains with increased values of η, despite the cost of increasing η in the transmission rates in DIP$^+$ HIV$^+$ individuals (interscale conflict, Fig. 4.22B). The opposing selection pressures on η from the presence of DIP$^+$ and DIP$^-$ individuals (intrascale conflict, Fig. 4.22B) is more significant for $P = 6$ than for $P = 2.5$, where it is only a very weak effect. (C) Initial expansion or contraction rate of a mutant HIV strain with $\eta = \eta_{mut}$ upon introduction into a population infected with both DIP and wild-type HIV, $\eta = \eta_{wt}$. Each HIV mutant and HIV wild-type combination is represented by a point (η_{mut}, η_{wt}) on the plane. At each point, the shade of the map shows the maximal eigenvalue of the Jacobian for the mutant strain, across all values of $P > 3$. Thus, the only mutants that can grow are those with $\eta_{mut} > \eta_{wt}$, and HIV cannot evolve away from DIP by starving it of protein, as long as $P > 3$. Based on Rast et al. (2016).

For an HIV mutant to spread through a host population, the fitness benefits should overcome the costs. Suppose, a mutant HIV variant with capsid-to-genome production ratio $\eta_{mut} = 1$, which is the most hopeful resistant mutant, competes against the initial (wild-type) HIV strain with $\eta_{wt} = 2$. Such a mutant increases the HIV viral load in a dually infected individual and hence, transmits better from him than the wild type

(Fig. 4.22B bottom). At the same time, it will lose to the wild type in transmission from HIV^+ DIP^- hosts and will also be at disadvantage within individual hosts (Fig. 4.22B). The two-strain model calculates the benefits and the costs and demonstrates that, at the population scale and in a host with two HIV strains, the mutant cannot spread, as long as DIPs are designed to have $P \sim 3$ or higher (Fig. 4.23B). See also original work (Rast et al. 2016, fig S4).

Increasing the designed value of P increases the transmission benefit of the HIV mutant in DIP^+ hosts, but does not change it in DIP^- hosts. Thus, the value of P controls the intrascale conflict. If coinfection did not occur, the interscale conflict were absent, but it occurs in high-risk patients and has strong effects on HIV evolution along the transmission chain (Rouzine and Coffin 1999c). Formally, turning coinfection “on” and “off” in the mathematical model controls the interscale conflict. When expression asymmetry P is decreased, or when coinfection is neglected, the mutant spreads better (Fig. 4.23B). Both high P and coinfection are required to prevent DIP-resistant mutant from spreading in a population.

To determine more general conditions on parameters η_{wt} and η_{mut}, the initial expansion rate of HIV mutants after introduction was calculated, and its largest value across interval $P > 3$ at was found (Fig. 4.23C). Generalizing the result in Fig. 4.23B, an HIV mutant with $\eta_{\mathrm{mut}} < \eta_{\mathrm{wt}}$ never propagates (Fig. 4.23C). This result is the key to the design criteria for engineering a resistant-proof DIP.

The result (Fig. 4.23C) demonstrates that HIV mutants increasing its prevalence cannot spread in a population at any parameter values. In addition, mutants increasing the viral load in a host cannot propagate if $\eta_{\mathrm{wt}} > \eta_c$. Such virus-load increasing mutants can propagate, at both biological scales, only when $\eta_{\mathrm{wt}} < \eta_c$, in which case all the selection pressures align (Rast et al. 2016, fig. S5). In this interval, an increase in η results in an increase in HIV load and transmission rate. However, this effect is unrelated to DIP and, hence, is not interesting to us. The selection pressure arises from the enhanced replication of HIV in a host due to the higher capsid production and only decreases when DIP is present (see transmission rates in Fig. 4.22B, right). In other words, at $\eta_{\mathrm{wt}} < \eta_c$, the population-level instability exists at the host level before DIP is introduced. Otherwise, the model predicts that DIP intervention is both dynamically and evolutionarily stable at any scale.

4.4.5 Discussion

A three-scale model of HIV dynamics is used to test whether interfering particles that steal critical HIV proteins can stably suppress rapidly evolving HIV in a high-risk population. In the absence of HIV evolution, the results predict that DIP will stably persist in the population and decrease HIV prevalence by transmitting from HIV^+ DIP^+ individuals to HIV^+ DIP^- individuals, provided HIV infection was sufficiently prevalent before the introduction of DIP (Fig. 4.21B). If DIP has the additional ability to transmit to HIV-negative individuals, DIP will be stable even if the initial HIV prevalence was low (Fig. 4.21C).

Critically, the results demonstrate that the spread of DIP-resistant HIV mutants, both those that enhance virus load in a host and those that enhance HIV prevalence, is prevented by their reduced transmission rates in DIP^- individuals and their reduced growth rates in both DIP^- and DIP^+ individuals (Fig. 4.22B). If DIP and HIV are evolutionarily stable within a host, these factors force HIV to evolve to become more, not less susceptible to DIP at the population scale (Fig. 4.23C). Combined, the study demonstrates that any preinfecting DIP that is dynamically and evolutionarily stable in a host, preserves this property at the population scale.

4.4.5.1 Cheaters and the "tragedy of the commons"

Competition of pathogen strains at multiple biological scales was modeled in previous studies [see Alizon (2013) for review]. Interestingly, these models often predict an increased pathogen virulence, because more virulent strains often have a higher replication ability (Nowak and May 1994; vanBaalen and Sabelis 1995; Alizon and van Baalen 2008). The previous results on the virulence evolution are analogous to our results on the direction of HIV evolution in the sense that HIV always evolves toward higher values of the capsid-to-genome production ratio, η. Biologically, however, our results are opposite: in the context of DIP therapy, the increase in η results in a decreased HIV load and, hence, a decreased virulence.

Decreased virulence was predicted in "public-goods" models in which selfish, but less virulent pathogens outcompete cooperative, but virulent strains (Alizon and Lion 2011). Paradoxically, DIPs can be viewed not only as HIV parasites, but also as "public goods" that the HIV strains coinfecting a host share. This is because an individual HIV strain, strangely enough, benefits from DIP in the long term. As demonstrated in previous sections, an increased capsid production forced by DIP in the course of strain evolution increases the relative fitness of the HIV strain. At the same time, an increased production of DIP due to the increase in capsids is, of course, deleterious to the entire HIV population, reducing the total load. Thus, the virus evolution toward the appeasement of DIP can be viewed as the enhancement of the "cheating" HIV strain at the expense of the other strains. We can interpret the overall HIV virulence reduction in time as "tragedy of the commons" (vanBaalen and Sabelis 1995; Rankin et al. 2007).

Because DIP production decreases HIV transmission in a population, the aim of this study was to find out whether the cheating virus strain would outcompete cooperative strains that produce smaller numbers of DIP. The outcome of the competition between cheaters and cooperators in a multi-scale evolutionary conflict depends on the relative strength of the lower-level and upper-level selection (Bonhoeffer and Nowak 1994; Mideo et al. 2008). The host-level benefits the cheater, and the population-level selects for the cooperator. The model analyzed in this section predicts the dominance of the host-level selection, because the population-level selection in favor of cooperators is absent or even inverted in DIP-negative individuals.

4.4.5.2 Model assumptions and limitations

(i) *Rare mutation:* In the population model, HIV mutants only introduced into patients via coinfection neglecting the generation of new HIV mutants within hosts. As easy to understand, this assumption is conservative, that is, it represents the best-case scenario for the evolution of DIP resistance at the population-scale. Indeed, if within-host mutations were to arise more frequently, there would be greater numbers of cheater HIV mutants that increase DIP production (i.e., HIV mutants with higher η values than the wild-type).

(ii) *Strong host selection:* When co-infecting a host, a higher-fitness strain with a higher η instantly replaces a lower-fitness strain with a smaller η. If the values of η are very close, this assumption may be not true. However, even for selection coefficient as small as 1%, the time scale of strain competition does not exceed a year (Rouzine and Coffin 1999c), which is much shorter than the time scales of epidemics. We do not need to consider smaller differences in η to test evolutionary stability.

(iii) *Route of escape:* HIV escapes DIP parasitism by reducing η, the production of protein (capsid) elements necessary for DIP. The choice of this parameter is obvious. Because a minimal design of DIP does not encode proteins, it would have no ability to restore η to high values. Because η is asymmetrically controlled, mutation affecting η is the most promising route of escape. Another possible single-cell parameter whose change offers a route of escape is the composite waste parameter, κ, which, in turn, depends on packaging coefficient k_{pck} and the gRNA production rate, θ (Tabs. 4.5 and 4.8), both depending on HIV genome. In Section 4.3, we demonstrated that the host-level selection favors small waste parameters κ, and that HIV cannot escape DIP by changing it. That fact, obviously, is not going to be changed by the presence of a population. Finally, HIV could try to escape by changing genome production rate θ, which directly affects the viral progeny number of a cell n and hence R_{00} (Tabs. 4.5 and 4.8). For example, mutations in could modify HIV Tat protein responsible for the transactivation of HIV and DIP genes. However, the change in θ would affect gRNA production of DIP and HIV in exactly the same way, which makes it a worse candidate than asymmetrically controlled η (Rast et al. 2016, *discussion*).

(iv) *The absence of epistasis* from the model remains the most important approximation of Sections 4.3 and 4.4, which may change the predictions for evolutionary costability significantly. Indeed, epistasis is omnipresent (Chapter II). For example, the presence of the obvious compensatory mutations in "kissing loop" makes genome-stealing DIP unstable (Section 4.3.2). Likewise, mutually compensating mutations in capsid and the capsid-binding domain of genomic RNA, could lead HIV astray from

DIP along a genetic path passing through a fitness valley (Rouzine and Coffin 1999c; Weissman et al. 2009; Pedruzzi and Rouzine 2021). Even though DIP could respond to the two mutations with a single mutation in its genomic RNA, such DIP-resistant double mutants are under a strong selection pressure to emerge (Rouzine 2020b). The presence of epistasis could also explain the slow progress with development of therapeutic DIP against any virus, despite many promising advances (Vignuzzi and Lopez 2019; Chaturvedi et al. 2021; Rezelj et al. 2021; Shirogane et al. 2021; Xiao et al. 2021). I hope that simulations of a Red Queen chase of HIV by DIP taking into account epistasis will be carried out in the future. For small modifications of the model and their effect on the predictions, see Section 4.4.8.

4.4.5.3 Frequency-dependent selection on the population scale

In general, selective forces acting on HIV at the population level cannot be described by a fixed fitness landscape. In other words, it is not possible to express the mutant expansion slope as the difference in log fitness w between two strains, $s = w(\eta_2) - w(\eta_1)$,for three following reasons:

(i) At the population level, the less-fit strain experiences a negative selection pressure with the strength depending on the frequency of individuals infected with fitter strains that can replace the less-fit strain on coinfection.

(ii) The expansion rate of a mutant has a term that does not depend on the magnitude of $\eta_2 - \eta_1$ but only on its sign (instant competitive exclusion);

(iii) even without host coinfection, s depends on the frequency of DIP^+ individuals in the populations, which is a dynamic parameter causing long-term oscillations (Rast et al. 2016, fig. S6).

Thus, natural selection at the population level is frequency-dependent. Fortunately, this complication can be safely neglected, when one calculates the fitness landscape due to incremental small changes in η, as we did. Although large jumps in parameter η would produce frequency-dependent corrections, these effects are important for selection magnitude, but not its direction, which the focus of the entire analysis.

4.4.5.4 DIP as a resistance-proof therapy?

Individuals with poor adherence to anti-retroviral therapy or undergoing suboptimal therapy regimens often evolve HIV resistance (Yerly et al. 2007; zur Wiesch et al. 2011). Resistant HIV strains can transmit from one individual to another, spreading in the population (Yerly et al. 2007). The difference between antivirals and DIP is that DIP action is based on the same biochemical processes as HIV replication. Hence, to develop resistance to DIP, HIV has to diminish its own fitness. In other words, the cost and benefit of a resistance mutation are tightly related. In contrast, for the escape from antiviral drugs, the benefit and the cost are not necessarily related, because drug binding to viral proteins has a different biochemistry. This explains why, at least in the simplest model without epistasis and fitness valleys, DIP works as a self-spreading drug from which there is no escape. While future modeling may find some escape routes for HIV, the design constraints predicted by these models may aid in the engineering of resistance-proof interventions against a range of viral and bacterial pathogens.

4.4.6 Derivation of DIP and HIV prevalence in a population of hosts

The remaining subsections of Section 4.4 serve as a mathematical appendix. The model of DIP and HIV dynamics comprises three linked scales of biological organization. The cell scale is identical to that in Section 4.3.6, and the host scale was analyzed in Section 4.3.6 without including the division of uninfected and DIP^{+} HIV^{-}cells in a host. For the analytic details related to that additional factor, the reader is referred to the original work (Rast et al. 2016, S1 text, section B). The method is the same as in Section 4.3.6. The evolutionary stability of HIV under DIP treatment on the individual-host scale was considered in Section 4.3.7. The updated results on an individual host are given in the text and figures above. The mathematical derivations for the dynamics at the host population level are presented in this subsection, and the evolutionary stability is considered in the next Section 4.4.7. The model variants are discussed in Section 4.4.8. Parentheses () are used for grouping, and brackets [] for an argument of a function.

Consider a well-mixed population with viral transmission during the chronic phase of HIV infection. According to the standard SIR approach, it is represented by the system of ODEs

$$\frac{dS}{dt} = \lambda - \frac{c}{N}\beta_1 SI - \frac{c}{N}\beta_H^{I_D} SI_D - \delta_S S \tag{4.116}$$

$$\frac{dI}{dt} = \frac{c}{N}\beta_1 SI + \frac{c}{N}\beta_H^{I_D} S\, I_D - \frac{c}{N}\beta_T^{I_D} I\, I_D - \delta_I\, I \tag{4.117}$$

$$\frac{dI_D}{dt} = \frac{c}{N}\beta_T^{I_D} I\, I_D - \delta_D\, I_D \tag{4.118}$$

The epidemiologic processes are described after eqs. (4.114). State variables S, I, I_D are the numbers of individuals that are susceptible uninfected, infected only by HIV, and dually infected with both HIV and DIP.

4.4.6.1 Link to the individual-host and single-cell scales

The model parameters in eqs. (4.116)–(4.118) are the frequency of contacts per unit time, c, the total population size, N, the death rates, $\delta_S, \delta_I, \delta_D$, and the dimensionless transmission coefficients β_1, $\beta_H^{I_D}, \beta_T^{I_D}$. These parameters were previously expressed in terms of HIV and DIP virus loads based on epidemiological data (Fraser et al. 2007; Metzger et al. 2011)

$$\beta_1 = F[V_H[\eta, P=0]] \tag{4.119}$$

$$\beta_H^{I_D} = F[V_H[\eta, P]] \tag{4.120}$$

$$\beta_T^{I_D} = F[V_T[\eta, P]] \tag{4.121}$$

$$F[V] = \frac{0.54V}{4.14 + V} \tag{4.122}$$

$$\delta_D = \left(\frac{25.4 \cdot (0.35)^{0.41}}{0.35^{0.41} + (V_H[\eta, P])^{0.41}} \right)^{-1} \tag{4.123}$$

$$\delta_I = \frac{1}{10}, \quad \delta_S = \frac{1}{35} \tag{4.124}$$

where $V_H[\eta, P]$ and $V_T[\eta, P]$ denote HIV and DIP loads, respectively, expressed in units of 10^5 RNA copy/mL blood. The case $P=0$ in eq. (4.119) corresponds to the absence of DIP. The virus loads, in turn, can be expressed in terms of single-cell parameters η and P (Rast et al. 2016, S1 text, section B). The average HIV load in the absence of DIP is $V_H(P=0) = 10^5$ RNA copies/mL, which corresponds to $\beta_H^{I_D} = \beta_1 = 0.105$.

4.4.6.2 Steady state

In an uninfected population, eq. (4.116) predicts a steady-state with

$$S^0 = \frac{\lambda}{\delta_S} \tag{4.125}$$

$$I = I_D = 0 \tag{4.126}$$

It is convenient to measure state variables in units of S^0, as given by

$$\hat{S} = \frac{S}{S^0}, \hat{I} = \frac{I}{S^0}, \hat{I}_D = \frac{I_D}{S^0}$$

In the presence of HIV, but before the introduction of DIP, the steady-state values of the new variables are

$$\hat{I} = \frac{1}{B}\left(1 - \frac{1}{R_0^{pop}}\right) \tag{4.127}$$

$$\hat{S} = \frac{1}{R_0^{pop}} \tag{4.128}$$

$$\hat{I}_D = 0 \tag{4.129}$$

$$R_0^{pop} \equiv \frac{\lambda\, c\beta_1}{N\delta_I\delta_S} \tag{4.130}$$

$$B \equiv \delta_I/\delta_S = 3.5 \tag{4.131}$$

One can express the reproduction ratio, R_0^{pop}, in terms of the steady-state prevalence of HIV infection $x = I/(I+S+D)$, as given by

$$R_0^{pop} = 1 + \frac{\delta_I x}{\delta_S(1-x)} \tag{4.132}$$

After introduction of DIP, for the steady state we obtain

$$\hat{S} = \frac{1}{\frac{B\tau}{\phi} + \mu R_0^{pop} B\hat{I}_D + 1} \tag{4.133}$$

$$\hat{I} = \frac{\tau}{R_0^{pop}\phi} \tag{4.134}$$

$$\hat{I}_D^{\,2} + \hat{I}_D \frac{1}{R_0^{pop}B}\left(\frac{B\tau}{\phi\mu} + \frac{B}{\phi} + \frac{1}{\mu}\right) + \frac{1}{\left(R_0^{pop}\right)^2 B\phi^2\mu}\left(B\tau - \phi\left(R_0^{pop}-1\right)\right) = 0 \tag{4.135}$$

where we used R_0^{pop} and B from eqs (4.130) and (4.131) and introduced three ratios, as follows

$$\mu \equiv \frac{\beta_H^{I_D}}{\beta_1} < 1 \tag{4.136}$$

$$\phi \equiv \frac{\beta_T^{I_D}}{\beta_1} > 1 \tag{4.137}$$

$$\tau \equiv \frac{\delta_D}{\delta_I} < 1 \tag{4.138}$$

To remind, parameters μ and ϕ are defined as the rates of HIV and DIP transmission from dually infected individuals relative to HIV transmission from HIV^+ DIP^- individuals; $1/\tau$ is the lifespan increase conferred by DIP. In the new notation, the starting eqs. (4.116)–(4.118) take the dimensionless form, eqs. (4.114).

To connect these results to the single-cell-level parameters, η, κ, P, population-level parameters μ, ϕ, τ are expressed in terms of HIV and DIP loads using eqs. (4.119)–(4.124). The virus loads, in the absence of HIV- cell division, are calculated in Section 4.3.6 as functions of η, κ, P. The updated expression including HIV- cell division are given in the original work (Rast et al. 2016, S1 text, section B). Steady-state HIV prevalence is then calculated as a function of η, P at fixed $\kappa = 0.01$ (Fig. 4.22A right). The reason for choosing a small κ is the result in Section 4.3 that HIV evolves toward small κ. A specific value is not important, as long as it is much less than unit.

4.4.6.3 Condition for DIP spread and stability in a population

To determine the general conditions for DIP spread, we return to the steady state. The biologically meaningful solution of quadratic eq. (4.135) has a form

$$\hat{I}_D = \frac{1}{2}\left(-X + \sqrt{X^2 + 4Y}\right) \tag{4.139}$$

$$X = \frac{1}{R_0^{pop}}\left(\frac{\tau}{\phi\mu} + \frac{1}{\phi} + \frac{1}{B\mu} - \frac{R_0^{pop}}{B\tau}\right)$$

$$Y = \frac{1}{\left(R_0^{pop}\phi\right)^2 B\mu}\left(R_0^{pop} - 1)\phi - 1) - B\tau\right)$$

At any sign of X, the dually infected population can exist and be positive, $\hat{I}_D > 0$, as long as $Y > 0$, which implies

$$R_0^{pop} > 1 + \frac{B\tau}{\phi} \tag{4.140}$$

Equation (4.140) is a necessary condition for the presence of DIP in a population. Note that it is stronger than the condition on the presence of HIV, which is, of course, $R_0^{pop} > 1$. To check whether it is also a sufficient condition, let us introduce a small number of dually infected individuals, $\hat{I}_D$, into a steady state in the absence of DIP, eqs. (4.127)–(4.129), The perturbations of the steady state to the presence of DIP are small, and we denote them as

$$\hat{S}[t] = \hat{S}^{ss}(1 + s[t])$$

$$\hat{I}[t] = \hat{I}^{ss}(1 + i[t])$$

$$\hat{I}_D[t] = i_D[t]$$

Substituting to the dynamic equations, one gets

$$\frac{i_D[t]}{dt} = \delta_S\left(\phi\left(R_0^{pop} - 1\right) - B\tau\right) i_D[t]$$

For DIP to propagate in a population, the right-hand side must be positive, which again leads to the inequality in eq. (4.140).

The condition given by eq. (4.140) is also related to the suppression of singly infected HIV prevalence by DIP introduction

$$\frac{\hat{I}[\phi = 0]}{\hat{I}[\phi]} = \frac{B\tau}{\phi\left(R_0^{pop} - 1\right)} < 1 \tag{4.141}$$

where the case $\phi = 0$ is the absence of DIP, eq. (4.127). In terms of the HIV prevalence in the absence of DIP x, defined as

$$x \equiv \hat{I}/\left(\hat{S}+\hat{I}\right) = \left(R_0^{pop}-1\right)/\left(B+R_0^{pop}-1\right) \tag{4.142}$$

the condition of DIP spread, eq. (4.140), can be written as

$$\phi > \tau\frac{1-x}{x} \tag{4.143}$$

Thus, the DIP stability threshold in the relative DIP transmission rate, ϕ, depends only on the initial HIV prevalence, x, and the lifespan increase by DIP, τ, but does not depend explicitly on the relative HIV transmission rate, μ, reproduction number in population R_0^{pop}, or HIV effect on lifespan B. As mentioned above after eq. (4.138), parameters ϕ and τ can be expressed in terms of the single-cell parameters η, κ, P. At fixed values of R_0^{pop}, x, and a low waste parameter $\kappa \ll 1$, the stability condition defines a region in the plane (η, P) (Fig. 4.21B, C). At high prevalence, $x > 0.5$, the population-level threshold is similar to that at the host-level.

4.4.7 Derivation of the evolutionary stability of HIV in the presence of DIP

HIV evolution within an individual host, in the absence of DIP$^+$ HIV$^-$ cell division, was discussed in Section 4.3.4 and analyzed in Section 4.3.7. The general expression for selection coefficient ∂s is given by eq. (4.105), and its specific expressions for mutation affecting cell-level parameters κ and η are given by eqs. (4.106)–(4.108). The presence of DIP$^+$ HIV$^-$ division introduces corrections (Rast et al. 2016, section C). Here we analyze the evolutionary stability of HIV in the presence of DIP on the host population scale.

In a single host, HIV evolves toward larger values of η (Fig. 4.22B top). In contrast, population-level selection is a balance between several opposing factors (Section 4.4.3). To analyze which factor is dominant, one considers two sets of population compartments representing two different HIV strains. Assume also that strain 1 has a higher fitness, so that when a host infected with strain 2 is superinfected with strain 1, strain 1 rapidly outcompetes strain 2 (Fig. 4.23A). Dropping caps over state variables, the equations have the form

$$\frac{1}{\delta_I}\frac{dS}{dt} = \frac{1}{B}(1-S) - R_{01}^{pop}(SI_1+\mu_1\, SI_{D1}) - R_{02}^{pop}(SI_2+\mu_2 SI_{D2})$$

$$\frac{1}{\delta_I}\frac{dI_1}{dt} = R_{01}^{pop}(SI_1+\mu_1 SI_{D1} - \phi_1\, I_1\, I_{D1} + c_0(I_2 I_1 + \mu_1 I_2 I_{D1})) - R_{02}^{pop}\phi_2\, I_1\, I_{D2} - \frac{\delta_{I1}}{\delta_I} I_1$$

$$\frac{1}{\delta_I}\frac{dI_2}{dt} = R_{02}^{pop}(SI_2+\mu_2 S\, I_{D2} - \phi_2 I_2 I_{D2}) - R_{01}^{pop}\ (\phi_1 I_2 I_{D1} + c_0(I_2 I_1 + \mu_1 I_2 I_{D1})) - \frac{\delta_{I2}}{\delta_I} I_2$$

$$\frac{1}{\delta_I}\frac{dI_{D1}}{dt} = R_{01}^{pop}\ (\phi_1 I_1 I_{D1} + c_0(I_{D2}I_1 + \mu_1 I_{D2}I_{D1})) + R_{02}^{pop}\ \phi_2 I_1 I_{D2} - \tau_1 I_{D1}$$

$$\frac{1}{\delta_I}\frac{dI_{D2}}{dt} = R_{02}^{pop}\ \phi_2 I_2 I_{D2} + R_{01}^{pop}[\phi_1 I_2 I_{D1} - c_0(I_{D2}\ I_1 + \mu_1 I_{D2}I_{D1})] - \tau_2 I_{D2} \qquad (4.144)$$

Here c_0 is the ratio of coinfection to single-infection rate. The model parameters are calculated in the same way as for the one-strain system. For example, to obtain $\mu_2 = \beta_H^{I_{D2}}/\beta_2$, one substitutes the steady-state viral load for strain 2 into eqs. (4.119) and (4.120).

Examples of the dynamics that follow after the introduction of a small mutant population are shown in Fig. 4.23B, where $c_0 = 0$ corresponds to "no coinfection" case and $c_0 = 1$ corresponds to "best coinfection" case (Section 4.4.3).

Analysis of the direction of virus evolution in η is similar that to the host-level analysis in Section 4.3, but with eqs. (4.144) as the starting dynamic equations. Suppose we start with population infected with strain 1 (wild type), which has a higher fitness in a host, and introduce into the population a small amount of mutant strain 2. The task is to determine the sign of the eigenvalues of the system. Denote the prevalence of wild-type compartments by S^{ss}, I^{ss}, and I_D^{ss} respectively, consider mutant variables small perturbations, and linearize the dynamic system. The changes in wild-type variables and S will only enter the mutant equations as higher-order terms. Hence, the eigenvalues can be obtained for the mutant and wild-type portions of the system separately. Because the system was stable before DIP introduction, only the eigenvalues from the mutant part of dynamic matrix have the potential to be positive. The mutant part is

$$\delta_I \begin{pmatrix} R_{02}^{pop}S^{ss} - R_{01}^{pop}\left(c_0\left(I^{ss} + \mu_1 I_D^{ss}\right) + \phi_1 I_D^{ss}\right) - \frac{\delta_{I2}}{\delta_I} & R_{02}^{pop}\mu_2 S^{ss} \\ R_{01}^{pop}\phi_1 I_D^{ss} & -R_{01}^{pop}c_0\left(I^{ss} + \mu_1 I_D^{ss}\right) - \tau_2 \end{pmatrix} \qquad (4.145)$$

For fully resistant mutants, there is no I_{D2}, and the mutant matrix is a scalar

$$\delta_I\left(R_{02}^{pop}\ S^{ss} - R_{01}^{pop}\ c_0\left(I^{ss} + \mu_1 I_D^{ss}\right) - \frac{\delta_{I2}}{\delta_I}\right) \qquad (4.146)$$

The results shown in the topographic map in Fig. 4.23C are obtained from the maximal eigenvalues of the matrices, eqs. (4.145), (4.146). The case for mutation from less-fit (in a host) strain 2 to more-fit strain 1 is done in the same way.

The next task is to obtain the selection coefficient for a mutant strain with a small change in η, denoted $\partial\eta$. This can be done using the method described in Section 4.3. Expanding eigenvalues in small changes in population-level parameters $\partial_\tau, \partial_{R0}, \partial_\mu$, we get

$$\partial s = \delta_I \frac{-\partial_\tau \left(R_0^{pop}\, S^{ss} - R_0^{pop}\, \phi_1 I_D^{ss} - 1\right) - \partial_{R0} \left(\tau_1 S^{ss} + R_0^{pop} \phi_1 \mu_1 I_D^{ss} S^{ss}\right) - \partial_\mu \left(\left(R_0^{pop}\right)^2 \phi_1 I_D^{ss}\right) S^{ss}}{R_0^{pop}\, S^{ss} - R_0 \phi_1 I_D^{ss} - 1 - \tau_1} \tag{4.147}$$

Here the asymmetry of genome production, P, is fixed, and all the small changes in $\partial_\tau, \partial_{R0}, \partial_\mu$ are proportional to $\partial\eta$ (Section 4.4.6).

4.4.8 Robustness to model variations

4.4.8.1 T-cell division and homeostasis

In the previous host-level model (Section 4.3), we assumed a fixed linear source of infectable CD4 T cells. In this section based on Rast et al. (2016), the host model includes division of all HIV-negative cells and its homeostatic shutdown. While this generalization makes the analysis more complex, only one parameter has been added, h_0. The main difference from the model without cell division is the expansion of the dynamic DIP stability region toward small η. Variations in the form of homeostatic shutdown function have weak effects (Rast et al. 2016, fig. S7).

4.4.8.2 DIP preinfects individuals

In the main model, DIP can superinfect only individuals infected with HIV, eqs. (4.116)–(4.118). As mentioned in Section 4.4.2, DIP is expected to infect HIV-negative individuals equally well and stay latent until infection with HIV. In this subsection, an alternative model is considered, in which DIP can only preinfect uninfected individuals. Thus, two opposite methods of DIP transmission are considered separately, for the sake of simplicity. The new model equations are

$$\frac{dS}{dt} = \lambda - \frac{c}{N}\beta_1 SI - \frac{c}{N}\left(\beta_H^{I_D} + \beta_T^{I_D}\right) SI_D - \delta_S S \tag{4.148}$$

$$\frac{dS_T}{dt} = \frac{c}{N}\beta_T^{I_D} SI_D - \frac{c}{N}\beta_1 S_T I - \frac{c}{N}\beta_H^{I_D} S_T I_D - \delta_S S_T \tag{4.149}$$

$$\frac{dI}{dt} = \frac{c}{N}\beta_1 SI + \frac{c}{N}\beta_H^{I_D} S\, I_D - \delta_I\, I \tag{4.150}$$

$$\frac{dI_D}{dt} = \frac{c}{N}\beta_1 S_T I + \frac{c}{N}\beta_H^{I_D} S_T I_D - \delta_D\, I_D \tag{4.151}$$

which replace eqs. (4.116)–(4.118) assuming DIP super-infection of HIV-infected individuals. A new state variable is the number of DIP-infected HIV-negative individuals, S_T.

As in the case of super-infection, eqs. (4.148)–(4.151) predict a steady state. In the absence of HIV, or in the presence of HIV but in the absence of DIP, the steady states levels for S, I and I_D are the same as in the super-infection model, eqs. (4.125)–(4.129). In either case, there is no individuals preinfected with DIP, $S_T = 0$.

In the presence of DIP, for the steady state, one finds

$$\hat{S} = \frac{\hat{I}}{R_0^{pop}(\hat{I} + \mu \hat{I}_D)}$$

$$\hat{I}_T = \frac{\tau \hat{I}_D}{R_0^{pop}(\hat{I} + \mu \hat{I}_D)}$$

$$\hat{I} = \frac{\tau}{\phi - \tau}\left(\frac{1}{R_0^{pop} B} + \mu \hat{I}_D\right)$$

$$(\hat{I}_D)^2\left(\frac{\mu}{\phi - \tau} + 1\right) + \hat{I}_D\left(-\frac{1}{B\tau} + \frac{1}{R_0^{pop} B}\left(\frac{1}{\phi - \tau} + \frac{1}{\mu} + \frac{\tau}{\phi(\phi - \tau)} + \frac{1}{\phi}\right)\right)$$
$$+ \frac{1}{(R_0^{pop})^2 B^2 \phi \mu}\left(-R_0^{pop} + \frac{\phi}{\phi - \tau}\right) = 0 \tag{4.152}$$

where state variables are measured in the units of uninfected population, and dimensionless parameters R_0^{pop}, B, ϕ, μ, and τ are defined in eqs (4.131), (4.132), and (4.136)–(4.138). If the last term in its left-hand side is negative, eq. (4.152) has a single positive solution. The resulting condition

$$\phi > \frac{R_0^{pop} \tau}{R_0^{pop} - 1} \tag{4.153}$$

is the new necessary condition for the presence of DIP in a population, which replaces eq. (4.140) obtained for the super-infection model. Equation (4.153) can also be shown to be the condition of the initial DIP spread in a population, in the same perturbative way as it was done for the superinfection model (Rast et al. 2016). The stability threshold in terms of the HIV prevalence x before DIP, eq. (4.142), now has the form

$$\frac{\phi}{\tau} > \frac{1 - x}{Bx} + 1 \tag{4.154}$$

Unlike in the superinfection model where the threshold in ϕ depends only on τ and x, eq. (4.143), now it depends also on the lifespan decrease by HIV, B. It still does not depend on R_0^{pop} or the DIP-induced decrease of HIV transmission rate. An important difference from the superinfection model is that, at moderately large values of B, the DIP stability threshold is not very sensitive to moderately low values of HIV prevalence, eq. (4.154) (Fig 4.21C).

4.4.8.3 Sensitivity to κ

Everywhere above, the waste parameter, κ, was assumed to be small and fixed at $\kappa = 0.01$. Indeed, in Section 4.3, it was shown that, regardless of the absence or presence or of DIP, HIV always evolves toward small κ. Some of the results are sensitive to the specific (low) value of κ, including the shape and height of the host instability region (Fig. 4.21B). Decreasing κ increases maximal P in that region to value $P = 3.5$, and moves $\eta_{max} \leq 1$ even closer toward 1 (Rast et al. 2016, figs S3A, C).

4.4.8.4 Timing of HIV transmission

The model assumed HIV and DIP transmission in the chronic stage when a steady state occurs. However, Wawer et al. (2005) suggested that a good part of HIV transmission occurs during the acute stage of infection. In this case, DIP-suppression of HIV transmission would be less effective, assuming that HIV is transmitted before the super-infection by the DIP. To test the robustness to this factor, Rast et al. (2016) allowed HIV transmission from DIP^+ individuals to be unsuppressed. In other words, they set $\mu = 1$. Although a major change in the model, it changed neither the prevalence of HIV^+ DIP^- hosts nor the predicted evolutionary behavior (Rast et al. 2016, fig. S8). The reason for this amazing robustness is that dually infected hosts become more frequent to compensate for the increased transmission of HIV.

4.4.8.5 Other approximations

Rate of super-infection: The superinfection rate with the second HIV strain may depend on the protection from the previous strain, which factor was neglected.

Single-resource stealing: The model considers stealing of a single protein and could be generalized to include stealing of several proteins.

Uniformity of wild-type population: The wild-type virus was assumed to be monomorphic to facilitate the comparison of two HIV strains. In real life, HIV is very diverse, the evolution at different sites is not independent, as it has strong Hill-Robertson effects (genetic background and clonal interference), which attracted a great body of work. The virus population represents a cloud of variants ("quasispecies") which moves as a traveling wave in fitness space (Rouzine and Weinberger 2013a; Rouzine 2020b). This work lumps it into a single variant, in order to focus on mutants in the relevant single-cell parameters.

References

Aaskov J, Buzacott K, Thu HM, Lowry K, Holmes EC. 2006. Long-term transmission of defective RNA viruses in humans and Aedes mosquitoes. Science 311:236–238.

Acevedo A, Andino R. 2014. Library preparation for highly accurate population sequencing of RNA viruses. Nat Protoc 9:1760–1769.

Acevedo A, Brodsky L, Andino R. 2014. Mutational and fitness landscapes of an RNA virus revealed through population sequencing. Nature 505:686–690.

Agol VI. 2006. Molecular mechanisms of poliovirus variation and evolution. Curr Top Microbiol Immunol 299:211–259.

Alizon S. 2013. Co-infection and super-infection models in evolutionary epidemiology. Interface Focus 3:20130031.

Alizon S, Lion S. 2011. Within-host parasite cooperation and the evolution of virulence. Proc Biol Sci 278:3738–3747.

Alizon S, Magnus C. 2012. Modelling the course of an HIV infection: insights from ecology and evolution. Viruses 4:1984–2013.

Alizon S, van Baalen M. 2008. Multiple infections, immune dynamics, and the evolution of virulence. Am Nat 172:E150–168.

Althaus CL, De Boer RJ. 2008. Dynamics of immune escape during HIV/SIV infection. PLoS Comput Biol 4: e1000103.

Alvarez-Castro JM, Le Rouzic A, Andersson L, Siegel PB, Carlborg O. 2012. Modelling of genetic interactions improves prediction of hybrid patterns – a case study in domestic fowl. Genet Res (Camb) 94:255–266.

An DS, Morizono K, Li QX, Mao SH, Lu S, Chen IS. 1999. An inducible human immunodeficiency virus type 1 (HIV-1) vector which effectively suppresses HIV-1 replication. J Virol 73:7671–7677.

Anderson B, May RM, Anderson RM. 1992. Infectious Diseases of Humans: Dynamics and Control Oxford Science Publications.

Archin NM, Vaidya NK, Kuruc JD, Liberty AL, Wiegand A, Kearney MF, Cohen MS, Coffin JM, Bosch RJ, Gay CL, et al. 2012. Immediate antiviral therapy appears to restrict resting $CD4^+$ cell HIV-1 infection without accelerating the decay of latent infection. Proc Natl Acad Sci U S A 109:9523–9528.

Arkin A, Ross J, McAdams HH. 1998. Stochastic kinetic analysis of developmental pathway bifurcation in phage lambda-infected Escherichia coli cells. Genetics 149:1633–1648.

Arnaout RA, Lloyd AL, O'Brien TR, Goedert JJ, Leonard JM, Nowak MA. 1999. A simple relationship between viral load and survival time in HIV-1 infection. Proc Natl Acad Sci U S A 96:11549–11553.

Asquith B, Edwards CT, Lipsitch M, McLean AR. 2006. Inefficient cytotoxic T lymphocyte-mediated killing of HIV-1-infected cells in vivo. PLoS Biol 4:e90.

Astier S. 2007. Principles of Plant Virology: Science Publishers.

Ayme V, Petit-Pierre J, Souche S, Palloix A, Moury B. 2007. Molecular dissection of the potato virus Y VPg virulence factor reveals complex adaptations to the pvr2 resistance allelic series in pepper. J Gen Virol 88:1594–1601.

Baggaley RF, Garnett GP, Ferguson NM. 2006. Modelling the impact of antiretroviral use in resource-poor settings. PLoS Medicine 3:e124.

Balaban NQ. 2011. Persistence: mechanisms for triggering and enhancing phenotypic variability. Curr Opin Genet Dev 21:768–775.

Balding DJ. 2006. A tutorial on statistical methods for population association studies. Nat Rev Genet 7:781–791.

Balfe P, Simmonds P, Ludlam CA, Bishop JO, Brown AJ. 1990. Concurrent evolution of human immunodeficiency virus type 1 in patients infected from the same source: rate of sequence change and low frequency of inactivating mutations. J Virol 64:6221–6233.

https://doi.org/10.1515/9783110697384-005

Bamdres JD, Shaw AS, Ratner L. 1994. HIV-1 nef protein downregulation of CD4 surface expression: Relevance of the lck binding domain of CD4. Virology 207:338–341.

Barlukova A, Rouzine IM. 2021. The evolutionary origin of the universal distribution of mutation fitness effect. PLoS Comput Biol 17:e1008822.

Barnes WM. 1992. The fidelity of Taq polymerase catalyzing PCR is improved by an N-terminal deletion. Gene 112:29–35.

Barre-Sinoussi F, Chermann JC, Rey F, Nugeyre MT, Chamaret S, Gruest J, Dauguet C, Axler-Blin C, Vezinet-Brun F, Rouzioux C, et al. 1983. Isolation of a T-lymphotropic retrovirus from a patient at risk for acquired immune deficiency syndrome (AIDS). Science 220:868–871.

Barrett AD, Dimmock NJ. 1986. Defective interfering viruses and infections of animals. Curr Top Microbiol Immunol 128:55–84.

Barton NH. 1995. Linkage and the limits to natural selection. Genetics 140:821–841.

Batada NN, Hurst LD. 2007. Evolution of chromosome organization driven by selection for reduced gene expression noise. Nat Genet 39:945–949.

Batorsky R, Kearney MF, Palmer SE, Maldarelli F, Rouzine IM, Coffin JM. 2011. Estimate of effective recombination rate and average selection coefficient for HIV in chronic infection. Proc Natl Acad Sci U S A 108:5661–5666.

Batorsky R, Sergeev RA, Rouzine IM. 2014. The route of HIV escape from immune response targeting multiple sites is determined by the cost-benefit tradeoff of escape mutations. PLoS Comput Biol 10: e1003878.

Bedford T, Rambaut A, Pascual M. 2012. Canalization of the evolutionary trajectory of the human influenza virus. BMC Biol 10:38.

Bedford T, Riley S, Barr IG, Broor S, Chadha M, Cox NJ, Daniels RS, Gunasekaran CP, Hurt AC, Kelso A, et al. 2015. Global circulation patterns of seasonal influenza viruses vary with antigenic drift. Nature 523:217–220.

Bedford T, Suchard MA, Lemey P, Dudas G, Gregory V, Hay AJ, McCauley JW, Russell CA, Smith DJ, Rambaut A. 2014. Integrating influenza antigenic dynamics with molecular evolution. Elife 3:e01914.

Behar DM, Yunusbayev B, Metspalu M, Metspalu E, Rosset S, Parik J, Rootsi S, Chaubey G, Kutuev I, Yudkovsky G, et al. 2010. The genome-wide structure of the Jewish people. Nature 466:238–242.

Bell JT, Timpson NJ, Rayner NW, Zeggini E, Frayling TM, Hattersley AT, Morris AP, M.I. M. 2011. Genome-wide association scan allowing for epistasis in type 2 diabetes. Annal Hum Genet 75:10–19.

Berg J, Lassig M, Wagner A. 2004. Structure and evolution of protein interaction networks: a statistical model for link dynamics and gene duplications. BMC Evol Biol 4:51.

Biebricher CK, Eigen M. 2005. The error threshold. Virus Res 107:117–127.

Biggerstaff M, Cauchemez S, Reed C, Gambhir M, Finelli L. 2014. Estimates of the reproduction number for seasonal, pandemic, and zoonotic influenza: a systematic review of the literature. BMC Infect Dis 14:480.

Bonhoeffer S, Chappey C, Parkin NT, Whitcomb JM, Petropoulos CJ. 2004. Evidence for positive epistasis in HIV-1. Science 306:1547–1550.

Bonhoeffer S, Nowak MA. 1994. Intra-host versus inter-host selection: Viral strategies of immune function impairment. Proc. Natl. Acad. Sci. U.S.A. 91:8062–8066.

Boutwell CL, Carlson JM, Lin TH, Seese A, Power KA, Peng J, Tang Y, Brumme ZL, Heckerman D, Schneidewind A, et al. 2013. Frequent and variable cytotoxic-T-lymphocyte escape-associated fitness costs in the human immunodeficiency virus type 1 subtype B Gag proteins. J Virol 87:3952–3965.

Brandin E, Thorstensson R, Bonhoeffer S, Albert J. 2006. Rapid viral decay in simian immunodeficiency virus-infected macaques receiving quadruple antiretroviral therapy. J Virol 80:9861–9864.

Brem RB, Storey JD, Whittle J, Kruglyak L. 2005. Genetic interactions between polymorphisms that affect gene expression in yeast. Nature 436:701.

Brenchley JM, Schacker TW, Ruff LE, Price DA, Taylor JH, Beilman GJ, Nguyen PL, Khoruts A, Larson M, Haase AT, et al. 2004. CD4$^+$ T cell depletion during all stages of HIV disease occurs predominantly in the gastrointestinal tract. J Exp Med 200:749–759.

Brunet E, Derrida B, Mueller AH, Munier S. 2007. Effect of selection on ancestry: An exactly soluble case and its phenomenological generalization. Physical Review E 76:041104–041101.

Brunet E, Rouzine IM, Wilke CO. 2008. The stochastic edge in adaptive evolution. Genetics 179:603–620.

Bukovsky AA, Song JP, Naldini L. 1999. Interaction of human immunodeficiency virus-derived vectors with wild-type virus in transduced cells. J Virol 73:7087–7092.

Bullock JM, Medway C, Cortina-Borja M, Turton JC, Prince JA, Ibrahim-Verbaas CA, Schuur M, Breteler MM, van Duijn CM, Kehoe PG, et al. 2013. Discovery by the Epistasis Project of an epistatic interaction between the GSTM3 gene and the HHEX/IDE/KIF11 locus in the risk of Alzheimer's disease. Neurobiol Aging 34:1309 e1301–1307.

Burch CL, Turner PE, Hanley KA. 2003. Patterns of epistasis in RNA viruses: a review of the evidence from vaccine design. J Evol Biol 16:1223–1235.

Burnett JC, Miller-Jensen K, Shah PS, Arkin AP, Schaffer DV. 2009. Control of stochastic gene expression by host factors at the HIV promoter. PLoS pathogens 5:e1000260.

Burns DP, Desrosiers RC. 1994. Envelope sequence variation, neutralizing antibodies, and primate lentivirus persistence. Curr Top Microbiol Immunol 188:185–219.

Burns DP, Desrosiers RC. 1991. Selection of genetic variants of simian immunodeficiency virus in persistently infected rhesus monkeys. J Virol 65:1843–1854.

Byarugaba DK, Erima B, Millard M, Kibuuka H, Lkwago L, Bwogi J, Mimbe D, Kiconco JB, Tugume T, Mworozi EA, et al. 2016. Whole-genome analysis of influenza A(H1N1)pdm09 viruses isolated in Uganda from 2009 to 2011. Influenza Other Respir Viruses 10:486–492.

Cale EM, Hraber P, Giorgi EE, Fischer W, Bhattacharya T, Leitner T, Yeh WW, Gleasner C, Green LD, Han CS, et al. 2011. Epitope-specific CD8$^+$ T lymphocytes cross-recognize mutant simian immunodeficiency virus (SIV) sequences but fail to contain very early evolution and eventual fixation of epitope escape mutations during SIV infection. J Virol 85:3746–3757.

Calvanese V, Chavez L, Laurent T, Ding S, Verdin E. 2013. Dual-color HIV reporters trace a population of latently infected cells and enable their purification. Virology 446:283–292.

Carlborg O, Jacobsson L, Ahgren P, Siegel P, Andersson L. 2006. Epistasis and the release of genetic variation during long-term selection. Nat Genet 38:418–420.

Carrat F, Vergu E, Ferguson NM, Lemaitre M, Cauchemez S, Leach S, Valleron AJ. 2008. Time lines of infection and disease in human influenza: a review of volunteer challenge studies. Am J Epidemiol 167:775–785.

Cave DR, Hendrickson FM, Huang AS. 1985. Defective interfering virus particles modulate virulence. J Virol 55:366–373.

Chahroudi A, Bosinger SE, Vanderford TH, Paiardini M, Silvestri G. 2012. Natural SIV hosts: showing AIDS the door. Science 335:1188–1193.

Chao L. 1990. Fitness of RNA virus decreased by Muller's ratchet. Nature 348:454–455.

Chattopadhyay SK, Morse HC, Makino M, Ruscetti SK, Hartley JW. 1989. Defective virus is associated with induction of murine retrovirus-induced immunodeficiency syndrome. Proc Natl Acad Sci U S A 86:3862–3866.

Chaturvedi S, Vasen G, Pablo M, Chen X, Beutler N, Kumar A, Tanner E, Illouz S, Rahgoshay D, Burnett J, et al. 2021. Identification of a therapeutic interfering particle – a single-dose SARS-CoV-2 antiviral intervention with a high barrier to resistance. Cell 184:6022–6036 e6018.

Chen CC, Schwender H, Keith J, Nunkesser R, Mengersen K, Macrossan P. 2011. Methods for identifying SNP interactions: a review on variations of Logic Regression, Random Forest and Bayesian logistic regression. IEEE/ACM Trans Comput Biol Bioinform 8:1580–1591.

Chen CJ, Banerjea AC, Harmison GG, Haglund K, Schubert M. 1992. Multitarget-ribozyme directed to cleave at up to nine highly conserved HIV-1 env RNA regions inhibits HIV-1 replication–potential effectiveness against most presently sequenced HIV-1 isolates. Nucleic Acids Res 20:4581–4589.

Chen J, Nikolaitchik O, Singh J, Wright A, Bencsics CE, Coffin JM, Ni N, Lockett S, Pathak VK, Hu WS. 2009. High efficiency of HIV-1 genomic RNA packaging and heterozygote formation revealed by single virion analysis. Proc Natl Acad Sci U S A 106:13535–13540.

Chun TW, Carruth L, Finzi D, Shen X, DiGiuseppe JA, Taylor H, Hermankova M, Chadwick K, Margolick J, Quinn TC, et al. 1997. Quantification of latent tissue reservoirs and total body viral load in HIV-1 infection. Nature 387:183–188.

Chun TW, Engel D, Mizell SB, Ehler LA, Fauci AS. 1998. Induction of HIV-1 replication in latently infected $CD4^+$ T cells using a combination of cytokines. The Journal of experimental medicine 188:83–91.

Chun TW, Stuyver L, Mizell SB, Ehler LA, Mican JA, Baseler M, Lloyd AL, Nowak MA, Fauci AS. 1997. Presence of an inducible HIV-1 latent reservoir during highly active antiretroviral therapy. Proc Natl Acad Sci U S A 94:13193–13197.

Clarke DK, Duarte EA, Elena SF, Moya A, Domingo E, Holland J. 1994. The red queen reigns in the kingdom of RNA viruses. Proc Natl Acad Sci U S A 91:4821–4824.

Clarke DK, Duarte EA, Moya A, Elena SF, Domingo E, Holland J. 1993. Genetic bottlenecks and population passages cause profound fitness differences in RNA viruses. J Virol 67:222–228.

Cleland A, Watson HG, Robertson P, Ludlam CA, Brown AJ. 1996. Evolution of zidovudine resistance-associated genotypes in human immunodeficiency virus type 1-infected patients. J Acquir Immune Defic Syndr Hum Retrovirol 12:6–18.

Cocco S, Feinauer C, Figliuzzi M, Monasson R, Weigt M. 2018. Inverse statistical physics of protein sequences: a key issues review. Rep Prog Phys 81:032601.

Coffin J, Swanstrom R. 2013. HIV pathogenesis: dynamics and genetics of viral populations and infected cells. Cold Spring Harb Perspect Med 3:a012526.

Coffin JM. 1995. HIV population dynamics in vivo: implications for genetic variation, pathogenesis, and therapy. Science 267:483–489.

Cohen D. 1966. Optimizing reproduction in a randomly varying environment. J Theor Biol 12:119–129.

Combarros O, Cortina-Borja M, Smith AD, Lehmann DJ. 2009. Epistasis in sporadic Alzheimer's disease. Neurobiol Aging 30:1333–1349.

Combarros O, van Duijn CM, Hammond N, Belbin O, Arias-Vasquez A, Cortina-Borja M, Lehmann MG, Aulchenko YS, Schuur M, Kolsch H, et al. 2009. Replication by the Epistasis Project of the interaction between the genes for IL-6 and IL-10 in the risk of Alzheimer's disease. J Neuroinflammation 6:22.

Condra JH, Holder DJ, Schleif WA, Blahy OM, Danovich RM, Gabryelski LJ, Graham DJ, Laird D, Quintero JC, Rhodes A, et al. 1996. Genetic correlates of in vivo viral resistance to indinavir, a human immunodeficiency virus type 1 protease inhibitor. J Virol 70:8270–8276.

Cong ME, Heneine W, Garcia-Lerma JG. 2007. The fitness cost of mutations associated with human immunodeficiency virus type 1 drug resistance is modulated by mutational interactions. J Virol 81:3037–3041.

Cook WJ, Green KA, Obar JJ, Green WR. 2003. Quantitative analysis of LP-BM5 murine leukemia retrovirus RNA using real-time RT-PCR. J Virol Methods 108:49–58.

Cooper TF. 2007. Recombination speeds adaptation by reducing competition between beneficial mutations in populations of Escherichia coli. PLoS Biol 5:e225.

Cordell HJ. 2009. Detecting gene-gene interactions that underlie human diseases. Nature reviews. Genetics 10:392–404.

Cordell HJ. 2002. Epistasis: what it means, what it doesn't mean, and statistical methods to detect it in humans. Hum Mol Genet 11:2463–2468.

Crawford H, Matthews PC, Schaefer M, Carlson JM, Leslie A, Kilembe W, Allen S, Ndung'u T, Heckerman D, Hunter E, et al. 2011. The hypervariable HIV-1 capsid protein residues comprise HLA-driven $CD8^+$ T-cell escape mutations and covarying HLA-independent polymorphisms. J Virol 85:1384–1390.

Crotty S, Cameron CE, Andino R. 2001. RNA virus error catastrophe: direct molecular test by using ribavirin. Proc Natl Acad Sci U S A 98:6895–6900.

D'Costa J, Mansfield SG, Humeau LM. 2009. Lentiviral vectors in clinical trials: Current status. Curr Opin Mol Ther 11:554–564.

Dahabieh MS, Ooms M, Simon V, Sadowski I. 2013. A double-fluorescent HIV-1 reporter shows that the majority of integrated HIV-1 is latent shortly after infection. Journal of virology.

Dapp MJ, Kober KM, Chen L, Westfall DH, Wong K, Zhao H, Hall BM, Deng W, Sibley T, Ghorai S, et al. 2017. Patterns and rates of viral evolution in HIV-1 subtype B infected females and males. PLOS ONE 12: e0182443.

Dar RD, Hosmane NN, Arkin MR, Siliciano RF, Weinberger LS. 2014. Screening for noise in gene expression identifies drug synergies. Science 344:1392–1396.

Dar RD, Razooky BS, Singh A, Trimeloni TV, McCollum JM, Cox CD, Simpson ML, Weinberger LS. 2012. Transcriptional burst frequency and burst size are equally modulated across the human genome. Proc Natl Acad Sci U S A 109:17454–17459.

Davenport MP, Ribeiro RM, Perelson AS. 2004. Kinetics of virus-specific $CD8^+$ T cells and the control of HIV infection. J. Virol. 78:10096–10103.

De Boer RJ. 2007. Understanding the failure of $CD8^+$ T-cell vaccination against simian/human immunodeficiency virus. Journal of virology 81:2838–2848.

De Boer RJ, Homann D, Perelson AS. 2003. Different dynamics of CD4 and CD8 T cell responses during and after acute lymphocytic choriomeningitis virus infection. J. Immunol. 171:3928–3935.

De Boer RJ, Perelson AS. 1998. Target cell limited and immune control models of HIV infection: a comparison. J Theor Biol 190:201–214.

Deeks SG. 2012. HIV: Shock and kill. Nature 487:439–440.

Deeks SG, Kitchen CM, Liu L, Guo H, Gascon R, Narvaez AB, Hunt P, Martin JN, Kahn JO, Levy J, et al. 2004. Immune activation set point during early HIV infection predicts subsequent $CD4^+$ T-cell changes independent of viral load. Blood 104:942–947.

Delwart EL, Sheppard HW, Walker BD, Goudsmit J, Mullins JI. 1994. Human immunodeficiency virus type 1 evolution in vivo tracked by DNA heteroduplex mobility assays. J Virol 68:6672–6683.

DePolo NJ, Giachetti C, Holland JJ. 1987. Continuing coevolution of virus and defective interfering particles and of viral genome sequences during undiluted passages: virus mutants exhibiting nearly complete resistance to formerly dominant defective interfering particles. J Virol 61:454–464.

DePolo NJ, Holland JJ. 1986. Very rapid generation/amplification of defective interfering particles by vesicular stomatitis virus variants isolated from persistent infection. J Gen Virol 67 (Pt 6):1195–1198.

Desai MM, Fisher DS. 2007. Beneficial mutation selection balance and the effect of linkage on positive selection. Genetics 176:1759–1798.

Desai MM, Fisher DS, Murray AW. 2007. The speed of evolution and maintenance of variation in asexual populations. Curr Biol 17:385–394.

Desai MM, Walczak AM, Fisher DS. 2013. Genetic diversity and the structure of genealogies in rapidly adapting populations. Genetics 193:565–585.

Dimmock NJ. 1985. Defective interfering viruses: modulators of infection. Microbiol Sci 2:1–7.

Doitsh G, Cavrois M, Lassen KG, Zepeda O, Yang Z, Santiago ML, Hebbeler AM, Greene WC. 2010. Abortive HIV infection mediates CD4 T cell depletion and inflammation in human lymphoid tissue. Cell 143:789–801.

Domingo E, Escarmis C, Sevilla N, Moya A, Elena SF, Quer J, Novella IS, Holland JJ. 1996. Basic concepts in RNA virus evolution. FASEB J 10:859–864.

Domingo E, Holland JJ. 1997. RNA virus mutations and fitness for survival. Annu Rev Microbiol 51:151–178.

Duarte E, Clarke D, Moya A, Domingo E, Holland J. 1992. Rapid fitness losses in mammalian RNA virus clones due to Muller's ratchet. Proc Natl Acad Sci U S A 89:6015–6019.
Dudley J, Johnson GR. 2009. Epistatic models improve prediction of performance in corn. Crop Sci 49:763–770.
Dull T, Zufferey R, Kelly M, Mandel RJ, Nguyen M, Trono D, Naldini L. 1998. A third-generation lentivirus vector with a conditional packaging system. J Virol 72:8463–8471.
Dutta RN, Rouzine IM, Smith SD, Wilke CO, Novella IS. 2008. Rapid adaptive amplification of preexisting variation in an RNA virus. J Virol 82:4354–4362.
Eigen M. 2002. Error catastrophe and antiviral strategy. Proc Natl Acad Sci U S A 99:13374–13376.
Eisele E, Siliciano RF. 2012. Redefining the viral reservoirs that prevent HIV-1 eradication. Immunity 37:377–388.
Elena SF, Gonzalez-Candelas F, Novella IS, Duarte EA, Clarke DK, Domingo E, Holland JJ, Moya A. 1996. Evolution of fitness in experimental populations of vesicular stomatitis virus. Genetics 142:673–679.
Elena SF, Sanjuan R. 2005a. Adaptive value of high mutation rates of RNA viruses: separating causes from consequences. J Virol 79:11555–11558.
Elena SF, Sanjuan R. 2005b. RNA viruses as complex adaptive systems. Biosystems 81:31–41.
Evans JT, Garcia JV. 2000. Lentivirus vector mobilization and spread by human immunodeficiency virus. Hum Gene Ther 11:2331–2339.
Felsenstein J. 1974. The evolutionary advantage of recombination. Genetics 78:737–756.
Finlay BB, McFadden G. 2006. Anti-immunology: evasion of the host immune system by bacterial and viral pathogens. Cell 124:767–782.
Finzi D, Blankson J, Siliciano JD, Margolick JB, Chadwick K, Pierson T, Smith K, Lisziewicz J, Lori F, Flexner C, et al. 1999. Latent infection of $CD4^+$ T cells provides a mechanism for lifelong persistence of HIV-1, even in patients on effective combination therapy. Nat Med 5:512–517.
Finzi D, Hermankova M, Pierson T, Carruth LM, Buck C, Chaisson RE, Quinn TC, Chadwick K, Margolick J, Brookmeyer R, et al. 1997. Identification of a reservoir for HIV-1 in patients on highly active antiretroviral therapy. Science 278:1295–1300.
Fischer W, Ganusov VV, Giorgi EE, Hraber PT, Keele BF, Leitner T, Han CS, Gleasner CD, Green L, Lo CC, et al. 2010. Transmission of single HIV-1 genomes and dynamics of early immune escape revealed by ultra-deep sequencing. PLOS ONE 5:e12303.
Fisher RA. 1930. The genetical theory of natural selection. Oxford, United Kingdom: Clarendon Press, 1958.
Fisher RA. 1958. The genetical theory of natural selection. Oxford, United Kingdom: Clarendon Press.
Fisher RA. 1990. On the dominance ratio. 1922. Bull Math Biol 52:297–318; discussion 201–297.
Fonville JM, Wilks SH, James SL, Fox A, Ventresca M, Aban M, Xue L, Jones TC, Le NMH, Pham QT, et al. 2014. Antibody landscapes after influenza virus infection or vaccination. Science 346:996–1000.
Frankel AD, Young JA. 1998. HIV-1: fifteen proteins and an RNA. Annu Rev Biochem 67:1–25.
Fraser C, Hollingsworth TD, Chapman R, de Wolf F, Hanage WP. 2007. Variation in HIV-1 set-point viral load: epidemiological analysis and an evolutionary hypothesis. Proc Natl Acad Sci U S A 104:17441–17446.
Fraser HB, Hirsh AE, Giaever G, Kumm J, Eisen MB. 2004. Noise minimization in eukaryotic gene expression. PLoS Biol 2:e137.
Friedrich TC, Dodds EJ, Yant LJ, Vojnov L, Rudersdorf R, Cullen C, Evans DT, Desrosiers RC, Mothe BR, Sidney J, et al. 2004. Reversion of CTL escape-variant immunodeficiency viruses in vivo. Nat Med 10:275–281.
Frost SD, Dumaurier MJ, Wain-Hobson S, Brown AJ. 2001. Genetic drift and within-host metapopulation dynamics of HIV-1 infection. Proc Natl Acad Sci U S A 98:6975–6980.
Frost SD, Nijhuis M, Schuurman R, Boucher CA, Brown AJ. 2000. Evolution of lamivudine resistance in human immunodeficiency virus type 1-infected individuals: the relative roles of drift and selection. J Virol 74:6262–6268.

Fryer HR, Frater J, Duda A, Roberts MG, Investigators ST, Phillips RE, McLean AR. 2010. Modelling the evolution and spread of HIV immune escape mutants. PLoS Pathog 6:e1001196.

Gallo RC, Salahuddin SZ, Popovic M, Shearer GM, Kaplan M, Haynes BF, Palker TJ, Redfield R, Oleske J, Safai B, et al. 1984. Frequent detection and isolation of cytopathic retroviruses (HTLV-III) from patients with AIDS and at risk for AIDS. Science 224:500–503.

Ganusov VV, De Boer RJ. 2006. Estimating Costs and Benefits of CTL Escape Mutations in SIV/HIV Infection. PLoS Comput Biol 2:e24.

Ganusov VV, Goonetilleke N, Liu MK, Ferrari G, Shaw GM, McMichael AJ, Borrow P, Korber BT, Perelson AS. 2011. Fitness costs and diversity of the cytotoxic T lymphocyte (CTL) response determine the rate of CTL escape during acute and chronic phases of HIV infection. J Virol 85:10518–10528.

García-Magariños M, López-de-Ullibarri I, Cao R, Salas A. 2009. Evaluating the Ability of Tree-Based Methods and Logistic Regression for the Detection of SNP-SNP Interaction. Annals of Human Genetics 73:360–369.

Génin E, Coustet B, Allanore Y, Ito I, Teruel M, Constantin A, Schaeverbeke T, Ruyssen-Witrand A, Tohma S, Cantagrel A, et al. 2013. Epistatic Interaction between BANK1 and BLK in Rheumatoid Arthritis: Results from a Large Trans-Ethnic Meta-Analysis. PLOS ONE 8:e61044.

Gerrish PJ, Colato A, Sniegowski PD. 2013. Genomic mutation rates that neutralize adaptive evolution and natural selection. J R Soc Interface 10:20130329.

Gerrish PJ, Lenski RE. 1998. The fate of competing beneficial mutations in an asexual population. Genetica 102–103:127–144.

Gheorghiu-Svirschevski S, Rouzine IM, Coffin JM. 2007. Increasing sequence correlation limits the efficiency of recombination in a multisite evolution model. Mol Biol Evol 24:574–586.

Giachetti C, Holland JJ. 1988. Altered replicase specificity is responsible for resistance to defective interfering particle interference of an Sdi-mutant of vesicular stomatitis virus. J Virol 62:3614–3621.

Gillespie JH. 1982. A Randomized Sas Cff Model of Natural-Selection in a Random Environment. Theoretical Population Biology 21:219–237.

Giorgi JV, Hultin LE, McKeating JA, Johnson TD, Owens B, Jacobson LP, Shih R, Lewis J, Wiley DJ, Phair JP, et al. 1999. Shorter survival in advanced human immunodeficiency virus type 1 infection is more closely associated with T lymphocyte activation than with plasma virus burden or virus chemokine coreceptor usage. J Infect Dis 179:859–870.

Gog JR, Rimmelzwaan GF, Osterhaus AD, Grenfell BT. 2003. Population dynamics of rapid fixation in cytotoxic T lymphocyte escape mutants of influenza A. Proc Natl Acad Sci U S A 100:11143–11147.

Goldstein S, Brown CR, Ourmanov I, Pandrea I, Buckler-White A, Erb C, Nandi JS, Foster GJ, Autissier P, Schmitz JE, et al. 2006. Comparison of simian immunodeficiency virus SIVagmVer replication and $CD4^+$ T-cell dynamics in vervet and sabaeus African green monkeys. J Virol 80:4868–4877.

Gonzalez-Ortega E, Ballana E, Badia R, Clotet B, Este JA. 2011. Compensatory mutations rescue the virus replicative capacity of VIRIP-resistant HIV-1. Antiviral Res 92:479–483.

Good BH, Desai MM. 2015. The impact of macroscopic epistasis on long-term evolutionary dynamics. Genetics 199:177–190.

Good BH, Rouzine IM, Balick DJ, Hallatschek O, Desai MM. 2012. Distribution of fixed beneficial mutations and the rate of adaptation in asexual populations. Proc Natl Acad Sci U S A 109:4950–4955.

Good BH, Walczak AM, Neher RA, Desai MM. 2014. Genetic diversity in the interference selection limit. PLoS Genet 10:e1004222.

Goodfellow IG, Kerrigan D, Evans DJ. 2003. Structure and function analysis of the poliovirus cis-acting replication element (CRE). RNA 9:124–137.

Goonetilleke N, Liu MK, Salazar-Gonzalez JF, Ferrari G, Giorgi E, Ganusov VV, Keele BF, Learn GH, Turnbull EL, Salazar MG, et al. 2009. The first T cell response to transmitted/founder virus contributes to the control of acute viremia in HIV-1 infection. J Exp Med 206:1253–1272.

Gordon SN, Dunham RM, Engram JC, Estes J, Wang Z, Klatt NR, Paiardini M, Pandrea IV, Apetrei C, Sodora DL, et al. 2008. Short-lived infected cells support virus replication in sooty mangabeys naturally infected with simian immunodeficiency virus: implications for AIDS pathogenesis. J Virol 82:3725–3735.

Gottlieb MS, Schroff R, Schanker HM, Weisman JD, Fan PT, Wolf RA, Saxon A. 1981. Pneumocystis carinii pneumonia and mucosal candidiasis in previously healthy homosexual men: evidence of a new acquired cellular immunodeficiency. N Engl J Med 305:1425–1431.

Goyal S, Balick DJ, Jerison ER, Neher RA, Shraiman BI, Desai MM. 2012. Dynamic mutation-selection balance as an evolutionary attractor. Genetics 191:1309–1319.

Gray RH, Wawer MJ, Brookmeyer R, Sewankambo NK, Serwadda D, Wabwire-Mangen F, Lutalo T, Li X, vanCott T, Quinn TC, et al. 2001. Probability of HIV-1 transmission per coital act in monogamous, heterosexual, HIV-1-discordant couples in Rakai, Uganda. Lancet 357:1149–1153.

Grenfell BT, Pybus OG, Gog JR, Wood JL, Daly JM, Mumford JA, Holmes EC. 2004. Unifying the epidemiological and evolutionary dynamics of pathogens. Science 303:327–332.

Grivel JC, Penn ML, Eckstein DA, Schramm B, Speck RF, Abbey NW, Herndier B, Margolis L, Goldsmith MA. 2000. Human immunodeficiency virus type 1 coreceptor preferences determine target T-cell depletion and cellular tropism in human lymphoid tissue. J Virol 74:5347–5351.

Groenink M, Andeweg AC, Fouchier RA, Broersen S, van der Jagt RC, Schuitemaker H, de Goede RE, Bosch ML, Huisman HG, Tersmette M. 1992. Phenotype-associated env gene variation among eight related human immunodeficiency virus type 1 clones: evidence for in vivo recombination and determinants of cytotropism outside the V3 domain. J Virol 66:6175–6180.

Grossman Z, Feinberg MB, Paul WE. 1998. Multiple modes of cellular activation and virus transmission in HIV infection: a role for chronically and latently infected cells in sustaining viral replication. Proc Natl Acad Sci U S A 95:6314–6319.

Grossman Z, Herberman R, Dimitrov DS. 1999. T cell turnover in SIV infection [comment]. Science 284:555a.

Guedj J, Neumann AU. 2010. Understanding hepatitis C viral dynamics with direct-acting antiviral agents due to the interplay between intracellular replication and cellular infection dynamics. J Theor Biol 267:330–340.

Haase AT. 2011. Early events in sexual transmission of HIV and SIV and opportunities for interventions. Annual review of medicine 62:127–139.

Haase AT. 1999. Population biology of HIV-1 infection: viral and $CD4^+$ T cell demographics and dynamics in lymphatic tissues. Annu Rev Immunol 17:625–656.

Haase AT, Henry K, Zupancic M, Sedgewick G, Faust RA, Melroe H, Cavert W, Gebhard K, Staskus K, Zhang ZQ, et al. 1996. Quantitative image analysis of HIV-1 infection in lymphoid tissue. Science 274:985–989.

Hallatschek O. 2011. The noisy edge of traveling waves. Proc Natl Acad Sci U S A 108:1783–1787.

Han Y, Wind-Rotolo M, Yang HC, Siliciano JD, Siliciano RF. 2007. Experimental approaches to the study of HIV-1 latency. Nature reviews. Microbiology 5:95–106.

Handel A, Regoes RR, Antia R. 2006. The role of compensatory mutations in the emergence of drug resistance. PLoS Comput Biol 2:e137.

Hartfield M, Otto SP, Keightley PD. 2010. The role of advantageous mutations in enhancing the evolution of a recombination modifier. Genetics 184:1153–1164.

Haseltine EL, Yin J, Rawlings JB. 2008. Implications of decoupling the intracellular and extracellular levels in multi-level models of virus growth. Biotechnol Bioeng 101:811–820.

Hazenberg MD, Otto SA, van Benthem BH, Roos MT, Coutinho RA, Lange JM, Hamann D, Prins M, Miedema F. 2003. Persistent immune activation in HIV-1 infection is associated with progression to AIDS. AIDS 17:1881–1888.

Henn MR, Boutwell CL, Charlebois P, Lennon NJ, Power KA, Macalalad AR, Berlin AM, Malboeuf CM, Ryan EM, Gnerre S, et al. 2012. Whole genome deep sequencing of HIV-1 reveals the impact of early minor variants upon immune recognition during acute infection. PLoS Pathog 8:e1002529.

Hill WG, Robertson A. 1968. Linkage disequilibrium in finite populations. Theor Appl Genet 38:226–231.

Hinkley T, Martins J, Chappey C, Haddad M, Stawiski E, Whitcomb JM, Petropoulos CJ, Bonhoeffer S. 2011. A systems analysis of mutational effects in HIV-1 protease and reverse transcriptase. Nat Genet 43:487–489.

Ho DD, Neumann AU, Perelson AS, Chen W, Leonard JM, Markowitz M. 1995. Rapid turnover of plasma virions and CD4 lymphocytes in HIV-1 infection. Nature 373:123–126.

Ho YC, Shan L, Hosmane NN, Wang J, Laskey SB, Rosenbloom DI, Lai J, Blankson JN, Siliciano JD, Siliciano RF. 2013. Replication-competent noninduced proviruses in the latent reservoir increase barrier to HIV-1 cure. Cell 155:540–551.

Hoh J, Ott J. 2003. Mathematical multi-locus approaches to localizing complex human trait genes. Nat Rev Genet 4:701–709.

Holland J. 1990. Generation and replication of defective viral genomes. In: BN F, DM K, editors. Virology. New York: Raven Press. p. 77–99.

Holland J, Spindler K, Horodyski F, Grabau E, Nichol S, VandePol S. 1982. Rapid evolution of RNA genomes. Science 215:1577–1585.

Holland JJ, Villarreal LP. 1975. Purification of defective interfering T particles of vesicular stomatitis and rabies viruses generated in vivo in brains of newborn mice. Virology 67:438–449.

Holmes EC. 2009. RNA virus genomics: a world of possibilities. J Clin Invest 119:2488–2495.

Holmes EC, Zhang LQ, Simmonds P, Ludlam CA, Brown AJ. 1992. Convergent and divergent sequence evolution in the surface envelope glycoprotein of human immunodeficiency virus type 1 within a single infected patient. Proc Natl Acad Sci U S A 89:4835–4839.

Hom N, Gentles L, Bloom JD, Lee KK. 2019. Deep Mutational Scan of the Highly Conserved Influenza A Virus M1 Matrix Protein Reveals Substantial Intrinsic Mutational Tolerance. J Virol 93.

Horodyski FM, Holland JJ. 1980. Viruses isolated from cells persistently infected with vesicular stomatitis virus show altered interactions with defective interfering particles. J Virol 36:627–631.

Hu WS, Temin HM. 1990. Retroviral recombination and reverse transcription. Science 250:1227–1233.

Hu Z, Li Y, Song X, Han Y, Cai X, Xu S, Li W. 2011. Genomic value prediction for quantitative traits under the epistatic model. BMC Genet 12:15.

Huang AS, Baltimore D. 1970. Defective viral particles and viral disease processes. Nature 226:325–327.

Huang Y, Wuchty S, Przytycka TM. 2013. eQTL Epistasis – Challenges and Computational Approaches. Front Genet 4:51.

Illingworth CJ, Mustonen V. 2012. Components of selection in the evolution of the influenza virus: linkage effects beat inherent selection. PLoS Pathog 8:e1003091.

Imhof M, Schlotterer C. 2001. Fitness effects of advantageous mutations in evolving Escherichia coli populations. Proc Natl Acad Sci U S A 98:1113–1117.

Jerison ER, Desai MM. 2015. Genomic investigations of evolutionary dynamics and epistasis in microbial evolution experiments. Curr Opin Genet Dev 35:33–39.

Jin X, Bauer DE, Tuttleton SE, Lewin S, Gettie A, Blanchard J, Irwin CE, Safrit JT, Mittler J, Weinberger L, et al. 1999. Dramatic rise in plasma viremia after $CD8^{(+)}$ T cell depletion in simian immunodeficiency virus-infected macaques. J Exp Med 189:991–998.

Johnson T, Barton NH. 2002. The effect of deleterious alleles on adaptation in asexual populations. Genetics 162:395–411.

Josefsson L, King MS, Makitalo B, Brannstrom J, Shao W, Maldarelli F, Kearney MF, Hu WS, Chen J, Gaines H, et al. 2011. Majority of $CD4^+$ T cells from peripheral blood of HIV-1-infected individuals contain only one HIV DNA molecule. Proc Natl Acad Sci U S A 108:11199–11204.

Joyce P, Rokyta DR, Beisel CJ, Orr HA. 2008. A general extreme value theory model for the adaptation of DNA sequences under strong selection and weak mutation. Genetics 180:1627–1643.

Jung A, Maier R, Vartanian JP, Bocharov G, Jung V, Fischer U, Meese E, Wain-Hobson S, Meyerhans A. 2002. Recombination: Multiply infected spleen cells in HIV patients. Nature 418:144.

Kadolsky UD, Asquith B. 2010. Quantifying the impact of human immunodeficiency virus-1 escape from cytotoxic T-lymphocytes. PLoS Comput Biol 6:e1000981.

Kahn JO, Walker BD. (98299385 co-authors). 1998. Acute human immunodeficiency virus type 1 infection. N Engl J Med 339:33–39.

Karlsson AC, Iversen AK, Chapman JM, de Oliviera T, Spotts G, McMichael AJ, Davenport MP, Hecht FM, Nixon DF. 2007. Sequential broadening of CTL responses in early HIV-1 infection is associated with viral escape. PLOS ONE 2:e225.

Kassen R, Bataillon T. 2006. Distribution of fitness effects among beneficial mutations before selection in experimental populations of bacteria. Nat Genet 38:484–488.

Kawashima Y, Pfafferott K, Frater J, Matthews P, Payne R, Addo M, Gatanaga H, Fujiwara M, Hachiya A, Koizumi H, et al. 2009. Adaptation of HIV-1 to human leukocyte antigen class I. Nature 458:641–645.

Ke R, Lloyd-Smith JO. 2012. Evolutionary analysis of human immunodeficiency virus type 1 therapies based on conditionally replicating vectors. PLoS Comput Biol 8:e1002744.

Kearney M, Maldarelli F, Shao W, Margolick JB, Daar ES, Mellors JW, Rao V, Coffin JM, Palmer S. 2009. Human immunodeficiency virus type 1 population genetics and adaptation in newly infected individuals. J Virol 83:2715–2727.

Keele BF, Giorgi EE, Salazar-Gonzalez JF, Decker JM, Pham KT, Salazar MG, Sun C, Grayson T, Wang S, Li H, et al. 2008. Identification and characterization of transmitted and early founder virus envelopes in primary HIV-1 infection. Proc Natl Acad Sci U S A 105:7552–7557.

Keightley PD, Eyre-Walker A. 2007. Joint inference of the distribution of fitness effects of deleterious mutations and population demography based on nucleotide polymorphism frequencies. Genetics 177:2251–2261.

Kelleher AD, Long C, Holmes EC, Allen RL, Wilson J, Conlon C, Workman C, Shaunak S, Olson K, Goulder P, et al. 2001. Clustered mutations in HIV-1 gag are consistently required for escape from HLA-B27-restricted cytotoxic T lymphocyte responses. J Exp Med 193:375–386.

Kessler D, Levine H, Ridgeway D, Tsimring L. 1997. Evolution on a smooth landscape. J. Stat. Phys. 87:519–544.

Kim CS, Lee SC, Kim YM, Kim BS, Choi HS, Kawada T, Kwon BS, Yu R. 2008. Visceral fat accumulation induced by a high-fat diet causes the atrophy of mesenteric lymph nodes in obese mice. Obesity (Silver Spring) 16:1261–1269.

Kimura M. 1965. Attainment of Quasi Linkage Equilibrium When Gene Frequencies Are Changing by Natural Selection. Genetics 52:875–890.

Kimura M. 1994. Population genetics, molecular evolution, and the neutral theory: selected papers. Chicago: University of Chicago Press.

Kirkegaard K, Baltimore D. 1986. The mechanism of RNA recombination in poliovirus. Cell 47:433–443.

Kirkwood TB, Bangham CR. 1994. Cycles, chaos, and evolution in virus cultures: a model of defective interfering particles. Proc Natl Acad Sci U S A 91:8685–8689.

Klatt NR, Shudo E, Ortiz AM, Engram JC, Paiardini M, Lawson B, Miller MD, Else J, Pandrea I, Estes JD, et al. 2010. $CD8^+$ lymphocytes control viral replication in SIVmac239-infected rhesus macaques without decreasing the lifespan of productively infected cells. PLoS Pathog 6:e1000747.

Klatzmann D, Barre-Sinoussi F, Nugeyre MT, Danquet C, Vilmer E, Griscelli C, Brun-Veziret F, Rouzioux C, Gluckman JC, Chermann JC, et al. 1984. Selective tropism of lymphadenopathy associated virus (LAV) for helper-inducer T lymphocytes. Science 225:59–63.

Klenerman P, Phillips RE, Rinaldo CR, Wahl LM, Ogg G, May RM, McMichael AJ, Nowak MA. 1996. Cytotoxic T lymphocytes and viral turnover in HIV type 1 infection. Proc Natl Acad Sci U S A 93:15323–15328.

Klimatcheva E, Planelles V, Day SL, Fulreader F, Renda MJ, Rosenblatt J. 2001. Defective lentiviral vectors are efficiently trafficked by HIV-1 and inhibit its replication. Mol Ther 3:928–939.

Koel BF, Burke DF, Bestebroer TM, van der Vliet S, Zondag GC, Vervaet G, Skepner E, Lewis NS, Spronken MI, Russell CA, et al. 2013. Substitutions near the receptor binding site determine major antigenic change during influenza virus evolution. Science 342:976–979.

Koldej RM, Anson DS. 2009. Refinement of lentiviral vector for improved RNA processing and reduced rates of self inactivation repair. BMC Biotechnol 9:86.

Kölsch H, Lehmann DJ, Ibrahim-Verbaas CA, Combarros O, van Duijn CM, Hammond N, Belbin O, Cortina-Borja M, Lehmann MG, Aulchenko YS, et al. 2012. Interaction of insulin and PPAR-α genes in Alzheimer's disease: the Epistasis Project. Journal of Neural Transmission 119:473–479.

Korboukh VK, Lee CA, Acevedo A, Vignuzzi M, Xiao Y, Arnold JJ, Hemperly S, Graci JD, August A, Andino R, et al. 2014. RNA virus population diversity, an optimum for maximal fitness and virulence. J Biol Chem 289:29531–29544.

Kouyos RD, Althaus CL, Bonhoeffer S. 2006. Stochastic or deterministic: what is the effective population size of HIV-1? Trends Microbiol 14:507–511.

Krakauer DC, Komarova NL. 2003. Levels of selection in positive-strand virus dynamics. J Evol Biol 16:64–73.

Kryazhimskiy S, Dushoff J, Bazykin GA, Plotkin JB. 2011. Prevalence of epistasis in the evolution of influenza A surface proteins. PLoS Genet 7:e1001301.

Kryazhimskiy S, Rice DP, Jerison ER, Desai MM. 2014. Microbial evolution. Global epistasis makes adaptation predictable despite sequence-level stochasticity. Science 344:1519–1522.

Kubo Y, Kakimi K, Higo K, Kobayashi H, Ono T, Iwama Y, Kuribayashi K, Hiai H, Adachi A, Ishimoto A. 1996. Possible origin of murine AIDS (MAIDS) virus: conversion of an endogenous retroviral p12gag sequence to a MAIDS-inducing sequence by frameshift mutations. J Virol 70:6405–6409.

Kuiken CL, Zwart G, Baan E, Coutinho RA, van den Hoek JA, Goudsmit J. 1993. Increasing antigenic and genetic diversity of the V3 variable domain of the human immunodeficiency virus envelope protein in the course of the AIDS epidemic. Proc Natl Acad Sci U S A 90:9061–9065.

Kuroda MJ, Schmitz JE, Charini WA, Nickerson CE, Lifton MA, Lord CI, Forman MA, Letvin NL. 1999. Emergence of CTL coincides with clearance of virus during primary simian immunodeficiency virus infection in rhesus monkeys. J Immunol 162:5127–5133.

Lambotte O, Boufassa F, Madec Y, Nguyen A, Goujard C, Meyer L, Rouzioux C, Venet A, Delfraissy JF, Group S-HS. 2005. HIV controllers: a homogeneous group of HIV-1-infected patients with spontaneous control of viral replication. Clin Infect Dis 41:1053–1056.

Lamers SL, Sleasman JW, She JX, Barrie KA, Pomeroy SM, Barrett DJ, Goodenow MM. 1993. Independent variation and positive selection in env V1 and V2 domains within maternal-infant strains of human immunodeficiency virus type 1 in vivo. J Virol 67:3951–3960.

Landau L, Lifshitz E. 1969. Statistical physics: Pergamon Press.

Lauring AS, Acevedo A, Cooper SB, Andino R. 2012. Codon usage determines the mutational robustness, evolutionary capacity, and virulence of an RNA virus. Cell Host Microbe 12:623–632.

Lech WJ, Wang G, Yang YL, Chee Y, Dorman K, McCrae D, Lazzeroni LC, Erickson JW, Sinsheimer JS, Kaplan AH. 1996. In vivo sequence diversity of the protease of human immunodeficiency virus type 1: presence of protease inhibitor-resistant variants in untreated subjects. J Virol 70:2038–2043.

Lee JM, Huddleston J, Doud MB, Hooper KA, Wu NC, Bedford T, Bloom JD. 2018. Deep mutational scanning of hemagglutinin helps predict evolutionary fates of human H3N2 influenza variants. Proc Natl Acad Sci U S A 115:E8276–E8285.

Leigh-Brown AJ. 1997. Analysis of HIV-1 env gene sequences reveals evidence for a low effective number in the viral population. Proc Natl Acad Sci U S A 94:1862–1865.

Leslie A, Kavanagh D, Honeyborne I, Pfafferott K, Edwards C, Pillay T, Hilton L, Thobakgale C, Ramduth D, Draenert R, et al. 2005. Transmission and accumulation of CTL escape variants drive negative associations between HIV polymorphisms and HLA. J Exp Med 201:891–902.

Letvin NL, Mascola JR, Sun Y, Gorgone DA, Buzby AP, Xu L, Yang ZY, Chakrabarti B, Rao SS, Schmitz JE, et al. 2006. Preserved $CD4^+$ central memory T cells and survival in vaccinated SIV-challenged monkeys. Science 312:1530–1533.

Levin BR, Perrot V, Walker N. 2000. Compensatory mutations, antibiotic resistance and the population genetics of adaptive evolution in bacteria. Genetics 154:985–997.

Levine BL, Humeau LM, Boyer J, MacGregor RR, Rebello T, Lu X, Binder GK, Slepushkin V, Lemiale F, Mascola JR, et al. 2006. Gene transfer in humans using a conditionally replicating lentiviral vector. Proc Natl Acad Sci U S A 103:17372–17377.

Levy DN, Aldrovandi GM, Kutsch O, Shaw GM. 2004. Dynamics of HIV-1 recombination in its natural target cells. Proc Natl Acad Sci U S A 101:4204–4209.

Levy JA, Hoffman AD, Kramer SM, Landis JA, Shimabukuro JM, Oshiro LS. 1984. Isolation of lymphocytopathic retroviruses from San Francisco patients with AIDS. Science 225:840–842.

Lewontin RC. 1964. The Interaction of Selection and Linkage. I. General Considerations; Heterotic Models. Genetics 49:49–67.

Li D, Lott WB, Lowry K, Jones A, Thu HM, Aaskov J. 2011. Defective interfering viral particles in acute dengue infections. PLOS ONE 6:e19447.

Li Q, Duan L, Estes JD, Ma ZM, Rourke T, Wang Y, Reilly C, Carlis J, Miller CJ, Haase AT. 2005. Peak SIV replication in resting memory $CD4^+$ T cells depletes gut lamina propria $CD4^+$ T cells. Nature 434:1148–1152.

Li Y, Kappes JC, Conway JA, Price RW, Shaw GM, Hahn BH. 1991. Molecular characterization of human immunodeficiency virus type 1 cloned directly from uncultured human brain tissue: identification of replication-competent and -defective viral genomes. J Virol 65:3973–3985.

Likhachev IV, Rouzine IM. 2023. Measurement of selection coefficients from genomic samples of adapting populations by computer modeling. STAR Protoc 4:101821.

Lin J, Andreasen V, Casagrandi R, Levin SA. 2003. Traveling waves in a model of influenza A drift. J Theor Biol 222:437–445.

Lippert C, Listgarten J, Davidson RI, Baxter S, Poon H, Kadie CM, Heckerman D. 2013. An exhaustive epistatic SNP association analysis on expanded Wellcome Trust data. Sci Rep 3:1099.

Liu MK, Hawkins N, Ritchie AJ, Ganusov VV, Whale V, Brackenridge S, Li H, Pavlicek JW, Cai F, Rose-Abrahams M, et al. 2013. Vertical T cell immunodominance and epitope entropy determine HIV-1 escape. J Clin Invest 123:380–393.

Liu SL, Schacker T, Musey L, Shriner D, McElrath MJ, Corey L, Mullins JI. 1997. Divergent patterns of progression to AIDS after infection from the same source: human immunodeficiency virus type 1 evolution and antiviral responses. J Virol 71:4284–4295.

Liu Y, McNevin JP, Holte S, McElrath MJ, Mullins JI. 2011. Dynamics of viral evolution and CTL responses in HIV-1 infection. PLOS ONE 6:e15639.

Liu Y, Mullins JI, Mittler JE. 2006. Waiting times for the appearance of cytotoxic T-lymphocyte escape mutants in chronic HIV-1 infection. Virology 347:140–146.

Lloyd-Smith JO, Schreiber SJ, Kopp PE, Getz WM. 2005. Superspreading and the effect of individual variation on disease emergence. Nature 438:355–359.

Lowry K, Woodman A, Cook J, Evans DJ. 2014. Recombination in enteroviruses is a biphasic replicative process involving the generation of greater-than genome length 'imprecise' intermediates. PLoS Pathog 10:e1004191.

Lu Q, Wei C, Ye C, Li M, Elston RC. 2012. A Likelihood Ratio-Based Mann-Whitney Approach Finds Novel Replicable Joint Gene Action for Type 2 Diabetes. Genetic epidemiology 36:583–593.

Luksza M, Lassig M. 2014. A predictive fitness model for influenza. Nature 507:57–61.

Macarthur RH. 1957. On the Relative Abundance of Bird Species. Proceedings of the National Academy of Sciences of the United States of America 43:293–295.
Maldarelli F, Kearney M, Palmer S, Stephens R, Mican J, Polis MA, Davey RT, Kovacs J, Shao W, Rock-Kress D, et al. 2013. HIV populations are large and accumulate high genetic diversity in a nonlinear fashion. J Virol 87:10313–10323.
Mansky LM, Temin HM. 1995. Lower in vivo mutation rate of human immunodeficiency virus type 1 than that predicted from the fidelity of purified reverse transcriptase. J Virol 69:5087–5094.
Marchi J, Lassig M, Walczak AM, Mora T. 2021. Antigenic waves of virus-immune coevolution. Proc Natl Acad Sci U S A 118.
Markowitz M, Louie M, Hurley A, Sun E, Di Mascio M, Perelson AS, Ho DD. 2003. A novel antiviral intervention results in more accurate assessment of human immunodeficiency virus type 1 replication dynamics and T-cell decay in vivo. J Virol 77:5037–5038.
Marriott AC, Dimmock NJ. 2010. Defective interfering viruses and their potential as antiviral agents. Rev Med Virol 20:51–62.
Masur H, Michelis MA, Greene JB, Onorato I, Stouwe RA, Holzman RS, Wormser G, Brettman L, Lange M, Murray HW, et al. 1981. An outbreak of community-acquired Pneumocystis carinii pneumonia: initial manifestation of cellular immune dysfunction. N Engl J Med 305:1431–1438.
Masur H, Ognibene FP, Yarchoan R, Shelhamer JH, Baird BF, Travis W, Suffredini AF, Deyton L, Kovacs JA, Falloon J, et al. 1989. CD4 counts as predictors of opportunistic pneumonias in human immunodeficiency virus (HIV) infection. Ann Intern Med 111:223–231.
Matano T, Shibata R, Siemon C, Connors M, Lane HC, Martin MA. 1998. Administration of an anti-CD8 monoclonal antibody interferes with the clearance of chimeric simian/human immunodeficiency virus during primary infections of rhesus macaques. Journal of virology 72:164–169.
Matthews PC, Koyanagi M, Kloverpris HN, Harndahl M, Stryhn A, Akahoshi T, Gatanaga H, Oka S, Juarez Molina C, Valenzuela Ponce H, et al. 2012. Differential clade-specific HLA-B*3501 association with HIV-1 disease outcome is linked to immunogenicity of a single Gag epitope. J Virol 86:12643–12654.
May RM. 2004. Uses and abuses of mathematics in biology. Science 303:790–793.
Mayers DL, McCutchan FE, Sanders-Buell EE, Merritt LI, Dilworth S, Fowler AK, Marks CA, Ruiz NM, Richman DD, Roberts CR, et al. 1992. Characterization of HIV isolates arising after prolonged zidovudine therapy. J Acquir Immune Defic Syndr (1988) 5:749–759.
Maynard Smith J. 1971. What use is sex? J. Theor. Bio. 30:319–335.
McCandlish DM, Shah P, Plotkin JB. 2016. Epistasis and the Dynamics of Reversion in Molecular Evolution. Genetics 203:1335–1351.
McCutchan FE, Artenstein AW, Sanders-Buell E, Salminen MO, Carr JK, Mascola JR, Yu XF, Nelson KE, Khamboonruang C, Schmitt D, et al. 1996. Diversity of the envelope glycoprotein among human immunodeficiency virus type 1 isolates of clade E from Asia and Africa. J Virol 70:3331–3338.
McKinney BA, Pajewski NM. 2011. Six Degrees of Epistasis: Statistical Network Models for GWAS. Front Genet 2:109.
McLain L, Armstrong SJ, Dimmock NJ. 1988. One defective interfering particle per cell prevents influenza virus-mediated cytopathology: an efficient assay system. J Gen Virol 69 (Pt 6):1415–1419.
McVean GA, Charlesworth B. 2000. The effects of Hill-Robertson interference between weakly selected mutations on patterns of molecular evolution and variation. Genetics 155:929–944.
Meher BR, Wang Y. 2012. Interaction of I50V mutant and I50L/A71V double mutant HIV-protease with inhibitor TMC114 (darunavir): molecular dynamics simulation and binding free energy studies. J Phys Chem B 116:1884–1900.
Messer PW, Neher RA. 2012. Estimating the strength of selective sweeps from deep population diversity data. Genetics 191:593–605.
Metzger VT, Lloyd-Smith JO, Weinberger LS. 2011. Autonomous targeting of infectious superspreaders using engineered transmissible therapies. PLoS Comput Biol 7:e1002015.

Michel N, Allespach I, Venzke S, Fackler OT, Keppler OT. 2005. The Nef protein of human immunodeficiency virus establishes superinfection immunity by a dual strategy to downregulate cell-surface CCR5 and CD4. Curr Biol 15:714–723.

Mideo N, Alizon S, Day T. 2008. Linking within- and between-host dynamics in the evolutionary epidemiology of infectious diseases. Trends Ecol Evol 23:511–517.

Miller CJ, Li Q, Abel K, Kim EY, Ma ZM, Wietgrefe S, La Franco-Scheuch L, Compton L, Duan L, Shore MD, et al. 2005. Propagation and dissemination of infection after vaginal transmission of simian immunodeficiency virus. J Virol 79:9217–9227.

Milush JM, Mir KD, Sundaravaradan V, Gordon SN, Engram J, Cano CA, Reeves JD, Anton E, O'Neill E, Butler E, et al. 2011. Lack of clinical AIDS in SIV-infected sooty mangabeys with significant $CD4^+$ T cell loss is associated with double-negative T cells. J Clin Invest 121:1102–1110.

Miralles R, Moya A, Elena SF. 1997. Is group selection a factor modulating the virulence of RNA viruses? Genet Res 69:165–172.

Mohri H, Bonhoeffer S, Monard S, Perelson AS, Ho DD. 1998. Rapid turnover of T lymphocytes in SIV-infected rhesus macaques. Science 279:1223–1227.

Moore JP, Cao Y, Leu J, Qin L, Korber B, Ho DD. 1996. Inter- and intraclade neutralization of human immunodeficiency virus type 1: genetic clades do not correspond to neutralization serotypes but partially correspond to gp120 antigenic serotypes. J Virol 70:427–444.

Moore MD, Fu W, Nikolaitchik O, Chen J, Ptak RG, Hu WS. 2007. Dimer initiation signal of human immunodeficiency virus type 1: its role in partner selection during RNA copackaging and its effects on recombination. J Virol 81:4002–4011.

Mostowy R, Kouyos RD, Fouchet D, Bonhoeffer S. 2011. The role of recombination for the coevolutionary dynamics of HIV and the immune response. PLOS ONE 6:e16052.

Mostowy R, Kouyos RD, Hoof I, Hinkley T, Haddad M, Whitcomb JM, Petropoulos CJ, Kesmir C, Bonhoeffer S. 2012. Estimating the fitness cost of escape from HLA presentation in HIV-1 protease and reverse transcriptase. PLoS Comput Biol 8:e1002525.

Mukherjee R, Plesa G, Sherrill-Mix S, Richardson MW, Riley JL, Bushman FD. 2010. HIV sequence variation associated with env antisense adoptive T-cell therapy in the hNSG mouse model. Mol Ther 18:803–811.

Muller H. 1932. Some genetic aspects of sex. Am. Nat. 66:118.

Muller V, Maree AF, De Boer RJ. 2001. Small variations in multiple parameters account for wide variations in HIV-1 set-points: a novel modelling approach. Proc Biol Sci 268:235–242.

Murphy K. 2011. Janeway's Immunobiology, Eighth Edition. London, New York: Garland Science.

Najera I, Holguin A, Quinones-Mateu ME, Munoz-Fernandez MA, Najera R, Lopez-Galindez C, Domingo E. 1995. Pol gene quasispecies of human immunodeficiency virus: mutations associated with drug resistance in virus from patients undergoing no drug therapy. J Virol 69:23–31.

Neher RA, Bedford T, Daniels RS, Russell CA, Shraiman BI. 2016. Prediction, dynamics, and visualization of antigenic phenotypes of seasonal influenza viruses. Proc Natl Acad Sci U S A 113:E1701–1709.

Neher RA, Hallatschek O. 2013. Genealogies of rapidly adapting populations. Proc Natl Acad Sci U S A 110:437–442.

Neher RA, Kessinger TA, Shraiman BI. 2013. Coalescence and genetic diversity in sexual populations under selection. Proc Natl Acad Sci U S A 110:15836–15841.

Neher RA, Leitner T. 2010. Recombination rate and selection strength in HIV intra-patient evolution. PLoS Comput Biol 6:e1000660.

Neher RA, Russell CA, Shraiman BI. 2014. Predicting evolution from the shape of genealogical trees. Elife 3.

Neher RA, Shraiman BI. 2009. Competition between recombination and epistasis can cause a transition from allele to genotype selection. Proc Natl Acad Sci U S A 106:6866–6871.

Neher RA, Shraiman BI. 2011. Statistical genetics and evolution of quantitative traits. Rev. Mod. Phys. 83:1283.

Neher RA, Shraiman BI, Fisher DS. 2010. Rate of adaptation in large sexual populations. Genetics 184:467–481.

Nelson GW, Perelson AS. 1995. Modeling defective interfering virus therapy for AIDS: conditions for DIV survival. Math Biosci 125:127–153.

Nguyen Ba AN, Cvijovic I, Rojas Echenique JI, Lawrence KR, Rego-Costa A, Liu X, Levy SF, Desai MM. 2019. High-resolution lineage tracking reveals travelling wave of adaptation in laboratory yeast. Nature 575:494–499.

Nielsen R, Slatkin M. 2013. An introduction to population genetics: Theory and applications: Sinaur Associates, Inc.

Nijhuis M, Schuurman R, de Jong D, Erickson J, Gustchina E, Albert J, Schipper P, Gulnik S, Boucher CA. 1999. Increased fitness of drug resistant HIV-1 protease as a result of acquisition of compensatory mutations during suboptimal therapy. AIDS 13:2349–2359.

Nishimura Y, Brown CR, Mattapallil JJ, Igarashi T, Buckler-White A, Lafont BA, Hirsch VM, Roederer M, Martin MA. 2005. Resting naive $CD4^+$ T cells are massively infected and eliminated by X4-tropic simian-human immunodeficiency viruses in macaques. Proc Natl Acad Sci U S A 102:8000–8005.

Norley S, Beer B, Holzammer S, zur Megede J, Kurth R. 1999. Why are the natural hosts of SIV resistant to AIDS? Immunol Lett 66:47–52.

Notton T, Sardanyes J, Weinberger AD, Weinberger LS. 2014. The case for transmissible antivirals to control population-wide infectious disease. Trends Biotechnol 32:400–405.

Novella IS. 2004. Negative effect of genetic bottlenecks on the adaptability of vesicular stomatitis virus. J Mol Biol 336:61–67.

Noviello CM, Lopez CS, Kukull B, McNett H, Still A, Eccles J, Sloan R, Barklis E. 2011. Second-site compensatory mutations of HIV-1 capsid mutations. J Virol 85:4730–4738.

Nowak MA. 2006. Evolutionary Dynamics: Exploring the Equations of Life. Cabmbridge, USA: Harvard University Press.

Nowak MA, Bangham CR. 1996. Population dynamics of immune responses to persistent viruses. Science 272:74–79.

Nowak MA, Lloyd AL, Vasquez GM, Wiltrout TA, Wahl LM, Bischofberger N, Williams J, Kinter A, Fauci AS, Hirsch VM, et al. 1997. Viral dynamics of primary viremia and antiretroviral therapy in simian immunodeficiency virus infection. J. Virol. 71:7518–7525.

Nowak MA, May R. 2000. Virus Dynamics: Mathematical Principles of Immunology and Virology. New York, NY, USA; Oxford, UK: Oxford University Press.

Nowak MA, May RM. 1994. Superinfection and the evolution of parasite virulence. Proceedings of the Royal Society of London. Series B: Biological Sciences 255:81–89.

O'Hara PJ, Horodyski FM, Nichol ST, Holland JJ. 1984. Vesicular stomatitis virus mutants resistant to defective-interfering particles accumulate stable 5′-terminal and fewer 3′-terminal mutations in a stepwise manner. J Virol 49:793–798.

Ogg GS, Jin X, Bonhoeffer S, Dunbar PR, Nowak MA, Monard S, Segal JP, Cao Y, Rowland-Jones SL, Cerundolo V, et al. 1998. Quantitation of HIV-1-specific cytotoxic T lymphocytes and plasma load of viral RNA. Science 279:2103–2106.

Ogg GS, Jin X, Bonhoeffer S, Moss P, Nowak MA, Monard S, Segal JP, Cao Y, Rowland-Jones SL, Hurley A, et al. 1999. Decay kinetics of human immunodeficiency virus-specific effector cytotoxic T lymphocytes after combination antiretroviral therapy. J Virol 73:797–800.

Okoye AA, Picker LJ. 2013. $CD4^{(+)}$ T-cell depletion in HIV infection: mechanisms of immunological failure. Immunol Rev 254:54–64.

Orr HA. 2003. The distribution of fitness effects among beneficial mutations. Genetics 163:1519–1526.

Otte A, Marriott AC, Dreier C, Dove B, Mooren K, Klingen TR, Sauter M, Thompson KA, Bennett A, Klingel K, et al. 2016. Evolution of 2009 H1N1 influenza viruses during the pandemic correlates with increased viral pathogenicity and transmissibility in the ferret model. Sci Rep 6:28583.

Otto SP, Barton NH. 1997. The evolution of recombination: removing the limits to natural selection. Genetics 147:879–906.

Paik SY, Banerjea A, Chen CJ, Ye Z, Harmison GG, Schubert M. 1997. Defective HIV-1 provirus encoding a multitarget-ribozyme inhibits accumulation of spliced and unspliced HIV-1 mRNAs, reduces infectivity of viral progeny, and protects the cells from pathogenesis. Hum Gene Ther 8:1115–1124.

Paillart JC, Shehu-Xhilaga M, Marquet R, Mak J. 2004. Dimerization of retroviral RNA genomes: an inseparable pair. Nat Rev Microbiol 2:461–472.

Palmer S, Kearney M, Maldarelli F, Halvas EK, Bixby CJ, Bazmi H, Rock D, Falloon J, Davey RT, Jr., Dewar RL, et al. 2005. Multiple, linked human immunodeficiency virus type 1 drug resistance mutations in treatment-experienced patients are missed by standard genotype analysis. J Clin Microbiol 43:406–413.

Pandrea I, Ribeiro RM, Gautam R, Gaufin T, Pattison M, Barnes M, Monjure C, Stoulig C, Dufour J, Cyprian W, et al. 2008. Simian immunodeficiency virus SIVagm dynamics in African green monkeys. J Virol 82:3713–3724.

Pang S, Vinters HV, Akashi T, O'Brien WA, Chen IS. 1991. HIV-1 env sequence variation in brain tissue of patients with AIDS-related neurologic disease. J Acquir Immune Defic Syndr (1988) 4:1082–1092.

Pang X, Wang Z, Yap JS, Wang J, Zhu J, Bo W, Lv Y, Xu F, Zhou T, Peng S, et al. 2013. A statistical procedure to map high-order epistasis for complex traits. Brief Bioinform 14:302–314.

Park AW, Daly JM, Lewis NS, Smith DJ, Wood JL, Grenfell BT. 2009. Quantifying the impact of immune escape on transmission dynamics of influenza. Science 326:726–728.

Patino-Galindo JA, Gonzalez-Candelas F. 2017. The substitution rate of HIV-1 subtypes: a genomic approach. Virus Evol 3: vex029.

Paun A, Shaw K, Fisher S, Sammels LM, Watson MW, Beilharz MW. 2005. Quantitation of defective and ecotropic viruses during LP-BM5 infection by real time PCR and RT-PCR. J Virol Methods 124:57–63.

Pearson JE, Krapivsky P, Perelson AS. 2011. Stochastic theory of early viral infection: continuous versus burst production of virions. PLoS Comput Biol 7:e1001058.

Pedruzzi G, Barlukova A, Rouzine IM. 2018. Evolutionary footprint of epistasis. PLoS Comput Biol 14: e1006426.

Pedruzzi G, Rouzine IM. 2019. Epistasis detectably alters correlations between genomic sites in a narrow parameter window. PLOS ONE in press.

Pedruzzi G, Rouzine IM. 2021. An evolution-based high-fidelity method of epistasis measurement: Theory and application to influenza. PLoS Pathog 17:e1009669.

Pennings PS, Kryazhimskiy S, Wakeley J. 2014. Loss and recovery of genetic diversity in adapting populations of HIV. PLoS Genet 10:e1004000.

Perelson AS, Essunger P, Cao Y, Vesanen M, Hurley A, Saksela K, Markowitz M, Ho DD. 1997. Decay characteristics of HIV-1-infected compartments during combination therapy. Nature 387:188–191.

Perelson AS, Neumann AU, Markowitz M, Leonard JM, Ho DD. 1996. HIV-1 dynamics in vivo: virion clearance rate, infected cell life-span, and viral generation time. Science 271:1582–1586.

Perelson AS, Ribeiro RM. 2013. Modeling the within-host dynamics of HIV infection. BMC Biol 11:96.

Peressin M, Proust A, Schmidt S, Su B, Lambotin M, Biedma ME, Laumond G, Decoville T, Holl V, Moog C. 2014. Efficient transfer of HIV-1 in trans and in cis from Langerhans dendritic cells and macrophages to autologous T lymphocytes. AIDS.

Perrault J, Holland JJ. 1972. Absence of transcriptase activity or transcription-inhibiting ability in defective interfering particles of vesicular stomatitis virus. Virology 50:150–170.

Pfeiffer JK, Kirkegaard K. 2003. A single mutation in poliovirus RNA-dependent RNA polymerase confers resistance to mutagenic nucleotide analogs via increased fidelity. Proc Natl Acad Sci U S A 100:7289–7294.

Pfutzner A, Dietrich U, von Eichel U, von Briesen H, Brede HD, Maniar JK, Rubsamen-Waigmann H. 1992. HIV-1 and HIV-2 infections in a high-risk population in Bombay, India: evidence for the spread of HIV-2 and presence of a divergent HIV-1 subtype. J Acquir Immune Defic Syndr (1988) 5:972–977.

Piana S, Carloni P, Rothlisberger U. 2002. Drug resistance in HIV-1 protease: Flexibility-assisted mechanism of compensatory mutations. Protein Sci 11:2393–2402.

Picker LJ, Hagen SI, Lum R, Reed-Inderbitzin EF, Daly LM, Sylwester AW, Walker JM, Siess DC, Piatak M, Jr., Wang C, et al. 2004. Insufficient production and tissue delivery of $CD4^+$ memory T cells in rapidly progressive simian immunodeficiency virus infection. J Exp Med 200:1299–1314.

Pieniazek D, Janini LM, Ramos A, Tanuri A, Schechter M, Peralta JM, Vicente AC, Pieniazek NK, Schochetman G, Rayfield MA. 1995. HIV-1 patients may harbor viruses of different phylogenetic subtypes: implications for the evolution of the HIV/AIDS pandemic. Emerg Infect Dis 1:86–88.

Pierson T, McArthur J, Siliciano RF. 2000. Reservoirs for HIV-1: mechanisms for viral persistence in the presence of antiviral immune responses and antiretroviral therapy. Annual review of immunology 18:665–708.

Poon AF, Swenson LC, Dong WW, Deng W, Kosakovsky Pond SL, Brumme ZL, Mullins JI, Richman DD, Harrigan PR, Frost SD. 2010. Phylogenetic analysis of population-based and deep sequencing data to identify coevolving sites in the nef gene of HIV-1. Mol Biol Evol 27:819–832.

Potter SJ, Lacabaratz C, Lambotte O, Perez-Patrigeon S, Vingert B, Sinet M, Colle JH, Urrutia A, Scott-Algara D, Boufassa F, et al. 2007. Preserved central memory and activated effector memory $CD4^+$ T-cell subsets in human immunodeficiency virus controllers: an ANRS EP36 study. J Virol 81:13904–13915.

Poulin R. 2007. Evolutionary Ecology of Parasites.: Princeton University Press.

Quagliarello V. 1982. The Acquired Immunodeficiency Syndrome: current status. Yale J Biol Med 55:443–452.

Quinones-Mateu ME, Arts EJ. 2002. Fitness of drug resistant HIV-1: methodology and clinical implications. Drug Resist Updat 5:224–233.

Racaniello VR, Baltimore D. 1981. Molecular cloning of poliovirus cDNA and determination of the complete nucleotide sequence of the viral genome. Proc Natl Acad Sci U S A 78:4887–4891.

Rainey GJ, Coffin JM. 2006. Evolution of broad host range in retroviruses leads to cell death mediated by highly cytopathic variants. J Virol 80:562–570.

Rambaut A, Pybus OG, Nelson MI, Viboud C, Taubenberger JK, Holmes EC. 2008. The genomic and epidemiological dynamics of human influenza A virus. Nature 453:615–619.

Rankin DJ, Bargum K, Kokko H. 2007. The tragedy of the commons in evolutionary biology. Trends Ecol Evol 22:643–651.

Rast LI, Rouzine IM, Rozhnova G, Bishop L, Weinberger AD, Weinberger LS. 2016. Conflicting Selection Pressures Will Constrain Viral Escape from Interfering Particles: Principles for Designing Resistance-Proof Antivirals. PLoS Comput Biol 12:e1004799.

Razooky BS, Pai A, Aull K, Rouzine IM, Weinberger LS. 2015. A hardwired HIV latency program. Cell 160:990–1001.

Razooky BS, Weinberger LS. 2011. Mapping the architecture of the HIV-1 Tat circuit: A decision-making circuit that lacks bistability and exploits stochastic noise. Methods 53:68–77.

Re F, Braaten D, Franke EK, Luban J. 1995. Human immunodeficiency virus type 1 vpr arrests the cell cycle in G2 by inhibiting activation of p34cdc2-cyclin B. J. Virol. 69:6859–6864.

Reimann KA, Li JT, Veazey R, Halloran M, Park IW, Karlsson GB, Sodroski J, Letvin NL. 1996. A chimeric simian/human immunodeficiency virus expressing a primary patient human immunodeficiency virus type 1 isolate env causes an AIDS-like disease after in vivo passage in rhesus monkeys. J Virol 70:6922–6928.

Reinhart TA, Rogan MJ, Amedee AM, Murphey-Corb M, Rausch DM, Eiden LE, Haase AT. 1998. Tracking members of the simian immunodeficiency virus deltaB670 quasispecies population in vivo at single-cell resolution. J Virol 72:113–120.

Rezelj VV, Carrau L, Merwaiss F, Levi LI, Erazo D, Tran QD, Henrion-Lacritick A, Gausson V, Suzuki Y, Shengjuler D, et al. 2021. Defective viral genomes as therapeutic interfering particles against flavivirus infection in mammalian and mosquito hosts. Nat Commun 12:2290.

Rhinn H, Fujita R, Qiang L, Cheng R, Lee JH, Abeliovich A. 2013. Integrative genomics identifies APOE epsilon4 effectors in Alzheimer's disease. Nature 500:45–50.

Ribeiro RM, Mohri H, Ho DD, Perelson AS. 2002a. In vivo dynamics of T cell activation, proliferation, and death in HIV-1 infection: why are $CD4^+$ but not $CD8^+$ T cells depleted? Proc Natl Acad Sci U S A 99:15572–15577.

Ribeiro RM, Mohri H, Ho DD, Perelson AS. 2002b. Modeling deuterated glucose labeling of T-lymphocytes. Bull. Math. Biol. 64:385–405.

Ribeiro RM, Qin L, Chavez LL, Li D, Self SG, Perelson AS. 2010. Estimation of the initial viral growth rate and basic reproductive number during acute HIV-1 infection. J Virol 84:6096–6102.

Rice DP, Good BH, Desai MM. 2015. The evolutionarily stable distribution of fitness effects. Genetics 200:321–329.

Rice S. 2004. Evolutionary Theory: Mathematical and Conceptual Foundations.: Sinauer Associated.

Rice WR. 2002. Experimental tests of the adaptive significance of sexual recombination. Nat Rev Genet 3:241–251.

Richman DD, Havlir D, Corbeil J, Looney D, Ignacio C, Spector SA, Sullivan J, Cheeseman S, Barringer K, Pauletti D, et al. 1994. Nevirapine resistance mutations of human immunodeficiency virus type 1 selected during therapy. J Virol 68:1660–1666.

Richman DD, Margolis DM, Delaney M, Greene WC, Hazuda D, Pomerantz RJ. 2009. The challenge of finding a cure for HIV infection. Science 323:1304–1307.

Ritchie MD. 2011. Using biological knowledge to uncover the mystery in the search for epistasis in genome-wide association studies. Ann Hum Genet 75:172–182.

Robertson DL, Sharp PM, McCutchan FE, Hahn BH. 1995. Recombination in HIV-1. Nature 374:124–126.

Rodriguez B, Sethi AK, Cheruvu VK, Mackay W, Bosch RJ, Kitahata M, Boswell SL, Mathews WC, Bangsberg DR, Martin J, et al. 2006. Predictive value of plasma HIV RNA level on rate of CD4 T-cell decline in untreated HIV infection. JAMA 296:1498–1506.

Rong L, Perelson AS. 2009a. Modeling HIV persistence, the latent reservoir, and viral blips. Journal of Theoretical Biology 260:308–331.

Rong L, Perelson AS. 2009b. Modeling latently infected cell activation: viral and latent reservoir persistence, and viral blips in HIV-infected patients on potent therapy. PLoS computational biology 5:e1000533.

Rouzine I, Coffin JM. 1999a. Interplay between experiment and theory in development of a working model for HIV-1 population dynamics. In: Domingo E, Webster R, Holland J, editors. Origin and evolution of viruses. London, United Kingdom: Academic Press Ltd.

Rouzine IM. 2020a. An Evolutionary Model of Progression to AIDS. Microorganisms 8.

Rouzine IM. 2020b. Mathematical Modeling of Evolution: Volume 1: One-Locus and Multi-Locus Theory and Recombination De Gruyter, Berlin/Boston.

Rouzine IM. 2022. A role for $CD4^+$ helper cells in HIV control and progression. AIDS 36:1501–1510.

Rouzine IM, Brunet E, Wilke CO. 2008. The traveling-wave approach to asexual evolution: Muller's ratchet and speed of adaptation. Theor Popul Biol 73:24–46.

Rouzine IM, Coffin JM. 2005. Evolution of human immunodeficiency virus under selection and weak recombination. Genetics 170:7–18.

Rouzine IM, Coffin JM. 2007. Highly fit ancestors of a partly sexual haploid population. Theor Popul Biol 71:239–250.

Rouzine IM, Coffin JM. 1999b. Linkage disequilibrium test implies a large effective population number for HIV in vivo. Proc Natl Acad Sci U S A 96:10758–10763.
Rouzine IM, Coffin JM. 2010. Multi-site adaptation in the presence of infrequent recombination. Theor Popul Biol 77:189–204.
Rouzine IM, Coffin JM. 1999c. Search for the mechanism of genetic variation in the pro gene of human immunodeficiency virus. J Virol 73:8167–8178.
Rouzine IM, Coffin JM. 1999d. T cell turnover in SIV infection [comment]. Science 284:555b.
Rouzine IM, Coffin JM, Weinberger LS. 2014. Fifteen years later: hard and soft selection sweeps confirm a large population number for HIV in vivo. PLoS Genet 10:e1004179.
Rouzine IM, McKenzie FE. 2003. Link between immune response and parasite synchronization in malaria. Proc Natl Acad Sci U S A 100:3473–3478.
Rouzine IM, Murali-Krishna K, Ahmed R. 2005. Generals die in friendly fire, or modeling immune response to HIV. J. Comput. Appl. Math. 184:258–274.
Rouzine IM, Razooky BS, Weinberger LS. 2014. Stochastic variability in HIV affects viral eradication. Proc Natl Acad Sci U S A 111:13251–13252.
Rouzine IM, Rodrigo A, Coffin JM. 2001. Transition between stochastic evolution and deterministic evolution in the presence of selection: general theory and application to virology. Microbiol Mol Biol Rev 65:151–185.
Rouzine IM, Rozhnova G. 2018. Antigenic evolution of viruses in host populations. PLoS Pathog 14: e1007291.
Rouzine I.M. and Rozhnova G. Evolutionary implications of SARS-CoV-2 vaccination for the future design of vaccination strategies. Communications Medicine 3, 86 (2023) https://doi.org/10.1038/s43856-023-00320-x
Rouzine IM, Sergeev RA, Glushtsov AI. 2006. Two types of cytotoxic lymphocyte regulation explain kinetics of immune response to human immunodeficiency virus. Proc Natl Acad Sci U S A 103:666–671.
Rouzine IM, Wakeley J, Coffin JM. 2003. The solitary wave of asexual evolution. Proc Natl Acad Sci U S A 100:587–592.
Rouzine IM, Weinberger AD, Weinberger LS. 2015. An evolutionary role for HIV latency in enhancing viral transmission. Cell 160:1002–1012.
Rouzine IM, Weinberger L. 2013a. The quantitative theory of within-host viral evolution [review]. J. Stat. Mech.: Theory and Experiment: P01009.
Rouzine IM, Weinberger LS. 2013b. Design requirements for interfering particles to maintain co-adaptive stability with HIV-1. Journal of Virology 87:2081–2093.
Rouzine IM, Weinberger LS. 2013c. Reply to “Coadaptive stability of interfering particles with HIV-1 when there is an evolutionary conflict”. J Virol 87:9960–9962.
Roze D, Barton NH. 2006. The Hill-Robertson effect and the evolution of recombination. Genetics 173:1793–1811.
Runckel C, Westesson O, Andino R, DeRisi JL. 2013. Identification and manipulation of the molecular determinants influencing poliovirus recombination. PLoS Pathog 9:e1003164.
Russell CA, Jones TC, Barr IG, Cox NJ, Garten RJ, Gregory V, Gust ID, Hampson AW, Hay AJ, Hurt AC, et al. 2008. Influenza vaccine strain selection and recent studies on the global migration of seasonal influenza viruses. Vaccine 26 Suppl 4:D31–34.
Sabino EC, Shpaer EG, Morgado MG, Korber BT, Diaz RS, Bongertz V, Cavalcante S, Galvao-Castro B, Mullins JI, Mayer A. 1994. Identification of human immunodeficiency virus type 1 envelope genes recombinant between subtypes B and F in two epidemiologically linked individuals from Brazil. J Virol 68:6340–6346.
Salazar-Gonzalez JF, Salazar MG, Keele BF, Learn GH, Giorgi EE, Li H, Decker JM, Wang S, Baalwa J, Kraus MH, et al. 2009. Genetic identity, biological phenotype, and evolutionary pathways of transmitted/founder viruses in acute and early HIV-1 infection. J Exp Med 206:1273–1289.

Sanjuan R, Nebot MR, Chirico N, Mansky LM, Belshaw R. 2010. Viral mutation rates. J Virol 84:9733–9748.

Santiago ML, Range F, Keele BF, Li Y, Bailes E, Bibollet-Ruche F, Fruteau C, Noe R, Peeters M, Brookfield JF, et al. 2005. Simian immunodeficiency virus infection in free-ranging sooty mangabeys (Cercocebus atys atys) from the Tai Forest, Cote d'Ivoire: implications for the origin of epidemic human immunodeficiency virus type 2. J Virol 79:12515–12527.

Schiffels S, Szollosi GJ, Mustonen V, Lassig M. 2011. Emergent neutrality in adaptive asexual evolution. Genetics 189:1361–1375.

Schmitz JE, Kuroda MJ, Santra S, Sasseville VG, Simon MA, Lifton MA, Racz P, Tenner-Racz K, Dalesandro M, Scallon BJ, et al. 1999. Control of viremia in simian immunodeficiency virus infection by $CD8^+$ lymphocytes. Science 283:857–860.

Schneidewind A, Brockman MA, Sidney J, Wang YE, Chen H, Suscovich TJ, Li B, Adam RI, Allgaier RL, Mothe BR, et al. 2008. Structural and functional constraints limit options for cytotoxic T-lymphocyte escape in the immunodominant HLA-B27-restricted epitope in human immunodeficiency virus type 1 capsid. J Virol 82:5594–5605.

Schuldiner M, Collins SR, Thompson NJ, Denic V, Bhamidipati A, Punna T, Ihmels J, Andrews B, Boone C, Greenblatt JF, et al. 2005. Exploration of the function and organization of the yeast early secretory pathway through an epistatic miniarray profile. Cell 123:507–519.

Sedaghat AR, Siliciano JD, Brennan TP, Wilke CO, Siliciano RF. 2007. Limits on replenishment of the resting $CD4^+$ T cell reservoir for HIV in patients on HAART. PLoS Pathog 3:e122.

Sedaghat AR, Siliciano RF, Wilke CO. 2008. Low-level HIV-1 replication and the dynamics of the resting $CD4^+$ T cell reservoir for HIV-1 in the setting of HAART. BMC infectious diseases 8:2.

Segrè D, DeLuna A, Church GM, Kishony R. 2004. Modular epistasis in yeast metabolism. Nature Genetics 37:77.

Seibert CW, Rahmat S, Krammer F, Palese P, Bouvier NM. 2012. Efficient transmission of pandemic H1N1 influenza viruses with high-level oseltamivir resistance. J Virol 86:5386–5389.

Sella G, Hirsh AE. 2005. The application of statistical physics to evolutionary biology. Proc Natl Acad Sci U S A 102:9541–9546.

Sergeev RA, Batorsky RE, Coffin JM, Rouzine IM. 2010. Interpreting the effect of vaccination on steady state infection in animals challenged with Simian immunodeficiency virus. J Theor Biol 263:385–392.

Sergeev RA, Batorsky RE, Rouzine IM. 2010. Model with two types of CTL regulation and experiments on CTL dynamics. J Theor Biol 263:369–384.

Shah P, McCandlish DM, Plotkin JB. 2015. Contingency and entrenchment in protein evolution under purifying selection. Proc Natl Acad Sci U S A 112:E3226–3235.

Sharov V, Rezelj VV, Galatenko VV, Titievsky A, Panov J, Chumakov K, Andino R, Vignuzzi M, Brodsky L. 2021. Intra- and Inter-cellular Modeling of Dynamic Interaction between Zika Virus and Its Naturally Occurring Defective Viral Genomes. J Virol 95:e0097721.

Shirogane Y, Rousseau E, Voznica J, Xiao Y, Su W, Catching A, Whitfield ZJ, Rouzine IM, Bianco S, Andino R. 2021. Experimental and mathematical insights on the interactions between poliovirus and a defective interfering genome. PLoS Pathog 17:e1009277.

Shirreff G, Pellis L, Laeyendecker O, Fraser C. 2011. Transmission selects for HIV-1 strains of intermediate virulence: a modelling approach. PLoS computational biology 7:e1002185.

Siliciano RF, Greene WC. 2011. HIV latency. Cold Spring Harb Perspect Med 1:a007096.

Silvestri G, Sodora DL, Koup RA, Paiardini M, O'Neil SP, McClure HM, Staprans SI, Feinberg MB. 2003. Nonpathogenic SIV infection of sooty mangabeys is characterized by limited bystander immunopathology despite chronic high-level viremia. Immunity 18:441–452.

Simmonds P, Balfe P, Ludlam CA, Bishop JO, Brown AJ. 1990. Analysis of sequence diversity in hypervariable regions of the external glycoprotein of human immunodeficiency virus type 1. J Virol 64:5840–5850.

Small CB, Klein RS, Friedland GH, Moll B, Emeson EE, Spigland I. 1983. Community-acquired opportunistic infections and defective cellular immunity in heterosexual drug abusers and homosexual men. Am J Med 74:433–441.

Smith DJ, Lapedes AS, de Jong JC, Bestebroer TM, Rimmelzwaan GF, Osterhaus AD, Fouchier RA. 2004. Mapping the antigenic and genetic evolution of influenza virus. Science 305:371–376.

Smith MW, Dean M, Carrington M, Winkler C, Huttley GA, Lomb DA, Goedert JJ, O'Brien TR, Jacobson LP, Kaslow R, et al. 1997. Contrasting genetic influence of CCR2 and CCR5 variants on HIV-1 infection and disease progression. Hemophilia Growth and Development Study (HGDS), Multicenter AIDS Cohort Study (MACS), Multicenter Hemophilia Cohort Study (MHCS), San Francisco City Cohort (SFCC), ALIVE Study. Science 277:959–965.

Song H, Pavlicek JW, Cai F, Bhattacharya T, Li H, Iyer SS, Bar KJ, Decker JM, Goonetilleke N, Liu MK, et al. 2012. Impact of immune escape mutations on HIV-1 fitness in the context of the cognate transmitted/founder genome. Retrovirology 9:89.

Stafford MA, Corey L, Cao Y, Daar ES, Ho DD, Perelson AS. 2000. Modeling plasma virus concentration during primary HIV infection. Journal of Theoretical Biology 203:285–301.

Steinhauer DA, de la Torre JC, Meier E, Holland JJ. 1989. Extreme heterogeneity in populations of vesicular stomatitis virus. J Virol 63:2072–2080.

Steinhauer DA, Domingo E, Holland JJ. 1992. Lack of evidence for proofreading mechanisms associated with an RNA virus polymerase. Gene 122:281–288.

Steinhauer DA, Holland JJ. 1987. Rapid evolution of RNA viruses. Annu Rev Microbiol 41:409–433.

Stern A, Bianco S, Yeh MT, Wright C, Butcher K, Tang C, Nielsen R, Andino R. 2014. Costs and benefits of mutational robustness in RNA viruses. Cell Rep 8:1026–1036.

Strelkowa N, Lassig M. 2012. Clonal interference in the evolution of influenza. Genetics 192:671–682.

Sun Y, Schmitz JE, Buzby AP, Barker BR, Rao SS, Xu L, Yang ZY, Mascola JR, Nabel GJ, Letvin NL. 2006. Virus-specific cellular immune correlates of survival in vaccinated monkeys after simian immunodeficiency virus challenge. J Virol 80:10950–10956.

Sztuba-Solinska J, Urbanowicz A, Figlerowicz M, Bujarski JJ. 2011. RNA-RNA recombination in plant virus replication and evolution. Annu Rev Phytopathol 49:415–443.

Tajima F. 1989. Statistical method for testing the neutral mutation hypothesis by DNA polymorphism. Genetics 123:585–595.

Takahashi T, Song J, Suzuki T, Kawaoka Y. 2013. Mutations in NA that induced low pH-stability and enhanced the replication of pandemic (H1N1) 2009 influenza A virus at an early stage of the pandemic. PLOS ONE 8:e64439.

Tang W, Wu X, Jiang R, Li Y. 2009. Epistatic module detection for case-control studies: a Bayesian model with a Gibbs sampling strategy. PLoS Genet 5:e1000464.

Thompson KA, Yin J. 2010. Population dynamics of an RNA virus and its defective interfering particles in passage cultures. Virol J 7:257.

Tindall KR, Kunkel TA. 1988. Fidelity of DNA synthesis by the Thermus aquaticus DNA polymerase. Biochemistry 27:6008–6013.

Tomiyama H, Fujiwara M, Oka S, Takiguchi M. 2005. Cutting Edge: Epitope-dependent effect of Nef-mediated HLA class I down-regulation on ability of HIV-1-specific CTLs to suppress HIV-1 replication. J Immunol 174:36–40.

Troyer RM, McNevin J, Liu Y, Zhang SC, Krizan RW, Abraha A, Tebit DM, Zhao H, Avila S, Lobritz MA, et al. 2009. Variable fitness impact of HIV-1 escape mutations to cytotoxic T lymphocyte (CTL) response. PLoS Pathog 5:e1000365.

Tsimring LS, Levine H, Kessler DA. 1996. RNA virus evolution via a fitness-space model. Phys Rev Lett 76:4440–4443.

Turelli M, Barton NH. 2006. Will population bottlenecks and multilocus epistasis increase additive genetic variance? Evolution 60:1763–1776.

Turnbull EL, Wong M, Wang S, Wei X, Jones NA, Conrod KE, Aldam D, Turner J, Pellegrino P, Keele BF, et al. 2009. Kinetics of expansion of epitope-specific T cell responses during primary HIV-1 infection. J Immunol 182:7131–7145.

Turner AM, De La Cruz J, Morris KV. 2009. Mobilization-competent Lentiviral Vector-mediated Sustained Transcriptional Modulation of HIV-1 Expression. Mol Ther 17:360–368.

Ueki M, Cordell HJ. 2012. Improved statistics for genome-wide interaction analysis. PLoS Genet 8: e1002625.

van Deutekom HW, Wijnker G, de Boer RJ. 2013. The rate of immune escape vanishes when multiple immune responses control an HIV infection. J Immunol 191:3277–3286.

vanBaalen M, Sabelis MW. 1995. The dynamics of multiple infection and the evolution of virulence. American Naturalist 146:881–910.

Vieira J, Frank E, Spira TJ, Landesman SH. 1983. Acquired immune deficiency in Haitians: opportunistic infections in previously healthy Haitian immigrants. N Engl J Med 308:125–129.

Vignuzzi M, Lopez CB. 2019. Defective viral genomes are key drivers of the virus-host interaction. Nat Microbiol 4:1075–1087.

Vignuzzi M, Stone JK, Arnold JJ, Cameron CE, Andino R. 2006. Quasispecies diversity determines pathogenesis through cooperative interactions in a viral population. Nature 439:344–348.

Vingert B, Perez-Patrigeon S, Jeannin P, Lambotte O, Boufassa F, Lemaitre F, Kwok WW, Theodorou I, Delfraissy JF, Theze J, et al. 2010. HIV controller $CD4^+$ T cells respond to minimal amounts of Gag antigen due to high TCR avidity. PLoS Pathog 6:e1000780.

Voynow SL, Coffin JM. 1985. Truncated gag-related proteins are produced by large deletion mutants of Rous sarcoma virus and form virus particles. J. Virol. 55:79–85.

Wain-Hobson S. 1993. The fastest genome evolution ever described: HIV variation in situ. Curr Opin Genet Dev 3:878–883.

Walczak AM, Nicolaisen LE, Plotkin JB, Desai MM. 2012. The structure of genealogies in the presence of purifying selection: a fitness-class coalescent. Genetics 190:753–779.

Wan X, Yang C, Yang Q, Xue H, Fan X, Tang NLS, Yu W. 2010. BOOST: A Fast Approach to Detecting Gene-Gene Interactions in Genome-wide Case-Control Studies. The American Journal of Human Genetics 87:325–340.

Wang D, Hicks CB, Goswami ND, Tafoya E, Ribeiro RM, Cai F, Perelson AS, Gao F. 2011. Evolution of drug-resistant viral populations during interruption of antiretroviral therapy. J Virol 85:6403–6415.

Wang D, Salah El-Basyoni I, Stephen Baenziger P, Crossa J, Eskridge KM, Dweikat I. 2012. Prediction of genetic values of quantitative traits with epistatic effects in plant breeding populations. Heredity (Edinb) 109:313–319.

Waterson G. 1975. Theor. Pop. Bio. 7:256–276.

Wawer MJ, Gray RH, Sewankambo NK, Serwadda D, Li X, Laeyendecker O, Kiwanuka N, Kigozi G, Kiddugavu M, Lutalo T, et al. 2005. Rates of HIV-1 transmission per coital act, by stage of HIV-1 infection, in Rakai, Uganda. J Infect Dis 191:1403–1409.

Wei W-H, Hemani G, Haley CS. 2014. Detecting epistasis in human complex traits. Nature Reviews Genetics 15:722.

Wei X, Ghosh SK, Taylor ME, Johnson VA, Emini EA, Deutsch P, Lifson JD, Bonhoeffer S, Nowak MA, Hahn BH, et al. 1995. Viral dynamics in human immunodeficiency virus type 1 infection. Nature 373:117–122.

Weigt M, White RA, Szurmant H, Hoch JA, Hwa T. 2009. Identification of direct residue contacts in protein-protein interaction by message passing. Proc Natl Acad Sci U S A 106:67–72.

Weinberger LS, Burnett JC, Toettcher JE, Arkin AP, Schaffer DV. 2005. Stochastic gene expression in a lentiviral positive-feedback loop: HIV-1 Tat fluctuations drive phenotypic diversity. Cell 122:169–182.

Weinberger LS, Dar RD, Simpson ML. 2008. Transient-mediated fate determination in a transcriptional circuit of HIV. Nat Genet 40:466–470.

Weinberger LS, Schaffer DV, Arkin AP. 2003. Theoretical design of a gene therapy to prevent AIDS but not human immunodeficiency virus type 1 infection. J Virol 77:10028–10036.

Weinreich DM, Watson RA, Chao L. 2005. Perspective: Sign epistasis and genetic constraint on evolutionary trajectories. Evolution 59:1165–1174.

Weissman DB, Desai MM, Fisher DS, Feldman MW. 2009. The rate at which asexual populations cross fitness valleys. Theor Popul Biol 75:286–300.

Weissman DB, Hallatschek O. 2014. The rate of adaptation in large sexual populations with linear chromosomes. Genetics 196:1167–1183.

Whitney JB, Hill AL, Sanisetty S, Penaloza-MacMaster P, Liu J, Shetty M, Parenteau L, Cabral C, Shields J, Blackmore S, et al. 2014. Rapid seeding of the viral reservoir prior to SIV viraemia in rhesus monkeys. Nature 512:74–77.

Whitton JL, Cornell CT, Feuer R. 2005. Host and virus determinants of picornavirus pathogenesis and tropism. Nat Rev Microbiol 3:765–776.

Wodarz D. 2001. Helper-dependent vs helper-independent CTL responses in HIV infection: implications for drug therapy and resistance. J Theor Biol 213:447–459.

Wolfs TF, de Jong JJ, Van den Berg H, Tijnagel JM, Krone WJ, Goudsmit J. 1990. Evolution of sequences encoding the principal neutralization epitope of human immunodeficiency virus 1 is host dependent, rapid, and continuous. Proc Natl Acad Sci U S A 87:9938–9942.

Wolinsky SM, Korber BT, Neumann AU, Daniels M, Kunstman KJ, Whetsell AJ, Furtado MR, Cao Y, Ho DD, Safrit JT. 1996. Adaptive evolution of human immunodeficiency virus-type 1 during the natural course of infection. Science 272:537–542.

Wong JK, Strain MC, Porrata R, Reay E, Sankaran-Walters S, Ignacio CC, Russell T, Pillai SK, Looney DJ, Dandekar S. 2010. In vivo $CD8^+$ T-cell suppression of siv viremia is not mediated by CTL clearance of productively infected cells. PLoS Pathog 6:e1000748.

Woodhouse DE, Rothenberg RB, Potterat JJ, Darrow WW, Muth SQ, Klovdahl AS, Zimmerman HP, Rogers HL, Maldonado TS, Muth JB, et al. 1994. Mapping a social network of heterosexuals at high risk for HIV infection. AIDS 8:1331–1336.

Worobey M, Holmes EC. 1999. Evolutionary aspects of recombination in RNA viruses. J Gen Virol 80 (Pt 10):2535–2543.

Wrenbeck EE, Azouz LR, Whitehead TA. 2017. Single-mutation fitness landscapes for an enzyme on multiple substrates reveal specificity is globally encoded. Nat Commun 8:15695.

Wright S. 1931. Evolution in Mendelian Populations. Genetics 16:97–159.

Wu L, KewalRamani VN. 2006. Dendritic-cell interactions with HIV: infection and viral dissemination. Nat Rev Immunol 6:859–868.

Wu X, Dong H, Luo L, Zhu Y, Peng G, Reveille JD, Xiong M. 2010. A novel statistic for genome-wide interaction analysis. PLoS Genet 6:e1001131.

Wyand MS, Manson KH, Garcia-Moll M, Montefiori D, Desrosiers RC. 1996. Vaccine protection by a triple deletion mutant of simian immunodeficiency virus. J Virol 70:3724–3733.

Xiao Y, Lidsky PV, Shirogane Y, Aviner R, Wu CT, Li W, Zheng W, Talbot D, Catching A, Doitsh G, et al. 2021. A defective viral genome strategy elicits broad protective immunity against respiratory viruses. Cell 184:6037–6051 e6014.

Xiao Y, Rouzine IM, Bianco S, Acevedo A, Goldstein EF, Farkov M, Brodsky L, Andino R. 2016. RNA recombination enhances adaptability and is required for virus spread and virulence. Cell Host Microbe 19:493–503.

Yan L, Neher RA, Shraiman BI. 2019. Phylodynamic theory of persistence, extinction and speciation of rapidly adapting pathogens. Elife 8:e44205.

Yerly S, von Wyl V, Ledergerber B, Boni J, Schupbach J, Burgisser P, Klimkait T, Rickenbach M, Kaiser L, Gunthard HF, et al. 2007. Transmission of HIV-1 drug resistance in Switzerland: a 10-year molecular epidemiology survey. AIDS 21:2223–2229.

Yoshimura FK, Diem K, Learn GH, Jr., Riddell S, Corey L. 1996. Intrapatient sequence variation of the gag gene of human immunodeficiency virus type 1 plasma virions. J Virol 70:8879–8887.

Yu Y, Wang J, Shao Q, Shi J, Zhu W. 2015. Effects of drug-resistant mutations on the dynamic properties of HIV-1 protease and inhibition by Amprenavir and Darunavir. Scientific Reports 5:10517.

Zeyl C, Bell G. 1997. The advantage of sex in evolving yeast populations. Nature 388:465–468.

Zhang LQ, MacKenzie P, Cleland A, Holmes EC, Brown AJ, Simmonds P. 1993. Selection for specific sequences in the external envelope protein of human immunodeficiency virus type 1 upon primary infection. J Virol 67:3345–3356.

Zhang Y, Jiang B, Zhu J, Liu JS. 2011. Bayesian models for detecting epistatic interactions from genetic data. Ann Hum Genet 75:183–193.

Zhang Y, Liu JS. 2007. Bayesian inference of epistatic interactions in case-control studies. Nat Genet 39:1167–1173.

Zhang Z, Schuler T, Zupancic M, Wietgrefe S, Staskus KA, Reimann KA, Reinhart TA, Rogan M, Cavert W, Miller CJ, et al. (20027623 co-authors). 1999. Sexual transmission and propagation of SIV and HIV in resting and activated $CD4^{+}$ T cells. Science 286:1353–1357.

Zhang ZQ, Notermans DW, Sedgewick G, Cavert W, Wietgrefe S, Zupancic M, Gebhard K, Henry K, Boies L, Chen Z, et al. 1998. Kinetics of $CD4^{+}$ T cell repopulation of lymphoid tissues after treatment of HIV-1 infection. Proc Natl Acad Sci U S A 95:1154–1159.

Zhu Z, Tong X, Zhu Z, Liang M, Cui W, Su K, Li MD, Zhu J. 2013. Development of GMDR-GPU for Gene-Gene Interaction Analysis and Its Application to WTCCC GWAS Data for Type 2 Diabetes. PLOS ONE 8: e61943.

Zuk O, Hechter E, Sunyaev SR, Lander ES. 2012. The mystery of missing heritability: Genetic interactions create phantom heritability. Proc Natl Acad Sci U S A 109:1193–1198.

zur Wiesch PA, Kouyos R, Engelstadter J, Regoes RR, Bonhoeffer S. 2011. Population biological principles of drug-resistance evolution in infectious diseases. Lancet Infect Dis 11:236–247.

Index

https://doi.org/10.1515/9783110697384-006

De Gruyter Series in Mathematics and Life Sciences

Volume 8/1
Igor M. Rouzine
Mathematical Modeling of Evolution. Volume 1: One-Locus and Multi-Locus Theory and Recombination, 2020
ISBN 978-3-11-060789-5, ISBN (PDF) 978-3-11-061545-6, ISBN (EPUB) 978-3-11-060819-9

Volume 7
George Dassios, Athanassios S. Fokas
Electroencephalography and Magnetoencephalography, 2020
ISBN 978-3-11-054583-8, ISBN (PDF) 978-3-11-054753-5, ISBN (EPUB) 978-3-11-054578-4

Volume 6
Piotr Biler
Singularities of Solutions to Chemotaxis Systems, 2019
ISBN 978-3-11-059789-9, ISBN (PDF) 978-3-11-059953-4, ISBN (EPUB) 978-3-11-059862-9

Volume 5
S. V. Masiuk, A. G. Kukush, S. V. Shklyar, M. I. Chepurny, I. A. Likhtarov
Radiation Risk Estimation. Based on Measurement Error Models, 2016
ISBN 978-3-11-044180-2, ISBN (PDF) 978-3-11-043366-1, ISBN (EPUB) 978-3-11-043347-0

Volume 4
Sergey Vakulenko
Complexity and Evolution of Dissipative Systems: An Analytical Approach, 2013
ISBN 978-3-11-026648-1, e-ISBN 978-3-11-026828-7

Volume 3
Zoran Nikoloski, Sergio Grimbs
Network-Based Molecular Biology: Data-Driven Modeling and Analysis, 2013
ISBN 978-3-11-026256-8, e-ISBN 978-3-11-026266-7

Volume 2
Shair Ahmad, Ivanka M. Stamova (Eds.)
Lotka-Volterra and Related Systems: Recent Developments in Population Dynamics, 2013
ISBN 978-3-11-026951-2, e-ISBN 978-3-11-026984-0

Volume 1
Alexandra V. Antoniouk, Roderick V. N. Melnik (Eds.)
Mathematics and Life Sciences, 2012
ISBN 978-3-11-027372-4, e-ISBN 978-3-11-028853-7

www.degruyter.com

FSC
www.fsc.org
MIX
Papier | Fördert gute Waldnutzung
FSC® C083411

Kolibri 360 GmbH
Ferdinand-Jühlke-Straße 7
99095 Erfurt, Deutschland
produktsicherheit@kolibri360.de